Introduction to Geospatial Technologies

Bradley A. Shellito
Youngstown State University

W.H. Freeman and Company ⊙ New York

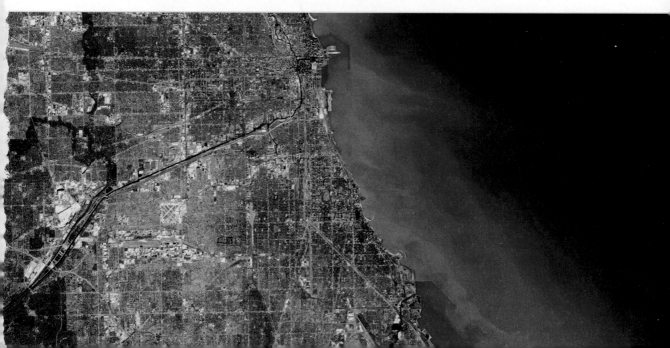

Executive Editor: Steven Rigolosi
Director of Marketing: John Britch
Media and Supplements Editor: Kerri Russini
Editorial Assistant: Stephanie Ellis
Art Director, Interior Design, and Cover Design: Diana Blume
Illustration Coordinator: Janice Donnola
Illustrations: Dragonfly Studio
Project Editor: Lisa Kinne
Photo Research Manager: Ted Szczepanski
Photo Editor and Researcher: Jacqui Wong
Director of Production: Ellen Cash
Production Coordinator: Paul Rohloff
Composition: MPS Limited, a Macmillan Company
Printing and Binding: R.R. Donnelley & Sons Company

ISBN-13: 978-1-4292-5528-8
ISBN-10: 1-4292-5528-5

Library of Congress Control Number: 2011930072

Printed in the United States of America

First printing

W.H. Freeman and Company
41 Madison Avenue
New York, NY 10010
Houndmills, Basingstoke RG21 6XS, England

www.whfreeman.com/geography

Dedication

This book is dedicated to my parents, David and Elizabeth, who taught me that with a lot of hard work—and a little luck—all things are possible.

Contents in Brief

Contents

Chapter 4
Finding Your Location with the Global Positioning System

GPS Origins, Position Measurement, Errors, Accuracy, GNSS around the World, Applications, and Geocaching 73

Part 2 Geographic Information Systems

Chapter 5
Working With Digital Spatial Data and GIS

Geographic Information Systems, Modeling the Real World, Vector Data and Raster Data, Attribute Data, Joining Tables, Metadata, Esri, and ArcGIS 99

Chapter 6
Using GIS for Spatial Analysis

Database Query and Selection, Buffers, Overlay Operations, Geoprocessing Concepts, and Modeling with GIS 147

Chapter 15

What's Next for Geospatial Technology?

New Frontiers in Geospatial Technology,
Geospatial Technologies in K-12 Education,
College and University Educational Programs,
How to Easily Get Data, a Geospatial World
Online, and Other Developments 439

Geospatial Lab Application 15.1

Thinking Critically with Geospatial Technology

Hands-on Applications

Preface

Why I Wrote *Introduction to Geospatial Technologies*

When people ask me what I teach, I say "geospatial technology." The usual response to this statement is a blank stare, a baffled "What?," or a variation on "I've never heard of that." However, if I say I teach "technologies like GPS, taking images from satellites, and using online tools like Google Earth or MapQuest," the response generally improves to "GPS is great" or "Google Earth is so cool," or even "Why do I get the wrong directions from that thing in my car?" Although geospatial technologies are everywhere these days—from software to Websites to cell phones—it seems that the phrase "geospatial technology" hasn't really permeated into common parlance.

I hope that *Introduction to Geospatial Technologies* will help remedy this situation. As its title implies, the goal of this book is to introduce several aspects of geospatial technologies—not only what they are and how they operate, but also how they are used in hands-on applications. In other words, the book covers a little bit of everything, from theory to application.

In a sense, the book's goal is to offer students an overview of several different fields and techniques and to provide a solid foundation on which further knowledge in more specialized classes (such as GIS or remote sensing) can be built. Whether the book is used for a basic introductory course, a class for non-majors, or as an introduction to widely used geospatial software packages, this book is aimed at beginners just starting out. At Youngstown State University (YSU), I teach an introductory class titled "Geospatial Foundations," but I've seen similar-sounding classes at other universities with names like "The Digital Earth," "Introduction to Geospatial Analysis," "Survey of Geospatial Technologies," "Introduction to GIS," or "Computer Applications in Geography." All of these courses seem aimed at the audience for which *Introduction to Geospatial Technologies* was written.

Organization

The book is divided into four main parts.

Part 1 focuses on geospatial technology as it relates to spatial measurements and data.

- Chapter 1, "It's a Geospatial World Out There" introduces some basic concepts and provides an overview of jobs, careers, and some key technologies and applications (such as Google Earth).

- Chapter 2, "Where in the Geospatial World Are You?," explains how coordinates for location-based data and measurements from a three-dimensional (3D) world are translated into a two-dimensional (2D) map on a computer screen.
- Chapter 3, "Getting Your Data to Match the Map," discusses reprojection and georeferencing, important information when you're using any sort of geospatial data.
- Chapter 4, "Finding Your Location with the Global Positioning System," introduces GPS concepts. Taking a handheld receiver outside, pressing a button, and then having the device specify your precise location and plot it on a map sounds almost like magic. This chapter demystifies GPS by explaining how the system works, why it's not always perfectly accurate, and how to get better location accuracy.

Part 2 focuses on Geographic Information Systems (GIS).

- Chapter 5, "Working with Digital Spatial Data and GIS," serves as an introduction to GIS, examining how real-world data can be modeled and how GIS data can be created and used.
- Chapter 6, "Using GIS for Spatial Analysis," covers additional uses of GIS, including querying a database, creating buffers, and geoprocessing.
- Chapter 7, "Using GIS to Make a Map," offers instruction on how to classify your data and how to transform GIS data into a professional-looking map.
- Chapter 8, "Getting There Quicker with Geospatial Technology," discusses concepts related to road networks, such as: How are streets, highways, and interstates set up and used in geospatial technology? How does the computer translate a set of letters and numbers into a map of an actual street address? How do programs determine the shortest route from point *a* to point *b*?

Part 3 examines issues related to remote sensing.

- Chapter 9, "Remotely Sensed Images from Above," focuses on aerial photography. It explains how the field started over 150 years ago (with a man, a balloon, and a camera) and continues today with unmanned aerial vehicles flying over Iraq and Afghanistan. This chapter also describes how to visually interpret features in aerial imagery and how to make accurate measurements from items present in an image.
- Chapter 10, "How Remote Sensing Works," delves into just what remote sensing is and how it works, and how exactly that image of a house is acquired by a sensor 500 miles away. The chapter also discussses all of the things that a remote sensing device can see that are invisible to the human eye.
- Chapter 11, "Images from Space," focuses on the field of satellite remote sensing and how satellites in orbit around Earth acquire images of the ground below.

- Chapter 12, "Studying the Environment from Space," discusses the Earth Observing System, a series of environmental observatories that orbit the planet and continuously transmit data back to Earth about the land, seas, and atmosphere.

Part 4 focuses on individual topics in geospatial technology that combine GIS and remote sensing themes and applications.

- Chapter 13, "Digital Landscaping," describes how geospatial technologies model and handle terrain and topographic features. Being able to set up realistic terrain, landscape features, and surfaces is essential in mapping and planning.

- Chapter 14, "See the World in 3D," delves into the realm of 3D modeling, shows how geospatial technologies create three-dimensional structures and objects, and then explains how to view or interact with them in programs like Google Earth.

- Chapter 15, "What's Next for Geospatial Technology?," wraps things up with a look at some recent developments in geospatial technologies as well as a look ahead to the future of the field.

Geospatial Lab Applications

Each chapter of *Introduction to Geospatial Technologies* covers one aspect of geospatial technology with an accompanying Geospatial Lab Application. Each lab application uses freely available software that can be downloaded from the Internet. These software packages include:

- ArcExplorer Java Edition for Educators (AEJEE)
- ArcGIS Explorer
- Google Earth
- Google SketchUp
- MapCruncher
- MICRODEM
- MultiSpec
- NASA World Wind
- Trimble Planning Software

The four chapters in Part 2 (Geographic Information Systems) offer two versions of the lab application. Instructors can choose to use either the free AEJEE or the desktop version of ArcGIS 10.

The labs provide hands-on application of the concepts and theories covered in each chapter. Clearly, it's one thing to read about how 3D structures can be created and placed into Google Earth, but it's another thing entirely to use Google SketchUp and Google Earth to do exactly that.

Each lab application has integrated questions that students must answer while working through the lab. These questions are designed to explore the various topics presented in the lab and also to keep students moving through the lab application.

Some labs use sample data that comes with the software when it's installed, while others require students to download sample data for use in the lab. Each lab provides links to a website from which you can download the software. (That same Website will also provide information regarding the necessary hardware or system requirements. Not all computers or lab facilities work the same, so be sure to check the software's Internet resources for help on installing the software.) The Instructors section of this book's companion Website also offers a "tech tips" section with some additional information related to installing or utilizing some of the software.

The lab applications for each chapter are set up as follows:

- Chapter 1: This lab introduces Google Earth as a tool for examining many facets of geospatial technology.
- Chapter 2: Students continue using Google Earth, investigating some other functions of the software as they relate to coordinates and measurements.
- Chapter 3: Students use Microsoft's MapCruncher program to match a graphic of a campus map with remotely sensed imagery and real-world coordinates.
- Chapter 4: This lab uses Trimble Planning Software and some other Internet resources to examine GPS planning and locations. It also provides suggestions for expanding the lab activities if you have access to a GPS receiver and want to get outside with it.
- Chapter 5: This lab introduces basic GIS concepts using Esri's ArcExplorer Java Edition for Educators (AEJEE). An alternate version of the lab uses ArcGIS 10.
- Chapter 6: This lab continues investigating the functions of AEJEE (or ArcGIS 10) by using GIS to answer some spatial analysis questions.
- Chapter 7: This lab uses AEJEE (or ArcGIS 10) to design and print a map of data.
- Chapter 8: This lab uses Google Maps, Google Earth, Internet resources, and GIS software (either AEJEE or ArcGIS 10) to match a set of addresses and investigate shortest paths between stops on a network.
- Chapter 9: This lab tests students' visual image interpretation skills by putting them in the role of high-tech detectives who are trying to figure out just what a set of aerial images are actually showing.
- Chapter 10: This lab is an introduction to MultiSpec, which allows users to examine various aspects of remotely sensed imagery obtained by a satellite.
- Chapter 11: This lab continues using MultiSpec by asking students to work with imagery from a Landsat satellite and investigate its sensor's capabilities.

- ⊚ Chapter 12: This lab uses NASA World Wind to examine phenomena such as hurricanes, fires, and pollution on a global scale.

- ⊚ Chapter 13: This lab uses Google Earth to examine how terrain is used in geospatial technology (and film a video of flying over 3D-style terrain). It also uses MICRODEM, which allows students to examine a digital terrain model and create a three-dimensional representation of it.

- ⊚ Chapter 14: In this lab, students work with 3D modeling. Starting from an aerial image of a building, they design a 3D version of it using Google SketchUp, and then look at it in Google Earth.

- ⊚ Chapter 15: This lab utilizes Esri's free ArcGIS Explorer program to wrap things up and look at many of the book's concepts combined in one package.

Additional Features

In addition to the lab applications, each chapter contains several **Hands-on Applications** that utilize free Internet resources to help students further explore the world of geospatial technologies and get directly involved with some of the chapter concepts. There's a lot of material out there on the Internet, ranging from interactive mapmaking to real-time satellite tracking, and these Hands-on Applications introduce students to it.

Each chapter also has one or more boxes titled **Thinking Critically with Geospatial Technology**. These boxes present questions to consider regarding potential societal, privacy, design, or ethical issues posed by geospatial technologies and their applications. These boxes are open-ended and are intended to stimulate discussion about geospatial technologies and how they affect (or could affect) human beings. For instance, how much privacy do you really have if anyone, anywhere, can obtain a clear image of your house or neighborhood (and directions how to drive there) with just a few clicks of a mouse?

Ancillary Materials and Student Website

www.whfreeman.com/shellito1e

The companion Website offers a set of valuable resources for both students and instructors.

For **students**, the Website offers a multiple-choice self-test for each chapter, as well as an extensive set of references, categorized by topic, to provide further information on a particular topic. The companion Website also provides a set of links to the free software packages needed to complete the lab activities, as well as the datasets required for specific lab applications. A set of world links is also provided.

For **instructors**, an instructor's manual provides (for each chapter) teaching tips on presenting the book's material, a set of "tech tips" related to software installation and usage, a set of key references for the chapter materials, and an answer key for all the lab activities. A test bank of questions is also provided.

Acknowledgments and Thanks

Books like this don't just spring out of thin air—I owe a great deal to the many people who have provided inspiration, help, and support for what would eventually become this book:

Steven Rigolosi of W.H. Freeman and Company, Editor Extraordinaire, for invaluable help, advice, patience, and guidance throughout this entire project. I would also very much like to thank Lisa Kinne, Kerri Russini, Stephanie Ellis, Ted Szczepanski, and Jacqui Wong at W.H. Freeman for their extensive "behind the curtain" work that shaped this book into a finished product. Thanks also go to the copy editor, Thomas Martin, and Anthony Calcara, who helped with the lab checking.

Neil Salkind, for great representation and advice. The students who contributed to the development of the YSU 3D Campus Model (Rob Carter, Ginger Cartright, Paul Crabtree, Jason Delisio, Sherif El Seuofi, Nicole Eve, Paul Gromen, Wook Rak Jung, Colin LaForme, Sam Mancino, Jeremy Mickler, Eric Ondrasik, Craig Strahler, Jaime Webber, Sean Welton, and Nate Wood). Many of the 3D examples presented in Chapter 14 wouldn't have existed without them. Jack Daugherty, for tech support, assistance with the labs, and help with the GeoWall applications. Lisa Curll, for assistance with the design and formatting of the lab applications. Grant Wilson, for his insightful technical reviews of the ArcGIS 10 lab applications. Margaret Pearce, for using an earlier draft of the manuscript with her students at the University of Kansas and for her extremely useful comments and feedback. Hal Withrow, for invaluable computer tech assistance.

I also offer very special thanks to all of my professors, instructors, colleagues, and mentors, past and present (who are too numerous to list), from Youngstown State University, the Ohio State University, Michigan State University, Old Dominion University, OhioView, and everywhere else, for the help, knowledge, notes, information, skills, and tools they've given me over the years. I am also deeply indebted to the work of Tom Allen, John Jensen, Mandy Munro-Stasiuk, and the members of SATELLITES for some methods used in some of the chapters and labs.

Finally, I owe a debt of gratitude to the colleagues who reviewed the original proposal and various stages of the manuscript. Thank you for your insightful and constructive comments, which have helped to shape the final product:

Robbyn Abbitt, *Miami University*
Amy Ballard, *Central New Mexico Community College*
Chris Baynard, *University of North Florida*
Robert Benson, *Adams State College*
Edward Bevilacqua, *SUNY College of Environmental Science and Forestry*
Julie Cidell, *University of Illinois*
Jamison Conley, *West Virginia University*
Kevin Czajkowski, *University of Toledo*
Adrienne Domas, *Michigan State University*
Christine Drennon, *Trinity University*
Jennifer Fu, *Florida International University*
Nandhini Gulasingam, *DePaul University*
Victor Gutzler, *Tarrant County College Southeast*
Jessica Kelly, *Millersville University*
Sara Beth Keough, *Saginaw Valley State University*
James Kernan, *SUNY Geneseo*
James Lein, *Ohio University*
Chris Lukinbeal, *Arizona State University*
Margaret Pearce, *University of Kansas*
Hugh Semple, *Eastern Michigan University*
Shuang-Ye Wu, *University of Dayton*
Donald Zeigler, *Old Dominion University*

As Chapter 15 points out, geospatial technology has become so widespread and prevalent that no book can cover every concept, program, or online mapping or visualization tool (as much as I'd like to). I hope that the students who use this book will view the concepts and applications presented herein as an introduction to the subject—and motivate them to take more advanced courses on the various aspects of geospatial technology.

One thing to keep in mind: In such a rapidly advancing field as geospatial technology, things can change pretty quickly. New satellites are being launched and old ones are ending their mission lives. Websites get updated and new updates for software and tools are released on a regular basis. As of the writing of this book, all of the Web data, software, and satellites were current, but if something's name has changed, or if a Website works differently, or if a satellite isn't producing any more data, there's probably something newer and shinier to take its place.

I'd very much like to hear from you regarding any thoughts or suggestions you might have for the book. You can reach me via email at bashellito@ysu.edu.

Bradley Shellito
Youngstown State University

Accessing Data Sets for Geospatial Lab Applications

Most of the Geospatial Lab Applications in this book use data that comes with the software, sample data that gets installed on your computer when you install the software itself, or data that you'll create during the course of the lab. However, the lab applications for Chapters 3, 8, 9, 10, and 11 require you to download a set of data that you'll use with those labs.

The lab applications will direct you to copy the dataset before beginning the lab. Each dataset is stored in its own folder online at

http://www.whfreeman.com/shellito1e

Under "Student Resources," click on "Lab Data Sets," then choose the relevant chapter.

1

It's a Geospatial World Out There

An Introduction to Geospatial Technology, Geospatial Data, Geospatial Jobs, and Google Earth

Have you ever done any of the following?

◉ Used an online mapping service like MapQuest, Google Maps, or Bing Maps to find directions (and the best route) to a destination or to print a map of an area?

◉ Used an in-car navigation system (like a device from companies like Garmin, Magellan, or Tom-Tom) to navigate to or from a destination?

◉ Used a Global Positioning System (GPS) receiver while hiking, jogging, hunting, fishing, golfing, or geocaching?

◉ Used an online resource to find a map of your neighborhood, compare nearby housing values, or to see where your property line ends and your neighbor's begins?

◉ Used a virtual globe program (like Google Earth) or an online map to look at photos or images of your home, street, school, or workplace?

If so, then congratulations—you've used geospatial technology applications. Anytime you're using any sort of technology-assisted information concerning maps, locations, directions, imagery, or analysis, you're putting this concept to use. Geospatial technology has become extremely widespread in society with a multitude of uses in both the private and public sectors. However, more often than not, if you tell someone you're using geospatial technology, you'll be asked "what's that?"

What Is Geospatial Technology?

Although geospatial technology is being used in numerous fields today, the term "geospatial technology" doesn't appear to have penetrated into everyday usage, despite the prevalence of the technology itself. Words like "satellite

geospatial technology a number of different high-tech systems that acquire, analyze, manage, store, or visualize various types of location-based data.

Geographic Information System (GIS) computer-based mapping, analysis, and retrieval of location-based data.

remote sensing acquisition of data and imagery from the use of satellites or aircraft.

satellite imagery digital images of the Earth acquired by sensors onboard orbiting spaceborne platforms.

aerial photography the acquisition of imagery of the ground taken from an airborne platform.

Global Positioning System (GPS) acquiring real-time location information from a series of satellites in Earth orbit.

images" or "Google Earth" and acronyms like "GIS" or "GPS" are growing increasingly commonplace in today's society, yet the phrase "geospatial technology" seems relatively unknown, while incorporating all of these things and more. **Geospatial technology** describes the use of a number of different high-tech systems and tools that acquire, analyze, manage, store, or visualize various types of location-based data. The field of geospatial technology encompasses several fields and techniques including:

- **Geographic Information System (GIS):** Computer-based mapping, analysis, and retrieval of location-based data.
- **Remote sensing:** Acquisition of data and imagery from the use of satellites (**satellite imagery**) or aircraft (**aerial photography**).
- **Global Positioning System (GPS):** Acquiring real-time location information from a series of satellites in Earth's orbit.

There are numerous related fields that utilize one or more of these types of technologies. For instance, an in-car navigation system already contains extensive road-network data mapped out and ready to use, which includes information about address ranges, speed limits, road connections, and special features of roads (such as one-way streets). It also requires the mapping of points of interest (such as the locations of gas stations or restaurants) and having a means of referencing new user-defined destinations. It also has to be able to plot the car's real-time position in relation to these maps and may even have a feature that shows a representation of the surrounding landscape as taken from an overhead viewpoint. As such, many of these types of systems combine different geospatial technologies to work together in one application.

Who Uses Geospatial Technology?

Geospatial technology is used in a wide variety of fields (Figure 1.1), including federal, state, and local government, forestry, law enforcement, public health, biology, and environmental studies (see *Hands-on Application 1.1: Industries Using Geospatial Technology* for a look at industries employing people in these fields). As long as the job field involves utilizing some sort of information or data that has a location associated with it, chances are that some sort of geospatial technology is being used. Geospatial technology has been heralded by the U.S. Department of Labor as one of the main emerging and evolving job fields in the United States with enormous growth potential. For instance, the O*NET utility from the Department of Labor contains job descriptions for fields such as "Geospatial Information Scientists and Technologists," "Remote Sensing Scientists and Technologists," and "Precision Agriculture Technicians" (see *Hands-on Application 1.2: Jobs in the Geospatial Field* on page 4 for more information about types of jobs).

FIGURE 1.1 Examples of geospatial technology in action on the job. (Source: (a) AP Photo/Wilfredo Lee (b) AP Photo/The Daily Press/Allison Williams (c) AP Photo/Gordon Hamilton/University of Maine (d) Bob Nichols/USDA NRCS (e) AP Photo/U.S. Geological Survey, Dr. Dan Dzurisin (F) Justin Sullivan/Getty Images)

Hands-on Application 1.1

Industries Using Geospatial Technology

Geospatial technology is being used in a variety of different applications in numerous different fields today. For a deeper look at some of these applications, open your Web browser and go to **http://www. esri.com/industries.html**, this is part of the Esri company's Website (we'll discuss more about Esri in Chapter 5, but the short version is that they're the market leader in GIS). This Website lists dozens of different fields that are using geospatial technology and describes how GIS (and Esri products) are being utilized in them. Examine a few of them that are connected to fields of interest of yours. For instance, if your interest is in criminal justice, examine some of the "Public Safety" applications. If you're involved in public or community health, examine some of the "Health and Human Services" applications, then describe how GIS is being utilized in some real-world, on-the-job applications.

Jobs in the Geospatial Field

Businesses are hiring in the geospatial field. For examples of current job openings, open your Web browser and visit some of the following Websites:

1. The GIS Jobs Clearinghouse: **http://www.gjc.org**
2. GIS Jobs.com: **http://www.gisjobs.com**
3. Geosearch: **http://www.geosearch.com**
4. Geocommunity (GIS Jobs and Careers): **http://careers.geocomm.com**
5. GIS Careers: **http://giscareers.com**
6. GIS Lounge: **http://jobs.gislounge.com**
7. GIS Connection: **http://www.gisconnection.com**

These are just a sampling of Websites where employers post job openings worldwide. Examine several jobs from areas near where you are (or where you'd like to go to). Describe the various types of jobs that are advertised, qualifications and skills employers are looking for, and salary ranges being offered.

The following are just a handful of examples of fields that utilize geospatial technology.

Archeology

Being able to pinpoint the location of artifacts uncovered on a dig, construct a map of the area, and then search for patterns on the site are all archeological techniques that can be rendered quickly and efficiently with geospatial technology (Figure 1.2). Archeologists can utilize historic maps, current aerial photography or satellite imagery, and location information obtained on the site throughout the course of their work.

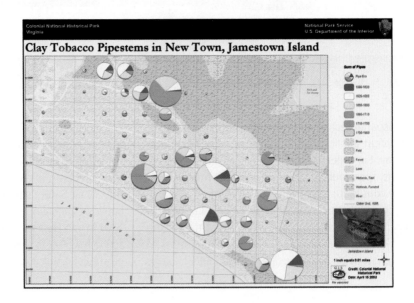

FIGURE 1.2 Using GIS for archeological mapping of Clay Tobacco Pipestems in Virginia. (Source: Courtesy of Colonial National Historical Park, National Park Service)

City Planning

Utilities, waste water, green space, traffic, roads, zoning, and housing are all topics that urban planners deal with. Geospatial technology provides a means of working with all of these entities together for planning purposes. Strategies for smart urban growth and managing and updating city resources can be examined through a variety of different applications.

Environmental Monitoring

Processes that affect the Earth's environment in a wide variety of ways can be tracked and assessed using geospatial technology. Information about land-use change, pollution, air quality, water quality, and global temperature levels is vital to environmental research ranging from monitoring harmful algae blooms to studies in climate change (see Figure 1.3).

Forestry

All manner of forest monitoring, management, and protection can be aided through the use of geospatial technology. Modeling animal habitats and the pressures placed upon them, examining the spatial dimensions of forest fragmentation, and managing fires are among the many different ways that geospatial technology is utilized within the field of forestry (Figure 1.4, page 6).

Homeland Security

Geospatial technology is a key component of examining vulnerable areas with regard to homeland security. Risk assessment of everything from

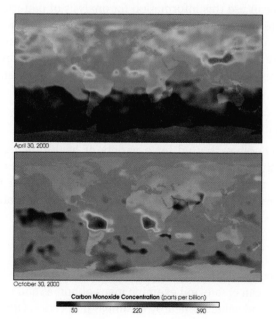

April 30, 2000

October 30, 2000

Carbon Monoxide Concentration (parts per billion)

50   220   390

FIGURE 1.3 Global carbon monoxide concentrations as monitored by NASA remote sensing satellites. (Source: NASA GSFC Scientific Visualization Studio, based on data from MOPITT (Canadian Space Agency and University of Toronto))

FIGURE 1.4 Geospatial
technology used for
assessing the risk of
wildfires in Virginia.
(Source: Courtesy Virginia
Department of Forestry)

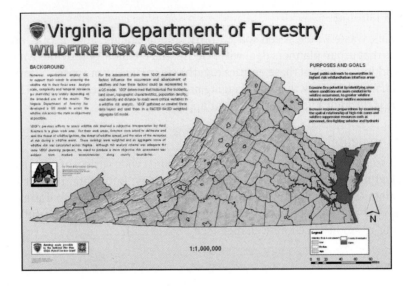

FIGURE 1.4 Geospatial technology used for assessing the risk of wildfires in Virginia. (Source: Courtesy Virginia Department of Forestry)

evacuation plans to smoke-plume modeling can be examined using geo-spatial technology. Disaster mitigation and recovery efforts can be greatly enhanced through the use of current satellite imagery and location-based capabilities.

Law Enforcement

The locations of various types of crimes can be plotted using GIS. Law enforcement officials can use this information for analysis of patterns (Figure 1.5) as well as determining potential new crime areas. Geospatial technology can be used in several other ways beyond mapping and analysis; for instance, high-resolution aerial photography of locations where police patrol can be gathered to provide further information about potentially dangerous areas.

Health and Human Services

Geospatial technology is used for a variety of health-related services. For example, monitoring of diseases, tracking sources of diseases, and mapping health-related issues (such as the spread of H1N1 or other influenza; see Figure 1.6) are all tasks that can be completed using geospatial technology applications.

Real Estate

Through geospatial technology, realtors and appraisers (as well as home buyers and sellers) can create and examine maps of homes and quickly compare housing prices and values of nearby or similar properties. Other features of a

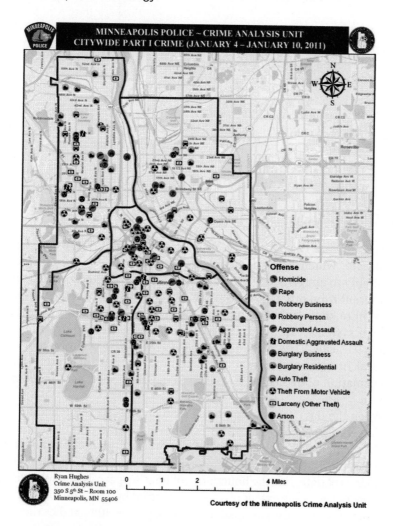

FIGURE 1.5 A GIS map showing crime locations in Minneapolis. (Source: Courtesy Minneapolis Police Crime Analysis Unit)

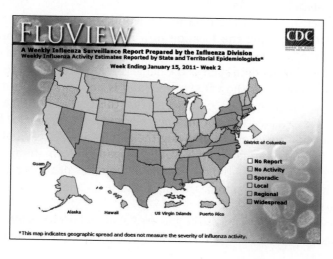

FIGURE 1.6 A CDC map examining the spatial distribution of influenza activity. (Source: Center for Disease Control and Prevention)

property can be examined by viewing high-resolution aerial images, the topography, the terrain, and even where it sits on a floodplain. You can also examine where a property is located in relation to schools, highways, wastewater treatment plants, and other urban features.

All of these areas and many, many more are reliant on the capabilities of geospatial technology. No matter the field, what distinguishes geospatial technology from other types of computer systems or technologies is that it explicitly handles geospatial data.

What Is Geospatial Data?

geospatial data items that are tied to a specific real-world location.

Geospatial data (also often referred to as spatial data) refers to location-based data, which is at the heart of geospatial technology applications. This ability to assign a location to data is what makes geospatial technology different from other systems. You're using geospatial concepts anytime you want to know "where" something is located. For instance, emergency dispatch systems can determine the location you're calling from if you make a 911 call. They also have access to information concerning where the local fire stations, ambulance facilities, and hospitals are. This type of information allows the closest emergency services to be routed to your location.

When the data you're using has a location that it can be tied to, you're working with geospatial data. This isn't just limited to point locations—the length and dimensions of a hiking trail (and the locations of comfort stations along the trail) are examples of real-world data with a spatial dimension. Other kinds of data, like the boundaries of housing parcels in a subdivision or a satellite image of the extent of an area impacted by a natural disaster would fall under this category. Geospatial technology explicitly handles these types of location-based concepts.

However, not all of the data in the world is geospatial data. For instance, data regarding the location of a residential parcel containing a house and the parcel's dimensions on the ground would be spatial information. However, other data such as the names of the occupants, the assessed value of the house, or the value of the land is not spatial information. A

non-spatial data data that is not directly linked to a spatial location (such as tabular data).

benefit of using geospatial technology is that this type of **non-spatial data** can be linked to a location. For instance, say you sell a lot of used books on eBay. You could make a map showing the destinations of each of the purchasers of your online sales—this would be spatial information. You could then link related non-spatial data (such as the name of the book that was sold or the price it was purchased for) to those locations, creating a basic database of sales.

Spatial information can also be gathered about other characteristics, such as the landscape, terrain, or land use. Remote sensing provides images of the

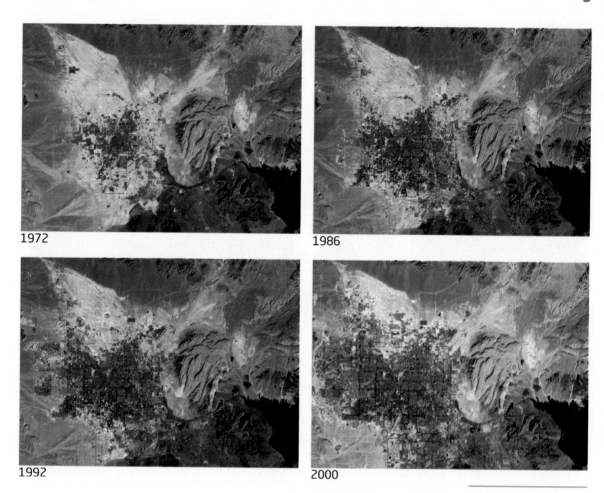

1972

1986

1992

2000

FIGURE 1.7 Satellite imagery (from 1972, 1986, 1992, and 2000) showing the growth of Las Vegas, Nevada over time. (Source: USGS)

ground that serve as a "snapshot" of a particular location at a specific time that can generate other forms of spatial information. For instance, a satellite image of Las Vegas, Nevada, can help denote the locations of new housing developments. Examining imagery from several dates can help track the location of where new houses are appearing (Figure 1.7). Similarly, geospatial technology can be used to determine where, spatially, new urban developments are most likely to occur, by analyzing different pressures of development on these areas.

Many types of decision making are reliant on useful geospatial information. For instance, when a new library is proposed to be built in an area, its choice of location is going to be important—the new site should maximize the number of people who can use it but minimize its impact on other nearby sites. The same type of spatial thinking applies to the placement of new schools, fire stations, or fast-food chain restaurants. Spatial information fuels other decisions that are location dependent. For instance, historic

Hands-on Application 1.3

Mapping Data Layers Online

Online mapping applications that combine several types of spatial data into a series of map layers are becoming increasingly common. A good, easy-to-use example of viewing spatial information online is MapWV, a utility for mapping various layers of data in West Virginia. Open your Web browser and go to **http://www.mapwv.gov** and then select the option for Create an Advanced Map. Numerous layers are available for you to create a map from—locations of cities and roads, tax district boundaries, historic places, shaded relief of the topography, and crisp aerial photographs. Select the layers you want to map in West Virginia. You can zoom to a major city (such as Morgantown), a county, or a particular place. Tools for zooming in and out of the map (a pair of magnifying glasses), panning around the map (the hand), getting further information about objects (the "i" in the circle), or resetting the view back to the entire state (the globe) are available on the toolbar in the middle of the screen. Examine various map layers for Morgantown, West Virginia, and see how they fit together in a spatial context (such as the hydrology data and the shaded relief or the aerial photography with the historic locations).

preservation efforts seek to defend historic Civil War battlefield areas from urban development. Knowing the location of the parcels of land that are under the greatest development pressure helps in channeling preservation efforts to have their greatest impact. With more and more geospatial technology applications becoming available, these types of geospatial problems are being taken into consideration (see *Hands-on Application 1.3: Mapping Data Layers Online* and *Hands-on Application 1.4: Examining Real Estate Values Online* for examples online).

Hands-on Application 1.4

Examining Real Estate Values Online

Zillow is an online tool used to examine housing prices in cities across the United States. You can examine the prices of houses for sale in the neighborhood and see—spatially—where housing prices change. In other words, Zillow provides online mapping related to real estate prices and appraisal. To check it out, open your Web browser and visit **http://www.zillow.com**. Start with the "Find Homes" search options, looking for houses in Virginia Beach, Virginia. Zoom in to some of the residential neighborhoods, and you'll see markers in yellow (recently sold homes) and red (homes for sale). Selecting a marker will give you the sale (or sold) price, along with the estimated value. When you're using this tool, you're selecting a location and examining the attributes (items like housing price or other characteristics) that accompany that place. Check your own local area for housing and examine local housing prices and their estimated values.

What Is Google Earth?

An example of a widespread geospatial technology application is the popular **Google Earth** software program. Google Earth is a **virtual globe**—a program that provides an interactive three-dimensional (3D) representation of the Earth. Google Earth brings together several aspects of geospatial technology into one easy-to-use program. You can examine satellite imagery or aerial photography of the entire globe, zoom from Earth orbit to a single street address, navigate through 3D cities or landscapes, measure global distances, and examine multiple layers of spatial data together, all with an intuitive interface. In many ways, the coolness and simplicity of use of Google Earth makes it feel like a video game of geospatial technology, but it's far more than that. Google Earth is able to handle vast amounts of geospatial data swiftly and easily, and enables users to specify and create their own location-based spatial data (see Figure 1.8 for a look at Google Earth and some of its layers).

Before Google Earth, there was Keyhole's "Earth Viewer," a program that enabled you to fly across a virtual Earth, zooming in and out of locations and imagery. Google purchased the Keyhole company and the first version of Google Earth was released in 2005. Google Earth is freely available for download off the Web and relies on a high-speed Internet connection to send the data from Google servers to your computer. For instance, when you use Google Earth to fly from space to your house, the imagery you see on the screen is being streamed across the Internet to you. The imagery on Google Earth is "canned," that is to say, it's fixed from one moment in time, rather than being a live view. For instance, if the imagery of your house used by Google was obtained in 2005, then that's the snapshot in time you'll be viewing, not a current image of your house. However, Google Earth has

Google Earth a freely available virtual globe program first released in 2005 by Google.

virtual globe a software program that provides an interactive three-dimensional map of the Earth.

FIGURE 1.8 A view of downtown Boston, Massachusetts from Google Earth. (Source: Gray Buildings © 2008 Sanborn, © 2010 Google)

Thinking Critically with Geospatial Technology 1.1

What Happens to Privacy In a Geospatial World?

Think about this type of scenario—if you own property, then that information is part of the public record with the county auditor's office and likely available online. Someone can type your address into an online GIS (or mapping Website such as MapQuest or Google Maps) and get a map showing how to travel to your property (house or business). In addition, the dimensions, boundaries, and a high-resolution image of your property can be easily obtained. It is possible that other information can be acquired, such as a close-up picture of your house taken from the street. All of this is free, publicly available information that can be had with just a few clicks of a mouse. Anonymity about where you live or work is suddenly becoming a thing of the past. What does the widespread availability (and access to) this type of spatial information mean about privacy? If anyone with an Internet connection can obtain a detailed image of the house you live in, has your personal privacy been invaded?

a feature so that you can examine past images. For instance, if imagery of your town was available in 2006, 2008, and today, you can choose which of these "canned" images you want to view. Many other data layers have been updated and are constantly being added, such as 3D buildings or other location-based application layers.

The *Geospatial Lab Application: Introduction to Geospatial Concepts and Google Earth* that accompanies this chapter walks you through several different usages of Google Earth. It assumes no prior experience with Google Earth, but even veteran users may find some new tricks or twists in how geospatial data is handled. You'll use a variety of Google Earth tools to investigate several of the avenues this book will explore, from aerial image interpretation, to 3D terrain and objects, to determining the shortest path between two destinations. Before you start using Google Earth, *Hands-on Application 1.5: The Google Earth Plug-In* will show you some of its very cool features and applications just using your Web browser.

Hands-on Application 1.5

The Google Earth Plug-In

The Google Earth Plug-in is a special add-on that will install into your Web browser and allow you to utilize some Google Earth features via the Web. Open your Web browser and go to **http://www. google.com/earth/explore/products/plugin.html** and follow the instructions for installing the Google Earth Plug-in for your Web browser. After installation, you'll see a preview window that will allow you to rotate, zoom, and tilt a virtual globe of the Earth.

The Google Earth Website also features several examples of how the plug-in is used for a variety of geospatial Web applications. For instance, the "Sports Stadiums" link allows you to zoom from space to a variety of 3D versions of stadiums across the United States, while the "Oil Spill" link allows you to interactively view the extent of the 2010 Deepwater Horizon oil spill in the Gulf of Mexico. Check out all of the above and also explore more examples of Google Earth to see its geospatial applications.

What's All This Have to do with Geography?

The first sentence for the entry "geography" in *The Dictionary of Human Geography* describes geography as "the study of the Earth's surface as the space within which the human population lives." To simplify a definition further into words of one syllable, geography is not just "where things are" but the study of "why things are where they are." Geography deals with concepts of the spatial characteristics of our planet and the spatial relationships and interactions of the people and features that populate it. The notion of "Geographic Information Science" has been acclaimed for years as the study of the multiple concepts that surround the handling of spatial data, and today the use of geospatial technologies is tightly interwoven with the discipline of geography.

At the higher education level, courses in GIS, remote sensing, GPS, and various applications of these technologies are key components of a Geography curriculum. Certificate programs in Geographic Information Science (GISci) or Geospatial Technologies are also becoming increasingly common to be offered by colleges and universities. Geospatial degree programs at the bachelors, masters, or doctoral level have been developed at numerous universities (see Chapter 15 for more information about geospatial technology in education). Even in the basic usages of collecting data in the field for a class project or producing a map of results, geospatial technologies have become an integral part of the geography discipline.

In a 2004 article, Jack Dangermond, CEO of the Environmental Systems Research Institute, Inc. (Esri; see Chapter 5), heralded GIS as a way of "Speaking the Language of Geography" and in a 2009 article noted that GIS was seen as "Geography in Action." Both of these are great descriptions—for example, when you use an in-vehicle navigation system to navigate your way through unfamiliar territory or use an online tool to examine housing values for your neighborhood, you're using basic geographic concepts of space and place through technology. Geospatial technology gives us the tools to more easily apply location-based principles to real-world situations. For example, how is the new freeway bypass going to affect the local community regarding traffic flows, new commercial development, and individual properties? Or what are the potential impacts of the site chosen for a new casino development in relation to the land use, economic and social effects, and the draw of gamers from far outside the local area? Throughout this book, you'll be dealing with geospatial technology concepts and applications, and using some sort of geospatial data, or measuring spatial characteristics, or examining spatial relations.

Chapter Wrapup

So what are you waiting for? Move on to this chapter's lab (see *Geospatial Lab Application: Introduction to Geospatial Concepts and Google Earth*) and start getting to work. This chapter's lab will have you doing all sorts of things with

Google Earth, while touching on aspects of all of this book's chapters. There are also questions throughout the lab for you to answer based on the tasks you'll be doing.

Important note: The references for this chapter are part of the online companion for this book and can be found at http://www.whfreeman.com/shellito1e.

Key Terms

geospatial technology (p. 2)

Geographic Information System (GIS) (p. 2)

remote sensing (p. 2)

satellite imagery (p. 2)

aerial photography (p. 2)

Global Positioning System (GPS) (p. 2)

geospatial data (p. 8)

non-spatial data (p. 8)

Google Earth (p. 11)

virtual globe (p. 11)

Introduction to Geospatial Concepts and Google Earth

This chapter's Geospatial Lab Application will introduce you to some basic concepts of geospatial technology through the use of the Google Earth software program. This lab also provides an introduction on how to use Google Earth and will help familiarize you with many of its features. Although the Geospatial Lab Applications in later chapters will use a variety of software packages, you'll also be using Google Earth in many of them. The investigations in this introductory lab may seem pretty basic, but labs in later chapters will be more in-depth and build on concepts learned here.

Objectives

The goals for you to take away from this lab are:

- Familiarizing yourself with the Google Earth (GE) environment and basic functionality and navigation using the software.
- Using different GE layers and features.

Obtaining Software

Google Earth (the current version, 6.0) is available for free download at http://earth.google.com.

Important note: Software and online resources sometimes change fast. This lab was designed with the most recently available version of the software at the time of writing. However, if the software or Websites have significantly changed between then and now, an updated version of this lab (using the newest versions) is available online at http://www.whfreeman.com/shellito1e.

Lab Data

There is no data to copy in this lab. All data comes as part of the GE data that is installed with the software or is streamed across the Internet when using GE.

Localizing This Lab

Although this lab visits popular locations in a tour around South Dakota, the techniques it uses can be easily adapted to any locale. Rather than using GE imagery and features of South Dakota, find nearby landmarks or popular spots and use GE to tour around your local area.

1.1 Starting to Use Google Earth

1. Start GE (the default install folder is called Google Earth). GE will usually open with a startup tip box, which you can close. Also, make sure the GE window is maximized, or else some features may not show up properly.

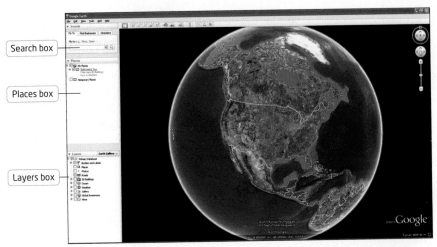

(Source: © 2011 Europa Technologies, US Dept of State Geographer, © East View Map Link, © Google)

A common use of GE is to view the ground (whether physical landscapes or structures) from above by using the aerial photography or satellite imagery streamed in from Google's servers. The Fly To option in the **Search Box** is used for this.

2. Select the **Fly To** tab and then type in **"Mitchell, SD."** Next, click on the **Begin Search** button (the magnifying glass next to where you type).

3. GE will rotate and zoom the view down to eastern South Dakota. One of the popular attractions in Mitchell is the Corn Palace, an event center that, each year, remakes the façade of the building into a new theme using corn. More information about the Corn Palace is available at http://www.cornpalace.com/index.php.

4. In the **Layers Box**, place a checkmark in the **Photos** box. This enables linking locations on the ground to photos that users have taken. The locations that have linked photos appear as blue squares over top of the imagery. You'll see numerous photo locations clustered together in one place—that would be the Corn Palace. Double-click on some of them to see some photos (taken from the ground) of the Corn Palace.

5. In GE, you can rotate and zoom your view to get a closer look at the imagery on the ground. Move the mouse to the upper right-hand side of the screen and a set of controls will appear. These fade out when not in use and reappear when the mouse rolls over them.

6. There are five controls. The first one (the ring with the N at the top) allows you to change the direction being faced. Grab the N with the mouse (by placing the pointer on N and holding down the left mouse button) and drag the N (that is, slowly rotate it) around the ring. (This is commonly called "grab and drag.") You'll see the view change (north will still be the direction in which the N is pointed). Double-clicking on the N will return the imagery so that the north direction is facing the top of the screen.

7. The second control is the **Look** control (the arrows surrounding the eye). By selecting one of the directional arrows and pressing it with the mouse, you can tilt your view around the terrain as if you were standing still and looking about. You can also grab the control and move it (by holding down the left button on your mouse), simulating looking around in multiple directions.

8. The third control is the **Move** control (the arrows surrounding the hand). By grabbing this control, you can pan the imagery around in whichever direction you choose.

9. The fourth control is the **Street View** icon (the yellow figure of a human—referred to as the "pegman"—standing on a green circle). This icon will appear when Google Street View imagery is available to see on the ground. We'll spend more time on Google Street View in Chapter 8, as it's a utility that allows you to see what places look like as if you were standing on the street in front of them. To use Street View, you would grab the icon from the controls with the mouse and drag it to a street that's visible in the view to enter the Street View mode. This control will only be visible if there are streets in the view that you can enter Street View mode with (which is why you won't initially see it when you start GE and are looking at the entire planet).

10. The last control is the **Zoom Slider**. By pushing forward (moving the bar toward the plus sign), you zoom closer, and by pulling backwards (moving the bar toward the minus sign), you zoom further out. By default, when GE zooms in, it tilts the terrain to give you a perspective view, rather than looking from the top directly down. To always use the top-down option rather than tilting the ground when zooming, select the **Tools** pull-down menu and choose **Options**. In the dialog box that appears, click on the **Navigation** tab, and then select the radio button that says **Do not automatically tilt when zooming**. To have GE tilt while zooming, select one of the other radio buttons.

11. You can also use the mouse for navigating around GE. Grab the imagery and drag it to pan across the map. To zoom in and out, roll the mouse

wheel back and forth. Pan and zoom over to the location of the Corn Palace. To confirm that you're in the right place, type the address of the Corn Palace into the Fly To box as follows: 604 North Main Street, Mitchell, SD 57301. (GE will automatically adjust to that location.)

12. For now, turn off the **Photos** options (remove the checkmark from the box).

13. Zoom in on the Corn Palace so that it fills most of the view and you can make out some details about it.

14. Grab the **Street View** icon from the controls and drag it to the street right in front of the Corn Palace. You'll see the street turn blue (this indicates that Street View imagery is available for this street). Release the icon on the street and the view will shift from an aerial view to imagery that looks like you're actually standing on the road in front of the Corn Palace.

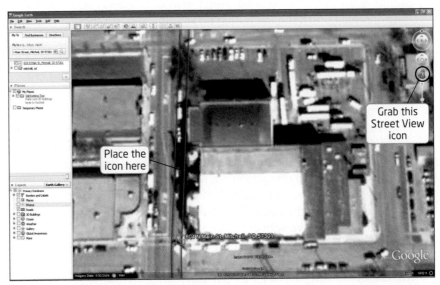

(Source: Image © 2011 Digital Globe, © 2011 Google)

15. In this Street View, use the mouse to pan around the image (you can see 360 degrees around your position) until you can see the entrance of the Corn Palace. You can pan and zoom with the mouse as necessary (and also click on the road—or elsewhere in the image—to advance your view).

16. Examine the Corn Palace and its surrounding area. Then answer Question 1.1. To return to the aerial view, click on the **Exit Street View** text in the upper right corner of the Street View. We'll do more with Street View (and street networks) in Chapter 8.

Question 1.1 By being able to view the Corn Palace from above and from standing on the ground in front of it, what details from the aerial view can help identify just what the building is? (You may need to switch back and forth between the aerial view and the street view to answer this question.)

17. With a number of remote sensing concepts, it's important to think of viewing the ground from above and that many features you can view from a street level are difficult to perceive from an aerial view. Chapter 9 delves further into interpreting aerial images and some strategies and tips for doing so, while Chapter 10 deals with how this type of remotely sensed imagery is acquired. Chapters 11 and 12 explore obtaining imagery from various satellites.

18. Here's another way of looking at the Corn Palace—a 3D representation. In the GE **Layers** options, select **3D Buildings**. A 3D representation of the Corn Palace will appear over top of the imagery of the building. Use the zoom and tilt controls and navigate around the building to get a better look at the 3D model of the Corn Palace, including its façade and turrets. Turn off the option for **3D Buildings** when you're done.

19. There are numerous buildings and objects that have 3D representations designed for use in GE. In Chapter 14, you'll investigate more about working with 3D versions of geospatial data, as well as constructing your own 3D buildings and viewing them in GE.

1.2 Finding Your Way Around with Google Earth

1. From the **Layers Box**, turn on the **Roads** layer. You'll see major roads (interstates, state highways) and their labels appear, as well as local roads when you're zoomed in close.

2. We're going to leave the Corn Palace behind and continue west through South Dakota to Wall Drug, a famous tourist location. More information on Wall Drug can be found at http://www.walldrug.com.

3. In the **Search Box**, click on the **Directions** tab.

4. In the **From** option, type in the Corn Palace's address as follows: 604 N Main St, Mitchell, SD 57301.

5. In the **To** option, type in Wall Drug's address as follows: 510 Main Street, Wall, SD 57790.

6. Click the **Begin Search** button. GE will zoom out to show you the path (in purple) it has calculated for driving distance between the Corn Palace and Wall Drug, while the Search Box will fill with directions featuring stops and turns along the way. At the bottom of the Search Box will be a checkbox marked "**Route**." By turning this on and off, you can view the route and the roads underneath it.

Question 1.2 By viewing the expanse of South Dakota between Mitchell and Wall, there are many possible roads between the two. Why do you suppose GE chose this particular path to go from the Corn Palace to Wall Drug?

Question 1.3 Based on the route that GE calculated, what is the driving distance (and approximate time equivalent) to get from the Corn Palace to Wall Drug?

7. This capability to take a bunch of letters and numbers and turn them into a mapped location, as well as being able to calculate the shortest driving distance between points is further developed in Chapter 8. Matching addresses and calculating shortest paths are key usages of geospatial technologies, and the Chapter 8 lab will go into more depth on how these types of operations are performed.

8. Also at the bottom of the **Search Box** is the **Play Tour** button. Click the **Play Tour** button. This will allow you to follow the path that GE has calculated for you.

9. A new set of controls will appear in the lower left corner of the view (if they don't, move your mouse over to that part of the view). These allow you to pause, fast-forward, rewind, and save the path that's been calculated.

10. Follow the path from Mitchell to Wall. It'll take a long time to watch the whole tour, so when you're ready, double-click on the final destination in the **Search Box** (that is, the "**Arrive at: 510 Main Street**" option) and you'll jump ahead to the end of the path—Wall Drug.

1.3 Using Google Earth to Examine Landscapes

1. There's a lot more to the South Dakota landscape than tourist attractions. Wall Drug is located in the town of Wall, named for the line of magnificent land formations nearby, and also lies directly adjacent to Badlands National Park. The Badlands are located to the south and southwest of Wall. For more information about the Badlands visit http://www.nps.gov/badl.

2. To see the boundaries of Badlands National Park in Google Earth, go to the **Layers Box** and expand the option for **More** (click on the plus button to the left of the name). In the options that appear under the expanded heading, put a checkmark in the **Parks/Recreation Areas** option. Zoom out from the boundaries of Wall and (to the south) you'll see the northernmost boundary of Badlands National Park highlighted in green.

3. Zoom out so that you can see the entire park in the view. New icons for the locations of Visitors Centers and other features will also appear.

4. Pan over to the eastern edge of the park and you will see a large Question Mark icon indicating a park entrance as well as a green arrow indicating an overlook.

(Source: Image USDA Farm Service Agency, Image © 2011 Digital Globe, © 2011 Google)

5. Zoom in to the point representing the overlook. At the bottom of the view you'll see numbers representing the latitude and longitude of the point, as well as the real-world elevation of that spot.

Question 1.4 What is the elevation for this particular overlook in Badlands National Park?

6. The imagery in GE is placed over top of a model of the Earth's terrain and landscape (hills, valleys, ridges, and so on). It's very difficult to make out the pseudo three-dimensional (3D) nature of the terrain model from above, so use the **Zoom Slider** to tilt the view down so that you can look around as if you were standing on the overlook point (in a perspective view of the planet). Once you've tilted all the way down, use the **Look** controls to examine the landscape. From here,

use the **Move** controls to "fly" over the Badlands from this overlook point. When you're down done cruising around the Badlands, answer Question 1.5.

Question 1.5   How does the terrain modeling (with the tilt function) aid in the visualization of the Badlands?

7.   This ability to model the peaks and valleys of the landscape with aerial imagery "draped" or "stretched" over the terrain for a more realistic appearance is often used with many aspects of geospatial technology. Chapter 13 greatly expands on this by getting into how this is actually performed as well as doing some hands-on terrain analysis.

1.4 Using Google Earth to Save Imagery

It's time to continue on the next leg of our South Dakota journey by heading to Mount Rushmore. Carved out of the side of a mountain in the Black Hills, Mount Rushmore is a monument featuring the faces of presidents George Washington, Thomas Jefferson, Theodore Roosevelt, and Abraham Lincoln. For more information about Mount Rushmore visit **http://www.nps.gov/ moru**.

1.   In GE's Fly To box, type **"Mount Rushmore."** GE will zoom around to an overhead view of the monument. Like the Badlands, Mount Rushmore is a national park—zoom out a little bit until you can see the extent of the park's boundaries (still outlined in green).

2.   To begin, you can save an image of what's being shown in the view. GE will take a "snapshot" and save it as a JPEG graphic file (.jpg), which is like a digital camera picture. Position the view to see the full outlined extent of Mount Rushmore, select the **File** pull-down menu, then select **Save**, and finally select **Save Image**. Name the image **"moru"** (GE will automatically add the .jpg file extension) and save it to your computer's hard drive.

3.   Minimize GE for now and go to the location on your computer where you saved the image, and then open it to examine it (for instance, with Microsoft Office Picture Manager).

Question 1.6 Note that even though the graphic contains the locations of Mount Rushmore, the outline of the park, and latitude/longitude and elevation information at the bottom, it doesn't have any spatial reference for measurements. Why is this?

4.   Close the image and return to Google Earth. Even though the saved image doesn't have any spatial reference, Chapter 3 describes how to take unreferenced imagery (or other unreferenced data) and transform it to match referenced data. In Chapter 3's lab, you do this in a hands-on fashion. You can also turn off the **Parks/Recreation** layer for now.

1.5 Using Google Earth to Make Placemarks and Measurements

While you're examining Mount Rushmore imagery, you can set up some points of reference to return to. GE allows you to create points of reference as Placemarks. There are three points to mark on the map—the top of the mountain, the amphitheater, and the parking area (see the graphic below for their locations).

(Source: © 2010 Google, Image © 2010 Digital Globe)

1. From the GE toolbar, select the **Add Placemark** button:

2. A yellow pushpin (labeled "Untitled Placemark") will appear on the screen. Using your mouse, click on the **pushpin** and drag it to the rear of the amphitheater so that the pin of the placemark is where the path meets the amphitheater.

3. In the Google Earth New Placemark dialog box, type the name "**Mount Rushmore Amphitheater.**"

4. By pressing the Placemark icon button next to where you typed the name, you can select a different icon rather than the yellow pushpin. Choose something more distinctive.

5. Click **OK** to close the dialog box.

6. Repeat the process by putting a new placemark at the top of the mountain where the monument is. Name this new placemark "**Top of the Monument.**"

7. Zoom GE's view in tightly, so you can prominently see the two placemarks in the view (with the two placemarks in diagonally opposite corners of the view).

8. Next, select the **Show Ruler** button from the GE toolbar:

9. In the Ruler dialog box that appears, select **Feet** from the **Length** pull-down menu.

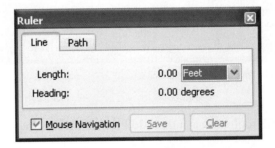

10. Use the options under the **Line** tab. This will let you compute the distance between two points. The Path option will let you find the total accumulated distance between more than two points.

11. Click the mouse on the point of the placemark in the amphitheater and drag it to the placemark on top of the mountain (your heading should be about 310 degrees or so). Click the mouse at the top of the mountain placemark.

12. Answer Question 1.7. When you're done, click on **Clear** in the **Ruler** dialog box to remove the drawn line from the screen, and then close the **Ruler** dialog box.

Question 1.7 What is the measured distance between the rear of the amphitheater and the top of the monument (keep in mind this is the ground distance, not a straight line between the two points)?

13. This ability to create points of reference (as well as lines and area shapes) and then compute distance between them is a very basic tool of GIS and can be used as the basis for multiple types of analysis. Chapter 5 introduces GIS concepts, Chapter 6 expands on how spatial analysis can be performed with GIS, and Chapter 7 demonstrates how to take your data and create a professional quality map of the results.

14. Use the tilt functions of GE to get a perspective view on Mount Rushmore (as you did with the Badlands) and take a look at the mountain from a pseudo-3D view. While you can see the height and dimensions of the mountain, the famous four presidents' faces can't be seen, despite zooming or rotating the model.

Question 1.8 Even with the terrain turned on and the view switching over to a perspective, why can't the presidents' faces on the side of the monument be seen?

1.6 Using Google Earth to Examine Coordinates

1. You'll notice at the bottom of the GE screen a set of coordinates for lat (Latitude) and lon (Longitude) next to the elevation value. Move the mouse around the screen and you'll see the coordinates change to reflect the latitude and longitude of whatever the mouse's current location is.

2. Zoom in closely to the road that enters the parking structure area of Mount Rushmore.

Question 1.9 What are the latitude and longitude coordinates of the entrance to the Mount Rushmore parking area?

3. You can also reference specific locations on the Earth's surface by their coordinates instead of by name. In the **Fly To** box type the following coordinates: **43.836584, -103.623403**. GE will rotate and zoom to this new location. Turn on the **Photos** to obtain more information on what you've just flown to. You can also turn on the **3D Buildings** layer to look at the location in a pseudo-3D view. Answer Questions 1.10 and 1.11, and then turn off the **Photos** (and the **3D Buildings** option) when you're done.

Question 1.10 What is located at the following geographic coordinates: latitude 43.836584, longitude -103.623403?

Question 1.11 What is located at the following geographic coordinates: latitude 43.845709, longitude -103.563499?

Chapter 2 deals with numerous concepts related to coordinate systems, reference systems, and projections of a three-dimensional Earth onto a two-dimensional surface.

1.7 Other Features of Google Earth

As noted before, you will be using GE for several other labs in this book and exploring some of these applications of geospatial technology. GE has a multitude of other features beyond those covered in this first lab. For instance, if you had a GPS receiver (say, one from Garmin or Magellan) you could import the data you've collected directly into GE.

1.   From the **Tools** pull-down menu, select **GPS**. You'll see several options for offloading information you collected in the field (such as waypoints) from the receiver into GE.

Chapter 4 will introduce you to several concepts related to GPS, as well as collecting data in the field and some pre-field work-planning options.

2.   Close the **GPS Import** dialog box now (unless you have a receiver you want to plug into the computer on which you can view your data in GE).

There are many other GE layers and features to explore, especially in relation to how geospatial data is visualized. To look at one of these features, return to the Badlands once more and do the following:

3.   In the **Layers** box, expand the options for **Gallery**, and then place a checkmark in the box next to **360 Cities**.

4.   Fly back to the Badlands and scroll to the western edge of the park and you'll see another symbol on the map—a red circle marked "360." Hover your cursor over it and the text should read "Mako Sica aka The Badlands."

5.   Double-click on the "360" circle. You will zoom down to see a sphere on the surface, and then the view should dive inside the sphere. The view will change to a new view of the Badlands (use the mouse to move about and look around a full 360 degrees). In essence, this tool simulates what the Badlands would look like from that position, all around you. You can look in any direction, as well as zooming in or out of places. Answer Question 1.12. When you're done looking around, click on the **"Exit Photo"** button in the view to return to regular GE. Chapter 15 will present more information about some other cool visualization techniques for geospatial data.

Question 1.12 How does the 360-degree image of the Badlands aid in visualizing the scenery (keep in mind this image has been tied to a particular location)?

Geospatial technology is such a rapidly changing field that new advances come quickly. Chapter 15 also explores some of these current developments on the frontiers of geospatial technology.

Google Earth is always changing, and new versions or upgrades are constantly being made available by Google for this program. Some of the newer options you may want to explore are Google Mars, Google Sky, and Flight Simulator.

Google Mars

From the **View** pull-down menu, select **Explore**, and then select **Mars**. The view will shift to the familiar-looking globe, but this time covered with imagery (from NASA and the USGS) from the surface of Mars. The usual controls work the same way as they do with GE and there's a lot of Martian territory to be explored. When you're ready to return to GE, from the **View** pull-down menu, select **Explore**, and then select **Earth**.

Google Sky

From the **View** pull-down menu, select **Explore**, and then select **Sky**. GE's view will change from instead of looking down on Earth, you're looking up to the stars—and instead of remotely sensed aerial or satellite imagery, you're looking at space-telescope (including the Hubble) imagery. There's plenty of imagery to be explored in this new mode. When you want to get back "down to Earth" again, select the **View** pull-down menu, select **Explore**, and then select **Earth**.

Flight Simulator

From the **Tools** pull-down menu, select **Enter Flight Simulator**. A new dialog box will appear asking if you want to pilot an F-16 fighter jet or an SR-22 propeller airplane. Select whichever plane you want to fly and GE will switch to the view as seen out the cockpit of your chosen plane and you can fly across South Dakota (and the rest of GE). Try not to crash (although you can easily reset if you do).

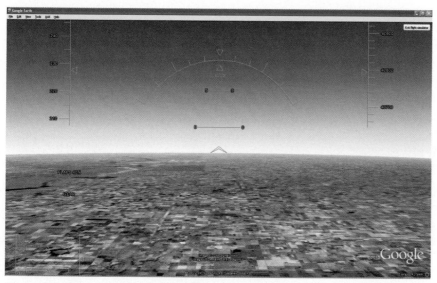

(Source: Image © 2011 Digital Globe, Image USDA Farm Service Agency)

Closing Time

This introductory lab was pretty straightforward but served to introduce you to the basic functions of Google Earth, as well as providing examples of many geospatial concepts you'll be working with over the course of this book. Chapter 2 starts looking at location concepts (like various coordinate systems) and returns to Google Earth to start applying some of the topics you've worked with in this chapter. You're all set, and now you can exit GE by selecting **Exit** from the **File** pull-down menu. There's no need to save any data (or Temporary Places) in this lab.

2

Where in the Geospatial World Are You?

Locations in a Digital World, Position Measurements, Datums, Coordinate Systems, GCS, Map Projections, UTM, and SPCS

Think about this—you're at the mall and your cell phone rings. It's a friend calling you with a really basic question: "Where are you?" For such a simple question, there's really a wide variety of answers. You could say "I'm at the mall," which gives the other person some basic location information, but only providing they know where the mall is. If they're calling from out of town and unfamiliar with the location of the mall, what you've told them is useless for them to know where you are. If you tell them "I'm in Columbus," then they have another type of information, but Columbus is a big city, and they still don't know where you are in the city. You could give them the name of the cross streets by the mall, the address of the mall, or the name of the store you're in at the mall, and even though all of the above contain different levels of information, your friend still won't have the information needed to know precisely where you are.

For instance, the answer of "I'm at the mall" effectively conveys your location only if the person you're talking to has some sort of knowledge of the mall's location. If she has a reference like "the mall's on the north side of town" or "the mall is just off exit 5 on the freeway," then she has some idea of your location. For the location information to make sense, she needs to have something (perhaps a map of the city you're in) to reference the data to. If you need specific information about your exact location, you'd have to be much more specific than saying you're at the mall. For example, in the case of a medical emergency, Emergency Medical Technicians (EMTs) would need to know your exact location (they couldn't afford to waste time searching the entire mall to find you). In this particular case, you would need to provide precise location information so that the EMTs would be able to find you, no matter where you are.

Geospatial technology works the same way. First, since it deals with geospatial data (that is linked to non-spatial data), there has to be a way of assigning location information to the data being handled. Every location on Earth has to be referenced in some way. In other words, every location has to be identified and measured. Second, these measurements require some sort of reference system so that locations at one point on Earth can be compared with locations at another area. Third, since we're dealing with points on a three-dimensional (3D) Earth but we only have a two-dimensional (2D) surface to examine (like a map), there has to be a way of translating real-world data to an environment we can use. This chapter examines how these concepts are treated in all forms of geospatial technology. If you're going to be able to precisely pin down the coordinates of a location, you'll need to first have a datum and a coordinate system.

What Is a Datum?

datum a reference surface of Earth.

ellipsoid a model of the rounded shape of Earth.

geoid a model of Earth using mean sea level as a base.

geodesy the science of measuring Earth's shape.

A **datum** is a reference surface, or model of Earth that is used for plotting locations across the globe. The datum represents the size and shape of Earth, which, contrary to popular belief, isn't perfectly round. Earth is actually an **ellipsoid** (or spheroid) shape, meaning that it's larger at its center than it is at its poles. Think of taking a basketball and squeezing the top and bottom of the ball to slightly compress it and you'll have a basic spheroid shape. However, the real form of Earth isn't a smooth ellipsoid—gravitational forces affect different parts of Earth in different ways, causing some areas to not be in synch with the ellipsoid. Another model of Earth, called a **geoid**, places Earth's surface at mean sea level to try and account for these differences (see Figure 2.1 for how the geoid and the ellipsoid stack up against each other as well as the variable topography of Earth). The science of measuring Earth's shape to develop these kinds of models and reference surfaces is referred to as **geodesy**.

Developing a datum means creating a mathematical model to reference locations and coordinates against. Thus, when you're plotting a point, you have a reference to measure this location to. The difficulty with datums in

FIGURE 2.1 How the real-world Earth, the ellipsoid, and the geoid match up with each other.

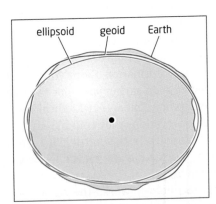

mapping arises because there isn't just one datum to use for all measurements of Earth's locations. In actuality, there are hundreds of datums being used. Some datums are global and measure the entire world, while some are more localized to a particular continent or region. Some of the more common datums you'll likely encounter with geospatial data are:

- ◉ **NAD27**: The North American Datum of 1927. This datum was developed for measurements of the United States and North America. It has its center point positioned at Meades Ranch in Kansas.

- ◉ **NAD83**: The North American Datum of 1983. This datum was developed by the National Geodetic Survey and Canadian agencies and is used as the datum for much data for the United States and the North American continent as a whole.

- ◉ **WGS84**: The World Geodetic System of 1984. This datum was developed by the U.S. Department of Defense and is used by the Global Positioning System (GPS) for locating points worldwide on Earth's surface (see Chapter 4 for more information about GPS).

Measurements made with one datum don't necessarily line up with measurements made from a second datum. Problems would arise if you take one set of data measured in one datum and your friend takes a second set of data measured from a different datum and you place both sets of data together. You have data of the same place, but your datasets wouldn't match up because you're both using different forms of reference for measurement. For instance, data measured from the NAD27 datum could be off by a couple hundred meters from data measured from the NAD83 datum. This problem is compounded with the availability of several local datums, or reference surfaces, designed to better fit a smaller geographic area. With the measurement differences that may arise, the best strategy to use when dealing with geospatial data is to keep everything in the same datum. A **datum transformation** is required to alter the measurements from one datum to another (for instance, changing the measurements made in NAD27 to NAD83). Datum transformation is a computational process, but a standard one in many geospatial software packages. A datum can be used to set up a geographic coordinate system (GCS) for measuring coordinates.

What Is a Geographic Coordinate System?

A **geographic coordinate system (GCS)** is a global reference system for determining the exact position of a point on Earth (see Figure 2.2 on page 32 for an example of GCS). GCS consists of lines of latitude and longitude that cover the entire planet and are used to find a position. Lines of **latitude** (also referred to as parallels) run in an east-to-west direction around the globe. The **Equator** serves as the starting point of zero, with measurements north of the Equator being numbers read as north latitude and measurements south of

NAD27 the North American Datum of 1927.

NAD83 the North American Datum of 1983.

WGS84 the World Geodetic System of 1984 datum (used with the Global Positioning System).

datum transformation changing measurements from one datum to measurements in another datum.

geographic coordinate system (GCS) a set of global latitude and longitude measurements used as a reference system for finding locations.

latitude imaginary lines on a globe north and south of the Equator that serve as a basis of measurement in GCS.

Equator the line of latitude that runs around the center of Earth and serves as the 0 degree line to make latitude measurements from.

FIGURE 2.2 The
geographic coordinate
system (GCS) covers the
globe and locations are
measured using latitude
and longitude. (Source:
© 2010 Google. Data SIO,
NOAA, U.S. Navy, NGA, GEBCO.
Image IBCAO. Image © 2010
Digital Globe. Image © 2010
TerraMetrics.)

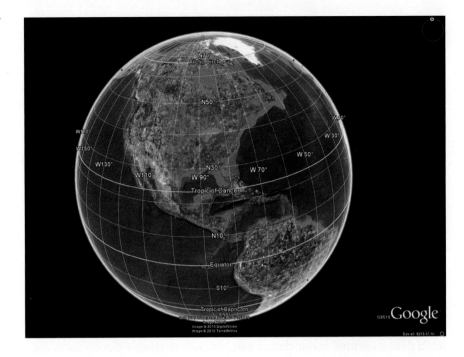

longitude imaginary
lines on a globe east
and west of the
Prime Meridian that
serve as a basis of
measurement in GCS.

Prime Meridian the
line of longitude
that runs through
Greenwich, England, and
serves as the 0 degree
line of longitude to base
measurements from.

**International Date
Line** a line of longitude
that follows a similar
path as the 180th
meridian (but changes
away from a straight
line to accommodate
geography).

**degrees, minutes, and
seconds (DMS)** the
measurement system
used in GCS.

the Equator being numbers read as south latitude. Lines of **longitude** (also called meridians) run north-to-south from the North Pole to the South Pole. The **Prime Meridian** (the line of longitude that runs through Greenwich, England) serves as the starting point of zero. Measurements made east of the Prime Meridian are numbers read as east longitude, while measurements made west of the Prime Meridian are numbers read as west longitude. The breaking point between east and west longitude is the 180th meridian, also referred to as the **International Date Line.**

Using this system (where measurements are made north or south of the Equator and east or west of the Prime Meridian), any point on the globe can be located. GCS measurements are not made in feet or meters or miles, but rather they are in **degrees, minutes, and seconds (DMS)**. A degree is a unit of measurement that can be broken into 60 subsections (each one referred to as a minute of distance). A minute can be subsequently broken down into 60 subsections (each one referred to as a second of distance). A single minute of latitude is equivalent to roughly one nautical mile (or about 1.15 miles). However, the distance between degrees of longitude varies across the globe as lines of longitude are closer together at the poles and further apart at the Equator.

When GCS coordinates are being written for a location, they are in terms of how far the distance (that is, in terms of degrees, minutes, and seconds) is from the Equator and the Prime Meridian. Latitude values run from 0 to 90 degrees north and south of the Equator, while longitude values run from 0 to 180 east and west of the Prime Meridian. For example, the GCS coordinates of a spot at the Mount Rushmore National Monument are 43 degrees, 52 minutes, 53.38 seconds north of the Equator (43°52′53.38″ N latitude) and 103 degrees, 27 minutes, 56.46 seconds west of the Prime Meridian (103°27'56.46" W longitude).

Negative values can also be used when expressing coordinates using latitude and longitude. When distance measurements are made west of the Prime Meridian, they are often listed with a negative sign to note the direction. For instance, the location at Mount Rushmore, being west longitude, could also be written as −103°27'56.46" longitude. The same use of negative values holds true when making measurements of south latitude (below the Equator).

GCS coordinates can also be expressed in their decimal equivalent, referred to as **decimal degrees (DD)**. In the decimal degrees method, values for minutes and seconds are converted to their fractional number of degrees. For instance, 1 degree and 30 minutes of latitude (in DMS) would equate to 1.5 degrees in DD (since 30 minutes is equal to 0.5 degrees). Using the Mount Rushmore coordinates as an example, the latitude measurement in decimal degrees would be 43.881494 north latitude (since 52 minutes and 53.38 seconds is equal to 0.881494 of one degree) and 103.465683 west longitude.

Using this system, any point on Earth's surface can be precisely measured and identified. Keep in mind that these locations are being made on a 3D globe (in reference to a datum), so measurements between points have to take this into account. When making measurements on a sphere, the shortest distance between two points is referred to as the **great circle distance**. Thus, if you wanted to find the shortest distance between the cities of New York and Paris, you would have to calculate the great circle distance between two sets of coordinates (see *Hands-on Application 2.1: Great Circle Distance Calculations* for more information about using the great circle distance).

The latitude and longitude system of GCS also serves as the basis for breaking Earth up into multiple **time zones** to account for difference in time around the world. Time zones set up the concept of a 24-hour day, with each zone having a different time as it relates to the current time in Greenwich, England (also called Greenwich Mean Time or GMT). The Earth spans 360 degrees of longitude, and dividing that by 24 hours equals 15 degrees of longitude for each hour. Thus, every 15 degrees away from the time zone containing the Prime Meridian marks the start of a new time zone (Figure 2.3 on pages 34–35).

decimal degrees (DD) the fractional decimal equivalent to coordinates found using degrees, minutes, and seconds.

great circle distance the shortest distance between two points on a spherical surface.

time zones a method of measuring time around the world, found by dividing the world into subdivisions and relating the time in that division to the time in Greenwich, England.

Hands-on Application 2.1

Great Circle Distance Calculations

To find the real-world distance around the planet between two cities—New York and Paris—you'll need the great circle distance between the two. To quickly calculate great circle distances, use the Surface Distance application online at **http://www.chemical-ecology.net/java/lat-long.htm**. Use the option for Great Circle Distance Between Cities to find the real-world distance between New York and Paris. Use it to find the city closest to you and calculate the great circle distance between sets of cities. Also, use a program like Google Earth to obtain the GCS coordinates (latitude and longitude) for a nearby location and find the surface distance from there to a nearby city.

FIGURE 2.3 The world time zones, as separated by lines of longitude (and geographic boundaries).

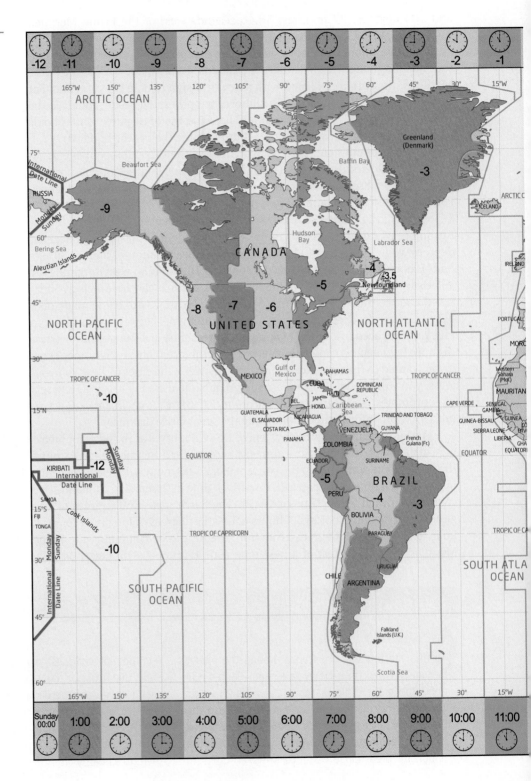

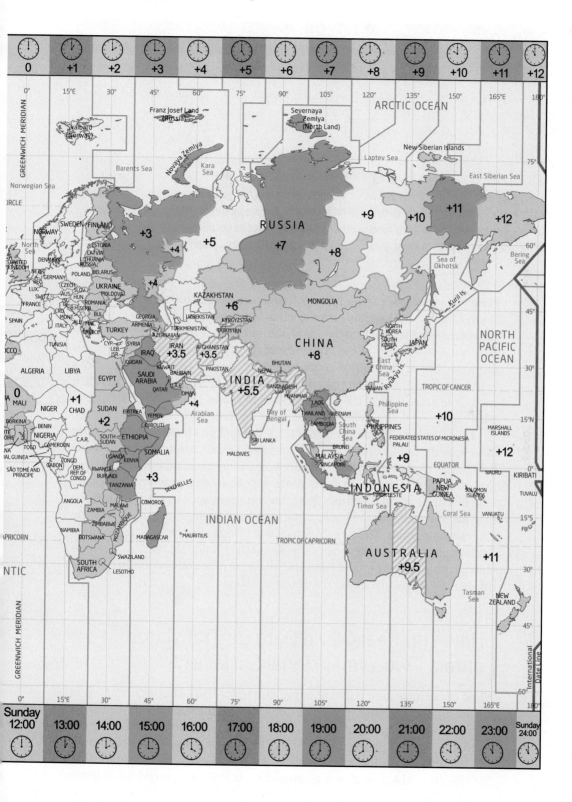

Thinking Critically with Geospatial Technology 2.1

Do You Really Need Printed Maps in a Digital World?

Let's face it, these days we're living in a digital world—road maps, road directions, and property maps are all available with a few clicks of a mouse in digital form. With everything available digitally (and more portable and accessible with wireless Internet access via laptops and phones), is there a need for printed maps? Do you really need to carry around a paper map version of the road network in Los Angeles, California, or will a mapping app on your smartphone suffice to find where your location is and get you

where you want to go? In the same vein, are paper map collections at universities or libraries relics of the past or still useful resources? When would having access to such a collection be necessary?

When driving a long distance is it necessary to have a printed map of the area you're traveling through when you've got GPS, satellite maps, and Internet mapping capabilities? If yes, what's a situation in which it would be necessary, and if no, why are printed maps still available?

From 7.5 degrees west longitude to 7.5 degrees east longitude is one time zone, 7.5 to 22.5 degrees longitude is a second time zone, and so on. Following these sections of longitude around the world, the east coast of the United States is five time zones to the west of the Prime Meridian, thus making the time in places like New York or Boston five hours behind the time in Greenwich. America's west coast lies in the region eight time zones to the west of the Prime Meridian (and three to the west of the United States east coast), thus making the time in San Francisco eight hours behind Greenwich (or three hours behind New York).

However, take a closer look at Figure 2.3 and you'll see that time zones don't exactly follow the lines of longitude. Some political or geographic boundaries cause some areas to entirely be in one time zone when the area may straddle two zones, or some areas may be in one zone or another despite how the lines of longitude lie. Other countries, such as China, adopt one time measurement for the entire country, despite China's geography crossing multiple time zones. The International Date Line marks the division between 24-hour periods and splits one day from another, but it also bends to accommodate some geography and islands to be in one time zone or another, rather than strictly follow the line of the 180th meridian.

How Can Real-World Data Be Translated onto a Two-Dimensional Surface?

map projection the translation of locations on a three-dimensional (3D) Earth to a two-dimensional (2D) surface.

All of these measurements are being done on a three-dimensional world but are somehow being translated onto a two-dimensional surface (like a piece of paper or a computer screen) for ease of use. A **map projection** is a translation of coordinates and locations from their places in the real world to a flat surface, and it is necessary to be able to use this geospatial data. The biggest problem with a map projection is that not everything is going

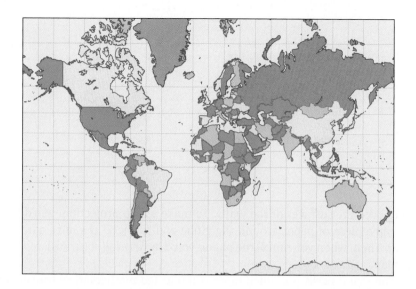

FIGURE 2.4 The Mercator projection, which retains the shapes of countries and continents, but greatly warps their sizes.

to perfectly translate from a three-dimensional object to a two-dimensional surface. In fact, the only representation of the world that would accurately capture all features would be a globe. Since it's hard to keep a globe in the glove compartment of the car, we're going to have to make due with a flat surface representation of the world (like a folded map) and realize that the map is going to have some sort of distortions built into it, simply because it's a two-dimensional version of a three-dimensional construct. One or more of the following will be distorted in a map projection: the shape of objects, the size of objects, the distance between objects, or the direction that objects are in. Some map projections may retain the shape of continents but warp their size. Other projections may keep the land area correct but throw the shapes out of place.

An example of this is the Mercator projection (Figure 2.4), a common type of map that was developed in 1569 for use as a navigation aid—it was designed so that every straight line on the map was a line of constant direction. The Mercator projection encompasses the whole globe and while it keeps the shapes of areas intact (that is, the continents look the way they should), the sizes (that is, the areas of land masses) are completely distorted, particularly the further you go from the Equator. For example, take a look at Greenland and Africa—they look to be about the same size on the map, even though Africa is really more than 13 times larger in area than Greenland. Similar problems can be seen with Antarctica or Russia which (while they have the right shape) look overwhelming in land area compared to their actual size.

Think of a projection like this—you have a see-through globe of Earth with the various landforms and lines of latitude and longitude painted on it. If you could place a light bulb inside the globe and turn it on, you'd see the shadows of the latitude, longitude, and continents projected onto the walls of the room you're in. In essence, you've "projected" the three-dimensional

FIGURE 2.5 The three
types of developable
surfaces in map
projections.

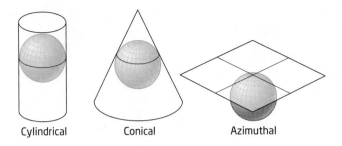

Cylindrical   Conical   Azimuthal

world onto the surfaces of the walls. To make a map projection, you'd have to wrap a piece of paper around the globe and "project" the lines onto that paper. Then, when you unroll the paper, you'd have the globe translated onto the map.

The real process is a much more complicated mathematical translation of coordinates, but the basic concept is similar. A map projection is created through the use of a developable surface, or a flat area onto which the coordinates from Earth can be placed. In this way, the locations on Earth's surface get translated onto the flat surface. The three main developable surfaces (**Figure 2.5**) that are used in this fashion are a cylinder (which creates a cylindrical projection), a cone (which creates a conical projection), and a flat plane (which creates an azimuthal surface). The place where Earth "touches" the surface is called the point of tangency, which is where the map's distortion is minimized.

Like datums, there are numerous map projections available for use, each having its own different uses and built-in distortions (see *Hands-on Application 2.2: Examining the Effects of Different Map Projections* for more information). Two of the more common projections that you may encounter with geospatial data are Lambert Conformal Conic and Transverse Mercator.

Hands-on Application 2.2

Examining the Effects of Different Map Projections

Different kinds of map projections will alter the map in various ways, depending on which parts are being distorted and which parts are being maintained. To interactively examine the effects of different projections on a map, open your Web browser and go to **http://projections.mgis.psu.edu**. The Website enables you to select different projections and view how the map looks with each of them. Switch the projection to view the world map as Mercator, Lambert Conformal Conic, and Transverse Mercator, and then select the option to redraw the map. How does each of these map projections alter the map as drawn? Try some of the other options and examine what types of distortions and changes enter the map with each projection.

Lambert Conformal Conic

This is a conic projection, in which a cone intersects Earth at two parallels. The choice of these tangency lines is important as the distortion on the map is minimized closer to them and maximized the farther you go from them. The Lambert Conformal Conic projection is commonly used for United States geospatial data (and other east-west trending areas).

Transverse Mercator

This is a cylindrical projection, with the tangency being the intersection between the cylinder and Earth. Measurements are most true at that point and distortions get worse the further east or west the areas displayed on the map are. It's considered "transverse" because the cylinder is wrapped around the poles, rather than the Equator. Transverse Mercator is used for more north-south trending areas.

GCS coordinates can find locations across the entire globe, down to precise measurements of degrees, minutes, and seconds. However, latitude and longitude is not the only coordinate system used today—there are several coordinate systems that use a map projection to translate locations on Earth's surface to a Cartesian grid system using x and y measurements. Two of the common grid systems used with geospatial data are UTM and SPCS.

What Is UTM?

The **Universal Transverse Mercator (UTM)** grid system works by dividing the world into a series of zones, then determining the x and y coordinates for a location in that zone. The UTM grid is set up by translating real-world locations to where their corresponding places would be on a two-dimensional surface (using the Transverse Mercator projection). However, UTM only covers Earth from between 84° N latitude and 80° S latitude (and thus couldn't be used for mapping the polar regions—a different system, the Universal Polar Stereographic Grid System, is used to find coordinates on the poles).

The first step in determining UTM coordinates is to find which **UTM zone** the location is in. UTM divides the world into 60 zones, with each zone being 6° of longitude wide (**Figure 2.6** on page 40), and each zone is set up as its own cylindrical projection. These zones are numbered 1 through 60 beginning at the 180th meridian and moving east. For instance, Mount Rushmore, outside of Keystone, South Dakota, is located in Zone 13. Beijing, China, is located in Zone 50. Sydney, Australia is in Zone 56. So, the first step in using UTM is to determine which UTM zone the point you want to locate is in.

The next step is to determine the x and y grid coordinates of the location. Unlike GCS, which uses degrees, minutes, and seconds, UTM utilizes

Universal Transverse Mercator (UTM) the grid system of locating coordinates across the globe.

UTM zone one of the 60 divisions of the world set up by the UTM system, with each zone being 6 degrees of longitude wide.

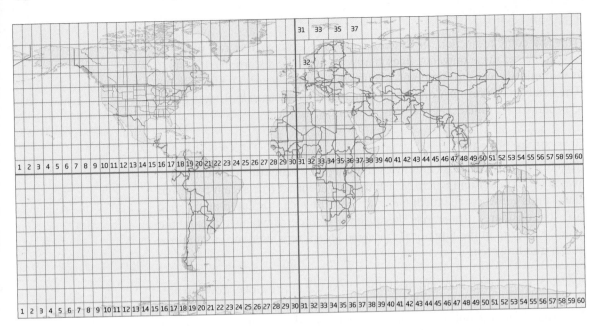

FIGURE 2.6 The 60
zones that comprise the
UTM system.

meters as its measurement unit. UTM's x and y coordinates are referred to as an **easting** and a **northing**. The UTM northing is found by counting the number of meters the point is north of the Equator. For instance, a particular location at Mount Rushmore is located 4,859,580 meters to the north of the Equator.

Locations in the southern hemisphere handle northing a little differently since UTM does not contain negative values. For southern hemisphere locations, count the number of meters south of the Equator to the point (as a negative value), then add 10,000,000 m to the number. For example, a location at the Opera House in Sydney, Australia (in the southern hemisphere), has a northing of 6,252,268 m. This indicates it would be approximately 3,747,732 meters south of the Equator ($-3,747,732 + 10,000,000$). Another way of conceptualizing this is to assume that (for southern hemisphere locations) rather than measuring north of the Equator, you're measuring north of an imaginary line drawn 10,000,000 meters south of the Equator. This extra 10 million meters is referred to as a **false northing** since it's a value set up by the system's designers to avoid negative values.

The next step is to measure the point's easting. Unlike GCS, there is no one Prime Meridian to make measurements from. Instead, each UTM Zone has its own central meridian which runs through the zone's center. This central meridian is given a value of 500,000 m, and measurements are made to the east (adding meters) or west (subtracting meters) from this value. For example, our specific spot at Mount Rushmore is located 123,733.22 m to the east of Zone 13's central meridian—so the easting for this location at Mount Rushmore is 623,733.22 m (500,000 + 123,733.22). A location in the town

easting a measurement of so many units east (or west) of some principal meridian.

northing a measurement of so many units north (or south) of a baseline.

false northing a measurement made north (or south) of an imaginary line such as used in measuring UTM northings in the southern hemisphere.

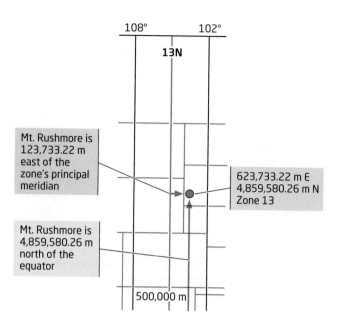

FIGURE 2.7 A simplified representation of determining the UTM Coordinates (easting, northing, and zone) of a location at Mount Rushmore.

of Buffalo, Wyoming, is also in Zone 13, and has an easting value of 365,568 m. This indicates this location is approximately 134,432 m to the west of Zone 13's central meridian—(500,000 – 134,432). No zone is more than 1,000,000 meters wide, so no negative values will be assigned to eastings in a zone. This value for easting is referred to as a **false easting** since the easting calculation is based on a central meridian for each zone.

When UTM coordinates are written, the false easting, northing, and zone must all be specified. If a false northing is used (that is, the coordinates are found in a location south of the Equator), then a notation that the coordinates are from the southern hemisphere is included. For example, the UTM coordinates of a location at Mount Rushmore would be written: 623733.22 E, 4859580.26 N, Zone 13. (See Figure 2.7 for a simplified graphical example of finding the location at Mount Rushmore's coordinates with UTM.) Also, see *Hands-on Application 2.3: Converting from Latitude/Longitude to UTM* on page 42 for converting latitude and longitude coordinates to UTM.

false easting a measurement made east (or west) of an imaginary meridian set up for a particular zone.

What Is SPCS?

A second grid-based method for determining coordinates is the **SPCS** (State Plane Coordinate System), a measurement system designed back in the 1930s. Today, SPCS is used as a coordinate system for United States data, especially city or county data and measurements. SPCS is a projected coordinate system that uses a translation method similar to UTM to set up a grid coordinate system to cover the United States. SPCS data was originally measured using the NAD27 datum, but there are now datasets using the NAD83 datum.

SPCS the State Plane Coordinate System used for determining coordinates of locations within the United States.

Hands-on Application 2.3

Converting from Latitude/Longitude to UTM

GCS and UTM are completely different systems, but there are utilities to make conversion from one system to another simple and easy. One of these is available at **http://www.cellspark.com/UTM.html**. This Web resource is used to convert from GCS to UTM or vice versa. You can specify your datum (reference ellipsoid), and then enter the latitude and longitude coordinates of a location (in either DD or DMS) and the utility will calculate the location's UTM coordinates. It can also calculate the latitude and longitude values when given UTM coordinates.

The first step is to obtain GCS coordinates of places before figuring out what their UTM equivalents are. A good source of coordinates can be found at **http://lat-long.com**. This Website allows you to input the name of an object (for instance, use the Empire State Building) and the state (New York) or county and it will return the lat/long (in both DD and DMS) for that place. Copy those coordinates into the UTM converter (use the WGS84 ellipsoid) to determine the UTM coordinates of the Empire State Building. Use the two Websites in conjunction with each other to determine the UTM coordinates of prominent locations around your area (public libraries, city hall, or your school).

SPCS zone one of the divisions of the United States set up by the SPCS.

Like UTM, SPCS requires you to specify a zone, an easting, and a northing in addition to the state the measurements are being made in.

SPCS divides the United States into a series of zones. Unlike UTM Zones, which follow lines of longitude, **SPCS zones** are formed by following state or county boundaries. A state is often divided into multiple zones—for example, Ohio has two zones: Ohio-North and Ohio-South. Nevada has three zones: Nevada-West, Nevada-Central, and Nevada-East. Texas has five zones, California has six zones, while all of Montana is represented by a single zone (Figure 2.8). To determine the coordinates for a point in SPCS, you must first know which state the point is in and then identify the zone in that state. For example, to map data from Miami, the state would be Florida and the zone would be Florida-East, while Cleveland would be in the state of Ohio and in the Ohio-North zone.

The next step is to determine the northing and easting. Like UTM, both of these are measured north and east from a pre-determined baseline and meridian, but in SPCS, measurements are made in feet (using the NAD27 datum) or in feet or meters (using the NAD83 datum). Each of the SPCS Zones has its own baseline and its own principal meridian. Two projected versions of SPCS are used—the Transverse Mercator version, which is usually used for states that have a north-south orientation (such as Indiana), and the Lambert Conformal Conic version, which is usually used for states that have a more east-west orientation (such as Ohio). Depending on the version used, the zones' baselines and meridians are positioned differently.

Baselines are placed a distance under the zone's southern border and measured north from there (to ensure all positive values). In the NAD27

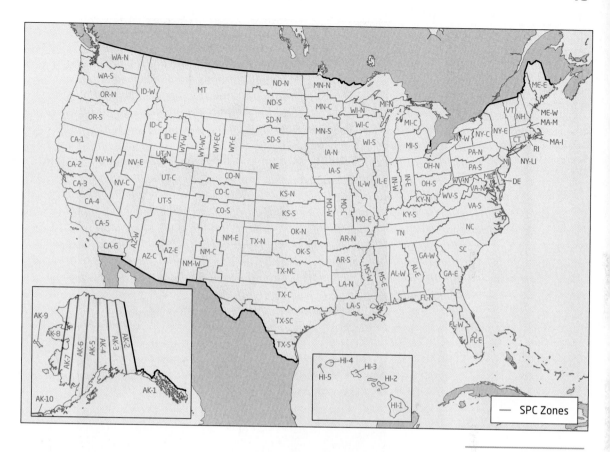

FIGURE 2.8 The SPCS zones for all states.

version, states using the Transverse Mercator format have their zone's origin 500,000 feet to the west of the principal meridian, while states using the Lambert Conformal Conic projection have their zone's origin 2,000,000 feet to the west of the meridian. Coordinates are determined by measuring distance using these false origin points, creating a false northing and a false easting. Thus, as each of the zones uses its own coordinate measurements, each zone has a different false northing and false easting from all other zones. Note that in the NAD83 version of SPCS, different values may be used for each zone's origin.

SPCS coordinates are referenced by listing the false easting, false northing, state, and zone. For example, the SPCS (NAD83 feet) coordinates of a location on the football field of Stambaugh Stadium in Youngstown, Ohio, are approximately: 2478424.96 E, 531216.31N, Ohio, N (North Zone; see Figure 2.9 on page 44 for a simplified graphical representation of SPCS measurements). Also, see *Hands-on Application 2.4: Using the State Plane Coordinate System* for a utility for converting latitude and longitude coordinates into SPCS values.

FIGURE 2.9 A simplified representation of SPCS measurements and coordinates for a location on the YSU football field in Youngstown, Ohio. (Source: Ohio Geographically Referenced Information Program (OGRIP), Ohio Statewide Imagery Program (OSIP), April 2006)

2478424.96 E, 531216.13N, Ohio, N

Hands-on Application 2.4

Using the State Plane Coordinate System

Like the UTM conversion tool in *Hands-on Application 2.3: Converting from Latitude/Longitude to UTM*, there are Web-based utilities used to convert GCS coordinates into their SPCS counterparts. A good one is set up by NOAA and available at **http:// www.ngs.noaa.gov/cgi-bin/spc_getpc.prl**. Similar to the utility used in *Hands-on Application 2.3*, this Web resource enables you to enter the latitude and longitude of a location and it will calculate the SPCS coordinates for that point. Note that the lat/long you give it have to be formatted in a specific way (see the example on the Website for more details) in order for it to work properly. Use Web resources or other mapping data you may have available to you to find lat/long coordinates of some national points of interest (such as the Transamerica Pyramid in California) and places more local to you and calculate the SPCS coordinates for them. Be careful to keep in mind the datum with which you're working for the lat/long coordinates and also that SPCS coordinates are available only for places within the United States.

Chapter Wrapup

This chapter provided a look at what goes into setting up coordinates and references for data used with geospatial technology. Whether you're reading coordinates from a GPS receiver or making measurements from a map, knowing how that geospatial information is put together and referenced has an impact on how you're able to make use of that data, especially if you're trying to fit several types of geospatial data together.

In the next chapter, we're going to switch gears and look at how you can take your data (whether it's a remotely sensed image, some information created using GIS, or a paper map of the local fairgrounds) and get it to match up with all of the other geospatial data you have. Before that, this chapter's lab (see *Geospatial Lab Application 2.1: Coordinates and Position Measurements* on page 46) will have you start applying the coordinate and measurement concepts described in this chapter with some more work with Google Earth.

Important note: The references for this chapter are part of the online companion for this book and can be found at http://www.whfreeman.com/shellito1e.

Key Terms

datum (p. 30)
ellipsoid (p. 30)
geoid (p. 30)
geodesy (p. 30)
NAD27 (p. 31)
NAD83 (p. 31)
WGS84 (p. 31)
datum transformation (p. 31)
geographic coordinate system
  (GCS) (p. 31)
latitude (p. 31)
Equator (p. 31)
longitude (p. 32)
Prime Meridian (p. 32)
International Date Line (p. 32)

degrees, minutes, and seconds
  (DMS) (p. 32)
decimal degrees (DD) (p. 33)
great circle distance (p. 33)
time zones (p. 33)
map projection (p. 36)
Universal Transverse Mercator
  (UTM) (p. 39)
UTM zone (p. 39)
easting (p. 40)
northing (p. 40)
false northing (p. 40)
false easting (p. 41)
SPCS (p. 41)
SPCS zone (p. 42)

Coordinates and Position Measurements

This chapter's lab will continue using Google Earth (GE), but only to specifically examine coordinate systems and the relationship between various sets of coordinates and the objects they represent in the real world. In addition, you'll be making some measurements using GE, as well as using some Web resources for comparing measurements made using different coordinate systems.

Note that this lab makes reference to things like "coordinates for the White House" or "coordinates for Buckingham Palace"—these represent measuring a set of coordinates at one specific location at these places and are used as simplifications of things for lab purposes.

Objectives

The goals for you to take away from this lab are:

◉ Setting up a graticule of lines in GE.

◉ Locating places and objects strictly by their coordinates.

◉ Making measurements across long and short distances, and then comparing measurements to surface distance calculations.

◉ Translating lat/long coordinates into UTM.

Obtaining Software

The current version of Google Earth (6.0) is available for free download at http://earth.google.com.

Important note: Software and online resources sometimes change fast. This lab was designed with the most recently available version of the software at the time of writing. However, if the software or Websites have significantly changed between then and now, an updated version of this lab (using the newest versions) is available online at http://www.whfreeman.com/shellito1e.

Lab Data

There is no data to copy in this lab. All data comes as part of the GE data that gets installed with the software or is streamed across the Internet through using GE.

Localizing This Lab

This lab flies around the world to numerous locations, including Washington, D.C., and London. However, it can easily be changed to examine the coordinates of nearby local areas. Rather than looking at the White House or Buckingham Palace, substitute the coordinates or locations for your local government offices (like city halls or courthouses).

2.1 Examining Coordinates and Distance Measurements in Google Earth

1. Start **GE**. Once Earth settles into view, scroll your mouse around Earth. You'll see a set of latitude and longitude coordinates appear at the bottom of the view representing the coordinates assigned to your mouse's location. By default, GE uses the GCS coordinate system and the WGS84 datum.

2. To examine the full graticule of latitude and longitude lines, from the **View** pull-down menu select **Grid**. Some key GCS lines will be highlighted in yellow amidst the Web of lat/long lines—the Equator, Prime Meridian, the Antimeridian, the Tropic of Cancer, and the Tropic of Capricorn.

3. We'll begin in Washington, D.C., so in the **Fly To** box, type **Washington, D.C. GE** will rotate and zoom to this area. You'll also see the spaces between the lat/long lines grow smaller and new values appear as GE zooms in.

4. Next, you'll go a specific location in Washington, D.C. In the Fly To box, type the following coordinates: **38.897631, −77.036566**. These are the decimal degree lat/long coordinates of the White House.

5. Press the **Placemark** button on the toolbar (see *Geospatial Lab Application 1.1* for more information on placemarks). The yellow pushpin will automatically be placed at the lat/long coordinates you had GE search for (so you don't have to manually place it anywhere). In the New Placemark dialog box, type **The White House** for the name of the placemark, and select another symbol for the yellow pushpin if you want.

6. The coordinates for the White House are in decimal degrees, but other methods of displaying coordinates are available in GE. From the **Tools** pull-down menu, select **Options**.

7. In the Show Lat/Long options, select the radio button next to **Degrees, Minutes, Seconds**, then click **Apply**, and then click **OK**.

8. From the Places box in GE, right-click on the **White House** placemark, and

then select **Properties**. The coordinates will be changed to Degrees, Minutes, and Seconds (DMS). Answer Question 2.1. Close the White House placemark dialog box.

Question 2.1 What are the coordinates for the White House in Degrees, Minutes, and Seconds?

9.   Decimal degree coordinates for the Lincoln Memorial are: **38.889257, −77.050137**. Use the **Fly To** box to zoom to these coordinates. Once GE arrives there, put a placemark at that spot, name it **Lincoln Memorial**, and answer Question 2.2.

Question 2.2 What are the coordinates for the Lincoln Memorial in Degrees, Minutes, and Seconds?

10. Adjust the view so you can see both placemarks at the edges of the view. Select the **Ruler** tool and compute the distance between the White House and the Lincoln Memorial (see *Geospatial Lab Application 1.1* for the use of the Ruler tool in measurements). Answer Question 2.3. Turn off the placemark for the Lincoln Memorial when you're done.

Question 2.3 According to GE, what is the distance between the White House and the Lincoln Memorial?

11. For such a short distance (as between the White House and the Lincoln Memorial), there should be little differences in measurements, so let's look at some larger distances.

12. Buckingham Palace (in London, England) is located at the following coordinates: **51.500821, −0.143089**. Use the **Fly To** box to zoom to these coordinates. Once GE arrives there, put a placemark at that spot, name it Buckingham Palace, and then answer Question 2.4.

Question 2.4 What are the coordinates for Buckingham Palace in Degrees, Minutes, and Seconds?

13. Zoom out and rotate the view so that you can see the placemarks for the White House and for Buckingham Palace in the same view. Use the **Ruler** to compute the distance between these two locations. Answer Questions 2.5 and 2.6. Close the Ruler dialog when you're done. Of course, this is a rough estimate of being able to place the start of the ruler and the end point of the ruler directly on top of the placemarks, given the scale of the view.

Question 2.5 According to your measurement in GE, what is the computed distance (in miles) between the White House and Buckingham Palace?

Question 2.6 Why is the line curved rather than a straight line? What kind of distance is being computed here?

14. We'll now check your measurements using a Web utility. Open your Web browser and go to **http://www.chemical-ecology.net/java/lat-long. htm**. This Website enables you to compute the surface (real-world)

distance between two sets of lat/long coordinates (and also between two cities) in DMS format.

15. Use your answer to Question 2.1 as the Degrees, Minutes, and Seconds for the Latitude and Longitude of Point one (you are using North latitude and West longitude).

16. Use your answer to Question 2.4 as the Degrees, Minutes, and Seconds for the Latitude and Longitude of Point two (you are again using North latitude and West longitude).

17. Click the **Distance Between** button. The surface distance between the two points will be computed in kilometers, miles, and nautical miles. Answer Question 2.7. Your answers to Questions 2.5 and 2.7 should be relatively similar given how they were both computed—if they're really way off, re-do your Question 2.5 measurement and make sure you're inputting the correct numbers for Question 2.7.

Question 2.7 According to this Website measurement, what is the computed surface distance (in miles) between the White House and Buckingham Palace?

2.2 Using UTM Coordinates and Measurements in Google Earth

Universal Transverse Mercator (UTM) is a projected coordinate system and GE can track coordinates using this UTM system as well as lat/long.

1. From GE's **Tools** pull-down menu, select **Options**. In the Show Lat/Long options, select the radio button for **Universal Transverse Mercator**. Click **Apply**, and then click **OK**.

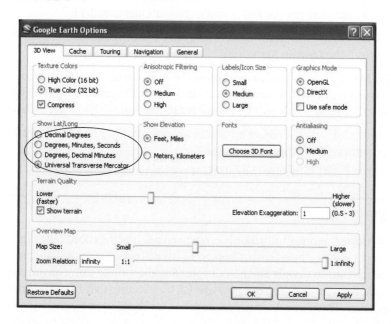

2. Zoom out to see the whole world—you'll also see that the graticule has changed from a Web of latitude/longitude lines to the UTM grid draped over Earth. You should see the UTM zones laid out across the world, numbered from 1 through 60.

Question 2.8 What UTM Zone is Buckingham Palace located in? What UTM Zone is the White House located in?

3. Double-click on the **White House** placemark (in the Places box) and GE will rotate and zoom back to the White House.

4. Scroll the mouse around the White House area. You'll see the new coordinates appear at the bottom of the view, but this time as the zone, easting, and northing measurements. Open the **Properties** of the **White House** placemark by right-clicking on it. The UTM easting, northing, and zone will appear.

Question 2.9 What are the UTM coordinates of the White House?

5. In the Places box, double-click on the **Lincoln Memorial** placemark and GE will zoom and rotate to it.

Question 2.10 What are the UTM coordinates of the Lincoln Memorial?

6. UTM coordinates are measured in meters, enabling an easier method of determining coordinates rather than degrees of latitude or longitude. If you move the mouse east, your easting will increase, and moving it west will decrease the easting. The same holds true for northing—moving the mouse north will increase the value of northing, while moving it south will decrease the northing value.

7. Change the view so that you can see both the White House and the Lincoln Memorial in the view. Using your answers for Questions 2.9 and 2.10 (and perhaps the **Ruler** tool), answer Question 2.11.

Question 2.11 How many meters away to the north and east is the White House from the Lincoln Memorial?

8. UTM can also be used to determine the location of other objects. Return to the **White House** placemark again. Determine what object is 912 meters south and 94.85 meters to the east of the White House.

Question 2.12 What object is located approximately 912 meters south and 94.85 meters east of the White House?

Question 2.13 What are the full UTM coordinates of this object?

Closing Time

This lab served as an introduction for using and referencing coordinates and measurements using Google Earth. Chapter 3's lab will use a new software program, but we'll return to Google Earth in Chapter 8 and continue using its functions in other chapters as well. You can now exit GE by selecting **Exit** from the **File** pull-down menu. There's no need to save any data (or Temporary Places) in this lab.

3

Getting Your Data to Match the Map

Reprojecting, Georeferencing, Control Points, and Transformations

Since there are all of these datums and map projections out there, it's important to know which one you're using when making measurements. Say, for instance, you have two sets of geospatial data that you need to bring together on one map of your town for analysis (that is, a street map and building footprints). Both of these pieces of data are in a different projection and coordinate system, and both use a different datum—the street map is in State Plane NAD27 feet and the buildings are in UTM NAD83 meters. When you put these things together, nothing is going to match up—the streets and buildings are going to be in different places in relation to one another since each one uses a different basis for its coordinates.

It is important to note when you're dealing with geospatial data that it can be constructed using a variety of different datums, coordinate systems, and projections. With all of the different variations involved with these items, it's best to try and work with all geospatial data on the same basis (that is, use the same datum, coordinate system, and projection). For instance, if you have multiple data layers in different map projections and you try to overlay them, then those pieces of data won't properly align with each other (Figure 3.1 on page 52).

How Can You Align Different Geospatial Datasets Together?

Back in Chapter 2, we discussed the concept of having to transform data measured from different datums to a common datum so that everything will align correctly. As such, it's best to **reproject** your data (or transform all of the datasets to match one, or transform them all to a common datum, coordinate

reproject changing a dataset from one map projection (or measurement system) to another.

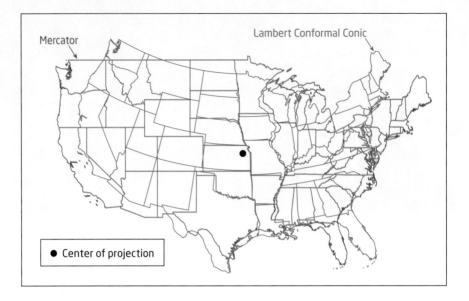

Mercator

Lambert Conformal Conic

● Center of projection

system, and projection) before moving forward. Reprojecting data will alter the coordinates, map projections, and measurements from one data source (for instance, coordinates measured in SPCS) to another system (such as UTM). Many geospatial software packages will give you the capability to take your initial dataset (such as your street map in State Plane NAD27) and reproject it to create a new dataset with new coordinates (so now you would have a new street map measured in UTM NAD83). This can also be done to change data from one map projection into another. In some cases, the software will be able to "project on the fly" or calculate the necessary transformations without actually altering the base dataset. Whatever the case, you'll usually avoid

Thinking Critically with Geospatial Technology 3.1

What Happens When Measurements Don't Match Up?

The scenario described in the text is not uncommon, having multiple data sources all with different projections, coordinate systems, and datums that all need to be used together in a project. Without everything aligning properly, you can end up with points being plotted in the wrong location with respect to the base map they're being located on, or street data not lining up with a satellite image of the same area. What kind of effects could this type of simple data mismatch have on real-world companies or government agencies? For instance,

what if the sewer data and the street data that a road construction work crew has for renovations of subdivision streets doesn't align properly (due to differing datums or projections)? What potential effects can this have? When the misalignment of data may cause one dataset to be off from its properly matched place with a map by a large margin, what could occur? Similarly, what types of effects could happen with things like shoreline construction, zoning maps, or species habitat mapping?

a lot of headaches later on by using a common basis for the various geospatial datasets you use. This is especially critical when getting several types of data to accurately match up with one another.

However, a problem's going to arise when one (or more) of your pieces of data doesn't have coordinates assigned to it, or you don't have any information as to what projection or datum was used to construct it. Think about this—you obtain an old map of your town, scan it into a computer, and want to add it to other layers (like a road network or parcel layer) in a GIS to see how things have changed since that old map was made. You add the layers into the GIS, but the scanned historic map won't match up with any of your data. This is because that image doesn't contain any **spatial reference** (such as a real-world coordinate system). Although it might show features like old roads or the dimensions of a building, the only coordinate system the old map possesses is a grid that begins at a zero point at the lower left-hand corner and extends to the dimensions of the scanned image. For instance, coordinates would be referenced by whatever units you're measuring with (such as 200 millimeters over from the left-hand corner and 40 millimeters up from the left-hand corner).

Items in the image can't be referenced by something like latitude and longitude since those types of coordinates haven't been assigned to anything in the image. Similarly, when you added this scanned image as a layer in a GIS, the software wouldn't know what to do with it when trying to match it up with other data layers (which have referenced coordinate systems). Without knowing "where" your scanned image matches up with the rest of the data, you'd be unable to use this image as geospatial data (or reproject it to match your other data). The same concept holds for other types of data used with geospatial technology. That is, if they all automatically had the correct spatial reference when they were created, things would be a lot easier. So what you'll have to do is match the data to the map, or assign a real-world set of coordinates to your data so it can synch up with other geospatial data. In other words, you're going to have to georeference your data.

> **spatial reference** the use of a real-world coordinate system for identifying locations.

| What Is Georeferencing?

Georeferencing is the process of aligning an unreferenced dataset with one that has spatial reference information. Other terms used for similar types of processes include "image to map rectification," "geometric transformation," or "registration." For example, say you're hosting a meeting at your school and want to make a map for the invitation to direct people not only to the school, but also to your specific building within the campus. You go to your school's Website and download a map that shows the campus boundaries and specific building locations on the campus. So far so good, but you want to place that downloaded map in the greater context of a map of the city in a software program to show where your campus building is in relation to the surrounding

> **georeferencing** a process whereby spatial referencing is given to data without it.

Hands-on Application 3.1

David Rumsey Historical Map Collection

An ongoing project for the collection and preservation of old historic maps on the part of David Rumsey can be seen at **http://www.davidrumsey.com**. The maps on the Website are part of a collection stretching from the seventeenth through the twentieth century. While there are many ways of viewing these maps, some of the historic maps have been georeferenced to make them match up with current geospatial imagery. By looking at georeferenced maps in relation to current data, you can see how areas have changed over time. On the Website, select the link for Google Earth under the Quicklinks section, then select the option to View Google Earth in your Web browser (you'll have to install a Google plug-in to make this work properly). Inside your Web browser, you'll be able

roads and neighborhoods, so people can get to your meeting. The problem is that while the data you have of the surrounding area (whether aerial photos, satellite imagery, or GIS data layers) contains real-world spatial references, your downloaded map is just a graphic or PDF without any spatial reference at all. Thus, the campus map won't match up with the spatially referenced data, and your plan of a snappy geospatial-style invitation is ruined. However, if you could just georeference that campus map, everything would fit together. The georeferencing procedure requires a few steps of work to proceed (see *Hands-on Application 3.1: David Rumsey Historical Map Collection* and **Figure 3.2** for an example of an online georeferenced map collection).

FIGURE 3.2 An example of a David Rumsey historical map (of Australia) georeferenced and overlaid in Google Earth. (Source: ©2010 Google. Data SIO, NOAA, U.S. Navy, NGA, GEBCO. ©2011 Cnes/Spot Image. Image IBCAO.)

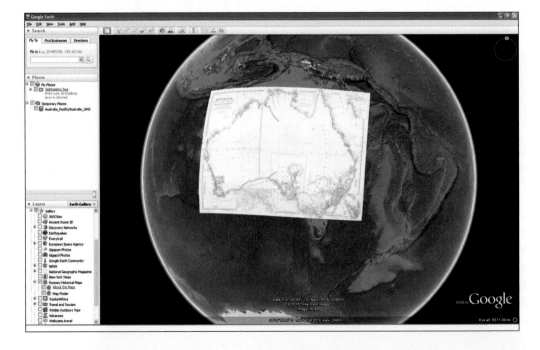

to interact with Google Earth just like in the last couple of labs, but you'll also see special symbols on the map representing the location of the georeferenced maps. Click on these symbols to see the maps appear—properly georeferenced—in relation to Google Earth's imagery. Check out some of the map icons for a variety of places around the globe and see how this historic information synchs up with modern boundaries and images.

The Rumsey Historical Map collection is also available as a layer in Google Earth itself. To use the maps as another Google Earth layer (like the roads or buildings), open the Gallery options and then open the Rumsey Historical Maps option, and click on the Map Finder icons. You'll see a new icon representing the maps appear in Google Earth that you can use to view the georeferenced maps. There's also an "About the Maps" pop-up where you can read more information about the map collection and how these maps were scanned and georeferenced for use in Google Earth.

How Can Data Be Georeferenced?

When data (such as an old scanned map or other unreferenced data) is georeferenced, each location in that data is aligned with real-world coordinates. So to start the whole process, you're going to need some type of source with real-world coordinates to align your unreferenced data against. The choice of a source to use for the matching is important since you have to be able to find the same locations on both the unreferenced image and the source. This is crucial to georeferencing, since the process needs some common locations to "tie" the unreferenced data to the source data. In the georeferencing procedure, these spots (where you can find the coordinates on the source data) are referred to as **control points** (sometimes called Ground Control Points or GCPs). Field work, such as surveying or GPS measurements (see Chapter 4) can provide sources for these control points, but without going into the field, you can also find common locations on the unreferenced image and the source to use in the process.

control points point locations where the coordinates are known—these are used in aligning the unreferenced image to the source.

Thus, the source should be something you can find the same locations on as the unreferenced data. For instance, a low-resolution satellite image isn't going to be a good match with the detailed map of your campus buildings. Trying to match individual locations like buildings to an image like this isn't going to work simply because the resolution is too low. A high-resolution image (where you can determine features of buildings) would be a more proper choice for this specific task. Datasets such as road networks (see Chapter 8) or orthophotos (see Chapter 9) are also good sources to use in finding these common points. Also, your source data should be a similar projection to your unreferenced data. If the projections are very different, then some features will be forced to match up well and others will end up distorted. Whatever the source, the georeferencing process will align an unreferenced image against the projection and coordinates of the source map (Figure 3.3 on page 56).

Where to select the control points is the key to the whole georeferencing procedure. Since control points are the objects that are going to connect

FIGURE 3.3 Comparing
the unreferenced
data (an older campus
map of Youngstown
State University)
to the reference
source (a referenced
image of campus).
(Sources: Mahoning County
Enterprise GIS, April 2004.)

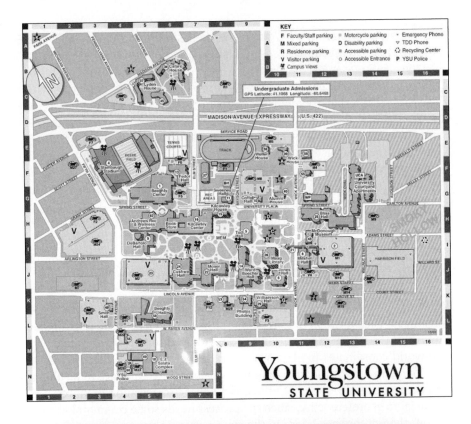

the unreferenced data to the source data, they have to be placed in the same location in both sets of data, and real-world coordinates for these points will be required. By finding these same locations on the source (that has real-world coordinates) and on the unreferenced data (without them), we can start to match the two datasets together. Thus, places where you can clearly find the same spot in the unreferenced data and the source data will be good choices for the placement of control points in that location. For instance, the intersections of roads or the corners of buildings are well-defined areas that can be found on both sets of data and would make good control point locations. These are also locations that aren't likely to change over time, so even if the unreferenced data and the source aren't from the same time period, they would still be valid as control point selections. Poor locations for control point selection spots are those that can't be accurately determined between the two sets of data, either because they're poorly defined or in spots that would vary. Examples of poor control point selection choices would be a spot on the top of a building, the center of a field, the shoreline of a beach, or a shrub on a landscaping display. See Figure 3.4 for some examples of good and poor control point location selections on the campus map.

Another thing to keep in mind when selecting control point locations is to not bunch them all together in one area of the data and ignore the rest of the map. Placing all of the control points at spots in one-quarter of the image is going to ensure that section of the map is well referenced while the rest of the map may not fit very well. Make sure to try and spread out the control points to representative locations throughout the data so that all areas fit well regarding their new referenced spots. You'll want to use several control points in the process—some georeferencing processes may only need a few, while others may require many more.

Three control points is the minimum number required to try to fit the unreferenced image to the source. With three control points, you'll be able have a rough fit of the image to the source. With this initial fit, some programs (for instance, MapCruncher, the program used in this chapter's lab)

FIGURE 3.4 Good (in red) and poor (in blue) locations for control point selection. (Sources: Mahoning County Enterprise GIS, April 2004. Esri® ArcGIS ArcMap ArcInfo graphical user interface Copyright© Esri.)

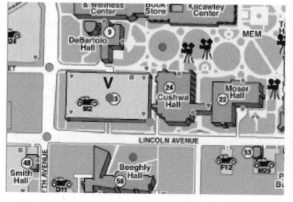

Hands-on Application 3.2

Old Maps Online

There's a research project in the Czech Republic dedicated to (in part) georeferencing old historic maps. To check out their work, open your Web browser and go to **http://help.oldmapsonline.org** to read about their current efforts. On the Website is a link to their Georeferencing Tool, an online utility for georeferencing historic data (also available at **http://www.georeferencer.org**). Click on the link to open the tool—you'll see a scanned historic map on the left-hand side and a source map on the right. Zoom in on both maps to try and find some common points, and double-click to create a control point.

will attempt to match up the next control point location with a spot you select on the source. As you properly add some more control points, the fit between the unreferenced data and the source should (hopefully) improve. Proper selection of control points is critical for georeferencing since the procedure needs to have some sort of real-world coordinates to match to, and it's these places that will be used to calculate coordinates for the other locations in the image. Often, you'll find that deleting a control point, adding another one, and tinkering with the placement of others is necessary to get the best fit of the data (see *Hands-on Application 3.2: Old Maps Online* for an online example of placing control points in relation to historic maps). Once your data is properly aligned using control points, you can then transform it to have the same coordinate system information as the source.

How Is Data Transformed to a Georeferenced Format?

rotated when the unreferenced image is turned during the transformation.

skewed when the unreferenced image is distorted or slanted during the transformation.

scaled when the scale of the unreferenced image is altered during the transformation.

translated when the unreferenced image is shifted during the transformation.

Take another look at Figure 3.3, which shows both the unreferenced campus map and the source image. In order to make the unreferenced map properly align with the source image, there are a number of actions that have to be taken with the map. First, check out the football stadium (Stambaugh Stadium) in the upper left part of the map and the upper portion of the image. In order to make it match from the map to the image, it would have to be turned about 45 degrees clockwise. Looking at the other features, the whole map would have to be rotated some amount (since the image is oriented to the north and the map is oriented to the northeast). Also, the map might have to be stretched or skewed to match up with the image.

When unreferenced data is transformed or "rectified", it's warped to make it match up with the source. It can be **rotated** (turned, like changing the orientation of the map), **skewed** (pulling the image, like at a slant), **scaled** (altering the entire map to match the image), and/or **translated** (altering

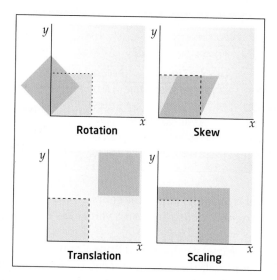

FIGURE 3.5 Four ways that the unreferenced data can be warped to align it with the source.

the location and placement of the map). See Figure 3.5 for examples of the effects of each of these transformations. This transformation of the data changes it to align with the new coordinate system. There are a number of transformation methods that are used, but a common one is the **affine transformation,** which will properly rotate, skew, translate, and scale the data.

> **affine transformation** a linear mathematical process by which data can be altered to align with another data source.

The affine transformation calculates real-world X and Y coordinates for each unreferenced x and y coordinate. The formulae used for this are:

$$X = Ax + By + C$$

and

$$Y = Dx + Ey + F$$

where $A, B, C, D, E,$ and F are the values calculated internally by the procedure and applied to the mathematical formulae as coefficients (and also control the properties like skewing and rotation). The affine transformation is a first-order transformation, and usually this type of transformation is what will be used in this procedure. The end result of a transformation is that the unreferenced data will now be spatially referenced.

So, once you select some control points, the software runs the transformation and the data is georeferenced. Everything appears to be working just fine. However, there are two questions that you should be asking yourself, and they are: "How do I know it worked?" and, more importantly, "How do I know the results are any good?" After all, just because the unreferenced data's been transformed, it doesn't necessarily mean that it's a good match with the source. Since you know the coordinates of each place you're putting a control point, you should be able to match up those locations and see

FIGURE 3.6 A sample of control point coordinate locations (in the unreferenced map and in the referenced image), and the computed error term for each control point. (Source: Clark Labs, Clark University, IDRISI Taiga)

Id	Include	Input X	Input Y	Output X	Output Y	Residual
1	Yes	176.834107	198.682058	2477637.6062	529915.11078	1.440036
2	Yes	207.211722	408.429262	2478335.4309	530863.61906	1.158569
3	Yes	329.441011	495.144545	2479158.6610	530964.75548	1.186549
4	Yes	520.071114	495.144545	2480081.0935	530473.22079	1.827131
5	Yes	519.775077	356.695021	2479698.6849	529784.98875	2.183062

Ground control points
Input reference: CAMPUSBW
Output reference: ysucampus_band1
Total RMS: 1.890850
Digitize GCP: Input / Output
Number of GCPs: 12
Retrieve GCP
Remove GCP
Save GCP as

how closely your selected control points match the source. Figure 3.6 shows a report from a software program measuring error values for each control point.

Root Mean Square Error (RMSE) an error measure used in determining the accuracy of the overall transformation of the unreferenced data.

A transformation procedure uses a value called **Root Mean Square Error (RMSE)** as an indicator of how well the now-referenced data matches with the source. After the initial set of control points has been chosen and the unreferenced data is aligned with the source, coordinates are computed for locations of the selected control points. The known coordinate location of each control point (in the source) is compared with the newly computed value for its corresponding location in the unreferenced data. If the control points were in the exact same location in both the unreferenced data and the source, then the difference in their coordinates will be zero. However, chances are, when placing coordinates—even if you selected good locations like road intersections—the two points aren't going to exactly match up, though they might be very close. The software will examine all error values for all control points and report back a value for the RMSE.

The lower the differences between where you put the control point on the unreferenced data (the estimated location) and the source data (the true location), the better the match between them will be. Thus, the lower the overall RMSE is, the better the transformation between the two datasets will be. Usually, the georeferencing program will allow you to also examine the difference between each point, allowing you to see how closely your control point selections matched each other. If the RMSE is too high, it's a good idea to check the difference in coordinates for each point—if a few of them are off too much, they can be adjusted, changed, or deleted (and new points selected instead) to try and get a better fit. If some parts of the data match well and others don't, perhaps the control points need to be moved to other places to better balance out the alignment, or other control points need to be added, or others may need to be removed.

When georeferencing an image, new locations for the image's pixels will need to be calculated and in some cases, new values for the pixels will need to be generated as well. This process is referred to as *resampling*. The end result of a good georeferencing process will have your unreferenced data properly aligned with the source data (Figure 3.7). Maps that were not referenced can now match up with current maps and datasets (see *Hands-on Application 3.3:*

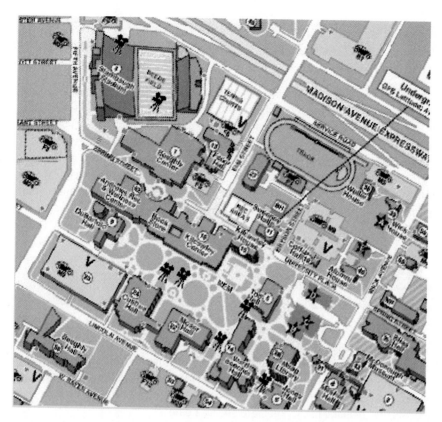

FIGURE 3.7 Detail from the end result of georeferencing the campus map. (Source: Clark Labs, Clark University, IDRISI Taiga)

Microsoft MapCruncher Gallery and *Hands-on Application 3.4: An Overview of the Georeferencing Process in ArcGIS*, both on page 62, for examples). Georeferencing can also be performed on other data created without a spatial reference such as unreferenced drawings and plans. A related procedure can also be used for remotely sensed images that need to be corrected.

Thinking Critically with Geospatial Technology 3.2

What Happens When the Georeferencing Is Wrong?

Think about this scenario—surveying is being done for a new beachfront development and a base map of the area will be created (using GIS) from current high-resolution remotely sensed imagery. Since the base map (and all layers created from it, such as property boundaries, utilities, and local land cover) will be derived from it, this initial layer is critical. However, something goes wrong in the georeferencing process of the image used to create the base map, and thus the initial base map has incorrect coordinates—and then each layer created from it has incorrect coordinates. Then on-site GPS coordinates and measurements are reprojected to match the new base map, and of course, they're now inaccurate since they use an incorrectly referenced base map. What's going to be the end result of all this for the land owners, developers, construction crews, and other stakeholders involved in the project? What kind of long-term or widespread problems can be caused by this type of "cascading error" when the initial referencing is incorrect or inaccurate?

Hands-on Application 3.3

Microsoft MapCruncher Gallery

In this chapter's lab, you'll be using a free software tool called MapCruncher that enables you to geo-reference your own images using Microsoft's Bing Maps as the source. There are several examples of MapCruncher's uses online. Open your Web browser and go to **http://research.microsoft.com/en-us/um/redmond/projects/mapcruncher/Gallery** to view some examples (including bike maps, campus maps, and park maps). Examine the georeferenced maps in the gallery, then locate and download some maps of places in your local area (such as parks, festivals, fairgrounds, tourist areas, or historical sites) in a format to use with MapCruncher. Once you go through the steps in this chapter's lab, you'll have an inventory of local maps on hand that you can apply your georeferencing skills to.

Hands-on Application 3.4

An Overview of the Georeferencing Process in ArcGIS

While you'll use the free MapCruncher to align un-referenced data in the lab, georeferencing tools are common in other GIS software such as ArcGIS (which we'll discuss further in Chapter 5). In order to demonstrate georeferencing in ArcGIS, its company has a video of how the steps are performed at **http://webhelp.esri.com/arcgisdesktop/9.3/index.cfm?TopicName=Georeferencing_video_sample**. Check out the video and see how GIS tools are used for georeferencing a dataset.

Chapter Wrapup

Proper georeferencing of data is a key element of working with geospatial technology and allows for all types of maps and imagery to work side by side with other forms of geospatial data. As noted in *Hands-on Application 3.3: Microsoft Map Cruncher Gallery*, this chapter's lab will have you apply all of this chapter's concepts using the MapCruncher program—you'll start with an unreferenced campus map, and by the end, it should be properly aligned with a real-world dataset.

   In the next chapter, we'll be looking at the Global Positioning System, an extremely useful (and freely available) system for determining your position on Earth's surface and providing you with a set of coordinates, so you can find where you are.

   Important note: The references for this chapter are part of the online companion for this book and can be found at http://www.whfreeman.com/shellito1e.

Key Terms

reproject (p. 51)
spatial reference (p. 53)
georeferencing (p. 53)
control points (p. 55)
rotated (p. 58)
skewed (p. 58)

scaled (p. 58)
translated (p. 58)
affine transformation (p. 59)
Root Mean Square Error (RMSE)
 (p. 60)

Georeferencing an Image

This chapter's lab will introduce you to some of the basics of working with georeferencing concepts by matching an unreferenced map image against a referenced imagery source. You'll be using the free Microsoft MapCruncher utility for this exercise.

Important note: MapCruncher will take a map (in an image or PDF format) and align it to its Virtual Earth (and transform it into a Mercator projection). The end result of the process is a series of "tiles" used by Virtual Earth to look at your map matched up with their data. As such, while you'll be using georeferencing concepts from the chapter, the end result will be that your map is turned into a "mashup" that could later be used on a Web page.

Objectives

The goals for you to take away from this lab are:

- Familiarizing yourself with the MapCruncher operating environment.
- Analyzing the unreferenced image and the source for appropriate control point locations.
- Performing a transformation on the unreferenced data.
- Evaluating the results of the procedure.

Obtaining Software

The current version of MapCruncher (3.2) is available for free download at http://www.microsoft.com/maps/product/mapcruncher.aspx.

Important note: Software and online resources sometimes change fast. This lab was designed with the most recently available version of the software at the time of writing. However, if the software or Websites have significantly changed between then and now, an updated version of this lab (using the newest versions) is available online at http://www.whfreeman.com/shellito1e.

Lab Data

Copy the Chapter 3 folder containing a PDF file (parkmap2010.pdf), which is a map—with no spatial reference attached to it, it's only a graphic—of the Youngstown State University (YSU) campus. This map of YSU can also be downloaded from the campus map Web page at http://web.ysu.edu/gen/ysu/Directions_Maps__Tour_m107.html.

Localizing This Lab

MapCruncher is capable of georeferencing a graphic or PDF map of an area. If you have a graphic or PDF of your school, it can be substituted for the YSU campus map and you can match your own local data instead. The same steps of the lab will still apply; you'll just be finding appropriate local control points rather than the ones in Youngstown, Ohio.

Alternatively, you could use a graphic or PDF of a map of a local park, hike or bike trail, fairgrounds, or local shopping district and georeference that instead.

3.1 Getting Started With MapCruncher

1. Start **MapCruncher**. It will open with the default screen referred to as "Untitled Mashup." Maximize the MapCruncher window to give yourself plenty of room to work with.

2. The left side of the screen will contain the unreferenced image you want to match, while the right side of the screen contains the referenced imagery from Microsoft's Virtual Earth (what you'll be matching your image to).

3. As the text box notes, select the **File** pull–down menu, then choose **Add Source Map**.

4. Navigate to the **Chapter 3** folder and select the **parkmap2010.pdf** file.

5. In the upper right corner of the MapCruncher screen is the Virtual Earth Position box.

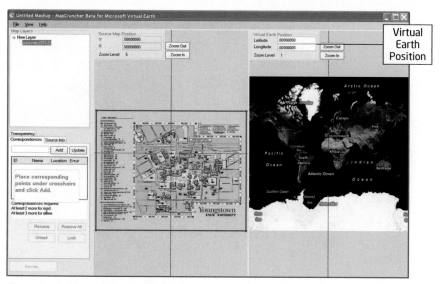

(Source: Used with permission from Microsoft)

6. Type the following coordinates and then click on **Zoom In** to move the view to YSU:

   ◉   Latitude: 41.10687400

   ◉   Longitude: –80.64670900

7. The crosshairs in the image will center in northeast Ohio. Press the Virtual Earth Position's **Zoom In** button several times to zoom in to YSU's campus (Zoom Level 17 is probably a good starting point).

8. Both of the views will now show the campus—the Source Map (left-hand view) has the PDF map image (or graphic) while the Virtual Earth (right-hand view) is the aerial imagery from Virtual Earth.

3.2 Selecting Correspondences

1. The next step is to choose the same positions on both images to use as control points (what MapCruncher refers to as "correspondences"). A good starting point is the intersection of Wick Avenue and Lincoln Avenue (a major road intersection that borders the southeastern portion of the campus).

2. Click on **Zoom In** on the Source Map until you reach Zoom Level 7.

3. The cursor will change to a hand—use this to pan around the map until the intersection of Wick and Lincoln Avenues is centered under the Source Map's crosshairs.

4. In the Virtual Earth view, pan around the map until that same intersection is in the Virtual Earth's crosshairs.

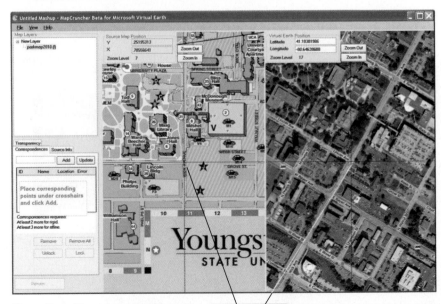

(Source: Used with permission from Microsoft)   Wick and Lincoln Intersection

5. When you have both crosshairs accurately positioned, click **Add** in the Correspondences box.

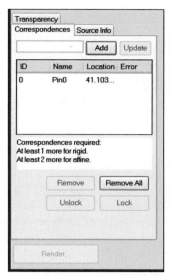

(Source: Used with permission from Microsoft)

6. The point will appear as ID 0. This is the first correspondence (what MapCruncher refers to as control points) being selected.

Question 3.1 Why is this location a good one for a correspondence?

Question 3.2 Just by visual examination of the two maps, what types of transformative operations will have to be performed to the Source Map to make it match the Virtual Earth?

7. Use the combination of the Source Map and the Virtual Earth to select another two correspondence points at clearly defined locations on both sources. Be sure to spread out the points geographically across the Source Map image, then click **Lock**. MapCruncher will perform an initial fit of the data, but pan around and you'll see that things don't match up just yet. Press **Unlock** and select a total of ten correspondences (you can use the Lock option to see how things are fitting together as you go, then Unlock to adjust the points).

Question 3.3 Which locations did you choose for the next nine correspondence points (and why)?

Question 3.4 Why would you want to spread the correspondence point selection across the image instead of clustering the points together?

8. Once you have about ten good locations chosen, click the **Lock** button in the Correspondences box.

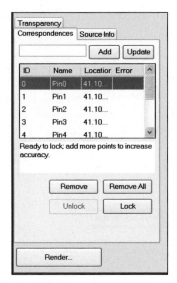

(Source: Used with permission from Microsoft)

9. A value for error will be computed for each of the points and they will be sorted from highest error value to lowest error value (rather than by ID number).

Question 3.5 What are the error values associated with your chosen control points?

10. Pan around the source image to the ID number with the highest error value. With the maps locked you'll see both the Source Map and the Virtual Earth move together.

   Important note: To zoom directly to one of the correspondence points, double-click on its ID in the box. MapCruncher will now be centered on that control point.

   If any of your error values are very high, you can change them by doing the following:

   a. Zoom in to the selected point. You'll see a green dot indicating where MapCruncher thinks the point should be and a line showing the distance from where you placed the correspondence point to where MapCruncher thinks it should be.

   b. Click **Unlock**.

   c. Move the crosshairs of the Source Map to the new spot.

   d. Click **Update**.

   e. Click **Lock** again to view the updated error terms.

11. Pan around the Source Map, viewing how well points on the interior and on the fringes match up between the Source Map and the Virtual

Earth representation. Add a few more correspondence points if you feel necessary to adjust for areas that don't match up well.

3.3 Matching the Source Map to the Virtual Earth

When you have the Source Map set up the way you want it, it's time to complete the process.

1. You may first want to save your work, by selecting the **File** pull–down menu and choosing **Save Mashup**. Give your work a file name and save it on your machine (it will save with the file extension ".yum").

2. Back in MapCruncher, click the **Render** button in the lower left–hand corner. The Render dialog box will open.

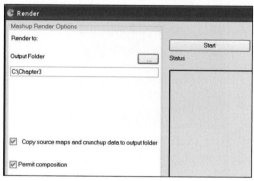

(Source: Used with permission from Microsoft)

3. Press the **"Pick"** button (the one with the "..." symbol on it) next to the Output Folder and then navigate to a folder on your machine to save the aligned file to.

4. Put checkmarks in the **Copy source maps and crunchup data to output folder** and the **Permit composition** boxes.

5. Click the **Start** button to begin the process.

    Important note: The render process may take several minutes and could fill up about 200 MB of space on your machine.

6. You'll see the new layer being "built" in the Status window of the Render dialog box.

7. After a couple minutes, the "Render Complete" box will appear. Click **OK** in this box.

8. Click on the **Preview rendered results** text at the bottom of the Render dialog box. The Mashup Viewer will open to see the map lined up with Virtual Earth.

9. From the Mashup Viewer's **Virtual Earth Background** pull-down menu, select the **Hybrid** option.

Question 3.6 How does the map match up with the roads leading in and out of campus? Are there any locations where the roads don't match?

10. Zoom in to the image—you'll see that the color white in the original map is now transparent and thus features like the roads in the aerial image can be seen through the map. Other colors can be made transparent as well, to better match up the new map with aerial imagery.

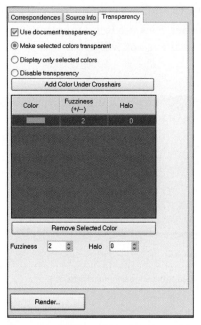

(Source: Used with permission from Microsoft)

11. Close the **Mashup Viewer** and back in MapCruncher, select the **Transparency** tab.

12. Put a checkmark in the **Use document transparency** box.

13. Choose the **Make selected colors transparent** option.

14. In the Source Map, pan the map so that the green color (for instance, the greenery in the campus center) is under the crosshairs, and then click the **Add Color Under Crosshairs** button. The green color will appear in the box, while the green on the Source Map should change to pink.

15. Click the **Render** button again.

16. In the Render dialog box, save the results to a new folder. Click **Start** and the image will re-render, this time with the color green being made transparent. Preview the rendered results again.

17. In the new Mashup Viewer, the map will be rendered, but will be different this time.

18. From the **Virtual Earth Background** pull–down menu, select the **Aerial Photos** option.

Question 3.7 How does the map match up with the aerial imagery around campus? Are there any locations where the images don't really match (that you can see with the greenery and roads made transparent)?

Closing Time

MapCruncher is a highly versatile tool that is used for making data (such as unreferenced images) match the map. Any type of map that is drawn to scale (such as a park trail map, a fairgrounds map, a theme park diagram, or an unreferenced image) can be aligned using MapCruncher. You can now exit MapCruncher by selecting **Exit** from the **File** pull–down menu.

4

Finding Your Location with the Global Positioning System

GPS Origins, Position Measurement, Errors, Accuracy, GNSS around the World, Applications, and Geocaching

Think about this—you're out hiking in a forest, a storm comes up and you need to quickly take shelter well off the trail. When things are all clear, you've gotten completely disoriented, can't find your location, and have gotten truly lost. Or consider this—you're driving in rural, unfamiliar territory late at night, you've missed an important turn, and none of the names or route numbers on the signs seem familiar at all. In both cases, you're out of cell phone range, and even though you have a printed map, you really have no idea where you are to start navigating yourself out of this mess.

A few years ago, these types of situations could have been potentially disastrous, but today there's a geospatial technology tool that's ready and available to help get you out of these situations. Using a small, hand-held, relatively inexpensive device, you can find the coordinates of your position on Earth's surface to within several feet of accuracy. Simply stand outside, turn the device on, and shortly your position will be found, and perhaps even plotted on a map stored on the device. Start walking around and it'll keep reporting where you are, and even take into account some other factors like the elevation you're at, the speed at which you're moving, and the compass direction in which you're going. Even though it all sounds like magic, it's far from it—there's an extensive network of equipment across Earth and in space continuously working in conjunction with your hand-held receiver to locate your position. This, in a nutshell, is the operability of the Global Positioning System (**GPS**).

GPS receivers have become so widespread that they seem to be used everywhere. In-car navigation systems have become standard items for sale on the shelves of an electronics store. GPS equipment is common at sporting goods stores, where it's used by campers, hikers, fishing enthusiasts, and golfers. Runners can purchase a watch containing a GPS receiver that will tell them their location, the distance they run, and their average speed.

GPS the Global Positioning System, a technology using signals broadcast from satellites for navigation and position determination on the Earth.

Smartphones will often have GPS capability as a standard function. Use of the GPS is free for everyone and available worldwide, no matter what remote area you find yourself lost in. So the first questions to ask are —who built this type of geospatial technology and why?

Who Made GPS?

The acronym "GPS" seems to be in the common vocabulary to describe any number of devices that use Global Positioning System technology for finding a location, but it's really an inaccurately used term. The term "GPS" originated with the initial designers and developers of the setup, the U.S. Department of Defense, and is part of the official name of the system, called **NAVSTAR GPS**. Other countries besides the United States have developed or are currently engaged in developing systems like the NAVSTAR GPS, so when we're describing "GPS," we're really discussing the United States system. Since the United States isn't the only nation with this technology, the more accurate way of describing things would be to call GPS one type of Global Navigation Satellite System (**GNSS**).

NAVSTAR GPS the United States Global Positioning System.

GPS isn't the first satellite navigation system. During the 1960s, the U.S. Navy used a system called Transit, which determined the location of sea-going vessels with satellites. The drawback of Transit was that it didn't provide continuous location information—you would have to wait a long time to get a fix on your position rather than always knowing where you were. The Naval Research Laboratory Timation program was another early satellite navigation program of the 1960s, which used accurate clock timings for ground position determination from orbit. The first GPS satellite was launched in 1978, and the twenty-fourth GPS satellite was launched in 1993, completing an initial full operating capacity of satellites of the system for use. Today, GPS is overseen and maintained by the 50th Space Wing, a division of the U.S. Air Force headquartered at Schriever Air Force Base in Colorado.

GNSS the Global Navigation Satellite System, an overall term for the technologies using signals from satellites for finding locations on the Earth's surface.

What Does the Global Positioning System Consist of?

Finding your position on the ground with GPS relies on three separate components, all operating together. They are as follows: a space segment, a control segment, and a user segment.

Space Segment

There are numerous GPS satellites (also referred to as "SVs" or "Space Vehicles") orbiting Earth in fixed paths, which make up the **space segment**. GPS satellites make two orbits around Earth every day (their orbit time is actually about 11 hours and 58 minutes) at an altitude of 20,200 kilometers (12,552 miles). The satellites are set in specific orbits, called a **constellation**,

space segment one of the three segments of GPS, consisting of the satellites and the signals being broadcast from space.

constellation the full complement of satellites comprising a GNSS.

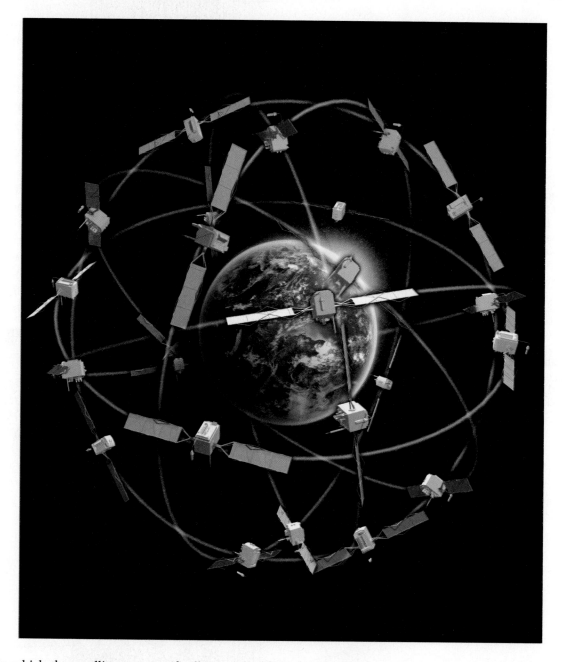

in which the satellites are specifically arranged for maximum coverage over Earth (**Figure 4.1**). The way the constellation is designed allows for a person to be able to receive enough GPS signals to find their location wherever they are on the planet. Twenty-four satellites is the minimum for a full constellation, and there are currently several additional operational GPS satellites in orbit to improve the global coverage.

The job of GPS satellites is to broadcast a set of signals down to Earth from orbit. These signals (we'll discuss this later) contain information about the

FIGURE 4.1 The constellation of GPS satellites orbiting Earth.

position of the satellite and the precise time at which the signal was transmitted from the satellite (GPS satellites measure time using extremely accurate atomic clocks onboard the satellites). The signals are sent on carrier frequencies: L1 (broadcast at 1575.42 MHz) and L2 (broadcast at 1227.60 MHz). There are also additional carrier frequencies planned for future GPS upgrades, including an L5 frequency to be used for safety-of-life functions. You need to have a direct line of sight to the satellites to receive these frequencies (which means that you have to be in the open-air, not indoors).

Control Segment

control segment one of the three segments of GPS, consisting of the control stations that monitor the signals from the GPS satellites.

The **control segment** of GPS represents a series of worldwide ground stations that track and monitor the signals being transmitted by the satellites. These ground stations are spread out to enable continuous monitoring of the satellites. The control stations collect the satellite data and transmit it to the master control station at Schriever Air Force Base. In the control segment, corrections and new data are uploaded to the satellites so that the satellites will be broadcasting correct data. The ground stations also monitor the satellites' position and relay updated orbit information to the satellites (**Figure 4.2**).

FIGURE 4.2 A control station monitors the signals from the satellites and sends correction data back to the satellite.

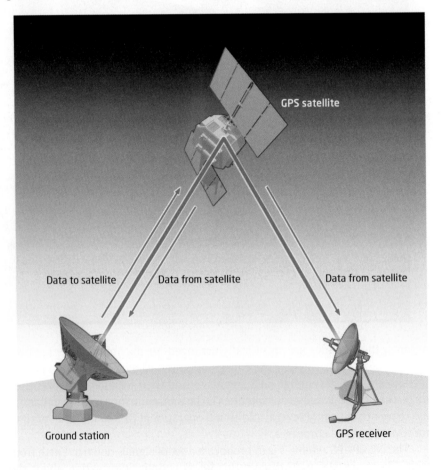

GPS satellite

Data to satellite   Data from satellite   Data from satellite

Ground station   GPS receiver

User Segment

The **user segment** represents a GPS unit somewhere on the Earth that is receiving the signals from the satellite. The receiver can then use this information to compute its position on the Earth. A key component of a GPS receiver is how many **channels** the receiver has—the number of channels reflects the number of satellites the receiver can obtain signals from at one time. Thus, a 12-channel receiver can potentially pick up signals from up to 12 satellites. Also, the type of unit affects which of the frequencies it can receive—it can either be a **single frequency** receiver (which can pick up the L1 frequency) or a **dual frequency** model (which can read both L1 and L2). One thing to keep in mind about the receiver is that no matter what size, capabilities, or number of channels it has, a receiver (like the name implies) can only receive satellite data, not transmit data back up to the satellites. As such, GPS is referred to as a one-way ranging system since the user on the ground cannot send any information to a satellite.

How Does GPS Find Your Position?

At this point, the three segments of GPS have satellites broadcasting signals, ground stations monitoring and correcting those signals, and users receiving those signals—and all three of these segments work together to determine your position on the Earth's surface. The signals being sent from the satellites are the key to the whole process—the signals contain information about the satellite's status, the orbit, and location (referred to as the **almanac**) that's sending them, as well as more precise data about satellite location (this information is referred to as the **ephemeris**).

The information is sent in one of two digital pseudo-random codes—the **C/A code** (coarse acquisition code) and the **P code** (precise code). The C/A code is broadcast on the L1 carrier frequency and is the information that civilian receivers can pick up. The P code is broadcast on the L1 and L2 carrier frequencies and contains more precise information, but a military receiver is required to pick up this signal. The **Y code** is an encrypted version of the P code and is used to prevent false P code information from being sent to a receiver by hostile forces. This "anti-spoofing" technique is commonly used to make sure that only the correct data is being received. Basically, the satellites are transmitting information in the codes about the location of the satellite, and they can also be used to determine the time when the signal was sent. By using this data, the receiver can find its position relative to that one satellite.

The signals are transmitted from space using high-frequency radio waves (the L1 and L2 carrier frequencies). These radio waves are forms of electromagnetic energy (to be discussed in Chapter 10) and will thus be moving at the speed of light. Your receiver can compute the time it took for the signal to arrive from the satellite. If you know how long the transmission time (t) was, and also that the radio waves are moving at the speed of light (c), then the distance between you and that one satellite can be calculated

user segment one of the three segments of GPS, consisting of the GPS receivers on the ground that pick up the signals from the satellites.

channels the number of satellite signals a GPS unit can receive.

single frequency a GPS receiver that can pick up only the L1 frequency.

dual frequency a GPS receiver that can pick up both the L1 and L2 frequency.

almanac data concerning the status of a GPS satellite, which is included in the information being transmitted by the satellite.

ephemeris data referring to the GPS satellite's position in orbit.

C/A code the digital code broadcast on the L1 frequency, which is accessible by all GPS receivers.

P code the digital code broadcast on the L1 and L2 frequencies, which is accessible by the military.

Y code an encrypted version of the P code.

pseudorange the calculated distance between a GPS satellite and a GPS receiver.

trilateration finding a location in relation to three other points of reference.

by multiplying t by c. This result will give you the **pseudorange** (or distance) between your receiver and the satellite transmitting the signal.

Unfortunately, this still doesn't tell us much—you know where you are on the ground in relation to the position of one satellite over 12,000 miles away, which gives very little information about where you are on the Earth's surface. It would be like waking up in an unknown location and a passerby tells you that you're 500 miles away from Boston—it gives you very little to go on in determining your real-world location.

The process of determining your position is referred to as **trilateration**, which means using three points of reference to find where you are. Trilateration in two dimensions is commonly used when plotting a location on a map. Take for example that you're on a trip to Michigan, you've driven all night, and arrived in an unknown location with no identifying information as to where you are. The people you're travelling with are nowhere to be found, and the first person you bump into tells you (somewhat unhelpfully) that you're 50 miles away from Ann Arbor, Michigan (see Figure 4.3). That puts you somewhere on a circle sweeping outward 50 miles from Ann Arbor, giving you far too many potential places for you to be.

The second person you run into tells you (again, not being overly helpful) that you're located 150 miles away from Traverse City, Michigan. This again puts your location somewhere on a circle 150 miles from Traverse City—but when you combine this with the information that places you 50 miles away from Ann Arbor, you can limit your options down considerably. There are only two cities in Michigan that are 50 miles from Ann Arbor and 150 miles away from Traverse City—you're either in Lansing or Flint (Figure 4.4).

Luckily, the third person you see tells you that you're 60 miles from Kalamazoo. Your first thought should be "Who are all these geographically minded people I keep running into?" while your second thought is that you now know exactly where you are. Lansing is the only option that fits all three of the distances from your reference points, so you can disregard Flint and be satisfied that you've found your location (see Figure 4.5 and *Hands-on Application 4.1: Trilateration Concepts*).

FIGURE 4.3 Measuring distance from one point.

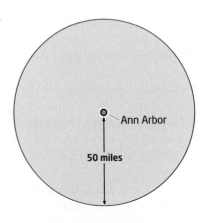

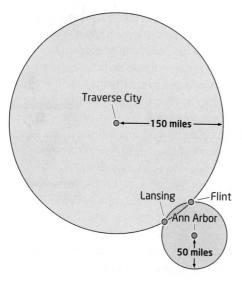

FIGURE 4.4 Using distances from two points in trilateration.

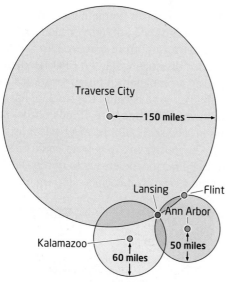

FIGURE 4.5 The results of trilateration (with distances from three points) to find a location.

Hands-on Application 4.1

Trilateration Concepts

Trilateration can be used to determine your location on a map. Open your Web browser and go to **http:// electronics.howstuffworks.com/gadgets/travel/ gps1.htm** to view the article by Marshall Brain and Tom Harris at *How Stuff Works*. Check out how three measurements are being made to determine a single location (like in the previous Michigan example), then start Google Earth and zoom to your location. Use Google Earth and its measuring tools to set up a similar scenario—find your location relative to three other cities or locations and calculate the measurements from each that are needed to trilaterate your position.

The same concept applies to finding your location using GPS, but rather than locating yourself relative to three other points on a map, your GPS receiver is finding its position relative to three satellites. Also, since a position on a three-dimensional (3D) Earth is being found with reference to positions surrounding it, a spherical distance is calculated rather than a flat circular distance. This process is referred to as trilateration in three dimensions (or **3D trilateration**). The concept is similar, though finding the receiver's position relative to one satellite means it's located somewhere on one sphere (similar to only having the information about being 50 miles from Ann Arbor). By the receiver finding its location in relation to two satellites, it's finding a location on a common boundary between two spheres. Lastly, by finding the location relative to three satellites, there are only two places those three spheres can intersect, and the way the geometry works out, one of them would be in outer space instead of on Earth's surface, so that one can be disregarded. That leaves one location where all three spheres intersect and thus one location on Earth's surface relative to all three satellites. That position would be the location of the GPS receiver (Figure 4.6).

However, there's a problem with all of this. Remember that what's being calculated is a pseudorange based on the speed of light (which is a constant) and the time it takes for the signal to transmit from space to Earth. If that time was off slightly, a different value for distance would be calculated, and your receiver's position could be placed at a completely different location away from its real position. This becomes a bigger issue because the satellite has a super-precise, but very expensive (somewhere in the range of tens of thousands of dollars) atomic clock, while the off-the-shelf GPS receiver has a less-precise (and inexpensive) quartz clock. Obviously, GPS receivers can't have atomic clocks in them or else they would be far too expensive to purchase. As such, the receiver's clock isn't as accurate as the atomic clock, and time differences

3D trilateration
finding a location on the Earth's surface in relation to the positions of three satellites.

FIGURE 4.6   The results of 3D trilateration to find a location using GPS.

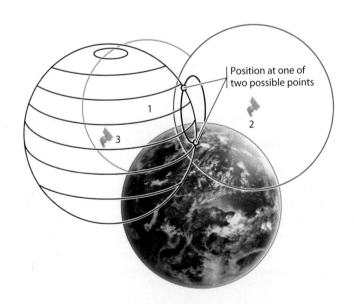

Position at one of two possible points

can cause inaccuracies in distances when it comes to finding a position. Fortunately, there's an easy solution to this mess, and that is, your receiver picks up information from a fourth satellite. By using a fourth measurement, the clock errors can be corrected for and your position can be accurately determined. The position you'll receive is measured using the WGS84 datum.

Why Isn't GPS Perfectly Accurate?

If GPS can be used to find a location on Earth's surface, the next question to ask is "just how accurate is GPS?" Through the use of pseudorange measurements from four satellites, it would seem the system should only give the user one accurately measured location. However, in average everyday use with an ordinary off-the-shelf civilian receiver, accuracy would be about 15 meters (or less) in the horizontal direction. This doesn't mean that the location is always off by 15 meters—it could be off by 5, 10, or 15 meters. This is due to a number of factors that can interfere with or delay the transmission or reception of the signal—and these delays in time can cause errors in position.

One of the biggest sources of error was actually a deliberate one worked into the system to intentionally make GPS less accurate. The United States introduced **Selective Availability** (SA) into GPS with the goal of intentionally making the C/A signal less accurate, presumably so that enemy forces couldn't use GPS as a tool against the United States military. Selective Availability introduced two errors into the signals being transmitted by the satellites—a delta error (which contained incorrect clock timing information) and an epsilon error (which contained incorrect satellite ephemeris information). The net effect of this was to make the accuracy of the C/A about 100 meters. As you can imagine, this level of accuracy certainly limited GPS in the civilian sector—after all, who wants to try to land an airplane with GPS when the accuracy of the runway location could be hundreds of feet off? In the year 2000, the U.S. Government turned Selective Availability off and has no announced plans to turn it back on (in fact, the U.S. Department of Defense has announced that new generations of GPS satellites will not have the capability for Selective Availability).

While Selective Availability is an intentional error originating from the satellites, other errors can naturally occur at the satellite level. Ephemeris errors can occur when the satellite broadcasts incorrect orbit position information. These types of errors indicate that the position of the satellite is not where it is supposed to be and a correction of ephemeris data hasn't been updated from a control station. Typically, these errors can introduce about 2 meters (or so) of error.

Even the arrangement of satellites in space (that is, how they are distributed) can have an effect on the position determination. The further the satellites are spread out from one another and the wider the angle between them, the more accurate a measurement will be obtained by the receiver. The error introduced by a poor geometric arrangement of the satellites is referred to as the Position Dilution of Precision or **PDOP**. Some receivers will calculate a value for PDOP for you

Selective Availability the intentional alteration of the timing and position information transmitted by a GPS satellite.

PDOP the Position Dilution of Precision—it describes the amount of error due to the geometric position of the GPS satellites.

FIGURE 4.7 The effects
of the ionosphere and
troposphere on GPS
signals.

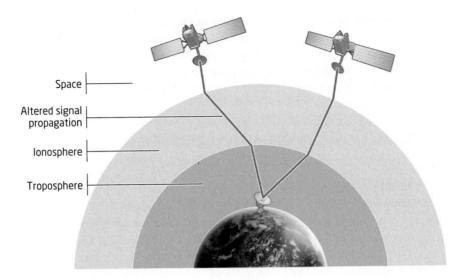

FIGURE 4.7 The effects of the ionosphere and troposphere on GPS signals.

to indicate what relative error could be introduced because of the arrangement of the satellites being used (basically, the lower the value the better).

Also keep in mind that the signals are being broadcast from space and have to pass through Earth's atmosphere before reaching the receiver. Though some of these atmospheric effects can be modeled and attempted to be accounted for, atmospheric problems can still introduce errors in accuracy by making changes to the speed of the signals (Figure 4.7). Atmospheric interference, especially in the ionosphere and troposphere, can cause delays in the signal before it gets to the ground. The ionosphere can alter the propagation speed of a signal, which can cause an inaccurate timing measurement, while the amount of water vapor in the troposphere can also interfere with the signals passing through it and also cause delays. It's estimated that the problems caused by the ionosphere can cause about 5 meters (or more) of position inaccuracy, while the troposphere can add about another 0.5 meters of measurement error.

The atmosphere is not the only thing that can interfere with the signal reaching the receiver. The receiver needs a good view of the sky to get an initial fix, so things like heavy tree canopy may interfere with the signal reception. In addition, the signals may be reflected off objects (such as tall buildings or bodies of water) before reaching the receiver rather than making a straight path from the satellite (Figure 4.8). This **multipath effect** can introduce additional delays into the reception of the signals since the signal is reaching the receiver later than it should (and add further error to your position determination).

multipath effect an error caused by a delay in the signal due to reflecting from surfaces before reaching the receiver.

All of these various types of error measurements can stack onto one another to possibly contribute several meters worth of errors into your position determination. When ephemeris introduces a few meters of error, the ionosphere and troposphere add a few more, and then multipathing and PDOP add a few more, before you know it they all start to add up to some significant

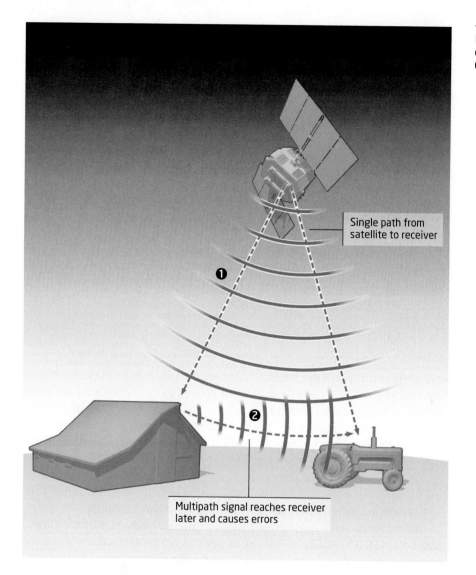

Single path from satellite to receiver

❶

❷

Multipath signal reaches receiver later and causes errors

FIGURE 4.8 The effect of multipathing on the GPS signal.

differences in where your position gets plotted. So, there should be some manner of solutions to improve the accuracy of GPS and get a better fix on your position.

How Can You Get Better Accuracy from GPS?

There are a few methods of obtaining a more accurate position measurement using GPS to try to overcome the error sources just described. The first of these is a process referred to as Differential GPS (or **DGPS**). This method uses a series of base stations at locations on the ground to provide a correction for GPS position determination. The base station is positioned at a place

DGPS differential GPS—a method using a ground-based correction in addition to the satellite signals in position determination.

FIGURE 4.9 The operation of DGPS.

where the coordinates are known and it receives signals from the satellites. Since it knows what its coordinates are supposed to be, it can calculate a correction to compensate for errors in where the position measurement from the satellites says it should be. The base station can then broadcast this correction out to receivers equipped to collect this differential correction. Thus, when you're using DGPS, your receiver is picking up the usual four satellite signals plus an additional correction from a nearby location on the ground (see Figure 4.9 for an example of the operation of DGPS). As a result of this, the position measurement will get reduced to less than 5 meters.

Today, there are a number of DGPS locations worldwide. In the United States, the U.S. Coast Guard has set up several reference stations along the coasts and waterways to aid ships in finding their location and navigation. The U.S. Department of Transportation operates the **NDGPS** (Nationwide Differential GPS) system of DGPS locations across the country, aimed at providing more accurate position information to drivers and travelers. NDGPS reduces position error down to one to two meters.

A related program, **CORS** (Continuously Operating Reference Stations), is operated by the National Geodetic Survey, and consists of numerous base stations located across the United States and other places throughout the world. CORS collects GPS information and makes it available to users of GPS data (such as surveyors and engineers) to make more accurate position measurements. CORS data provides a powerful tool for location information for a variety of applications (including being established in Iraq by the U.S. Army to aid in rebuilding there).

NDGPS the National Differential GPS—it consists of ground-based DGPS beacons around the United States.

CORS the Continuously Operating Reference Stations—a system operated by the National Geodetic Survey to provide a ground-based method of obtaining more accurate GPS positioning.

Satellite Based Augmentation Systems (**SBAS**) are additional resources to use in improving accuracy. SBAS works similar to DGPS in that a correction is calculated and picked up by your receiver in addition to the regular four satellites. However, in this case, the correction is being sent from an additional new satellite, not from an Earth-bound station. An example of a SBAS is called the Wide Area Augmentation System (**WAAS**), which was developed by the Federal Aviation Administration (FAA) to obtain more accurate position information for aircraft. WAAS has since spread to uses beyond aircraft as a quick and easy method of improving overall GPS position accuracy.

WAAS operates through a series of base stations spread throughout the United States that collect and monitor the signals sent by the GPS satellites. These base stations calculate position correction information (similar to the operation of DGPS) and transmit this correction to a geostationary WAAS satellite. These WAAS satellites then broadcast this correction signal back to Earth, and if your receiver can pick up this signal, it can use this correction information in its position calculation (Figure 4.10). WAAS reduces the position error to 3 meters or less with its use. A drawback to WAAS is that it will only function within the United States (including Alaska and Hawaii) and nearby portions of North America (although WAAS base stations have

SBAS Satellite Based Augmentation System—a method of using correction information sent from an additional satellite in the GPS position determination.

WAAS the Wide Area Augmentation System—a satellite based augmentation system that covers the United States and other portions of North America.

FIGURE 4.10 The setup and operation of the WAAS system.

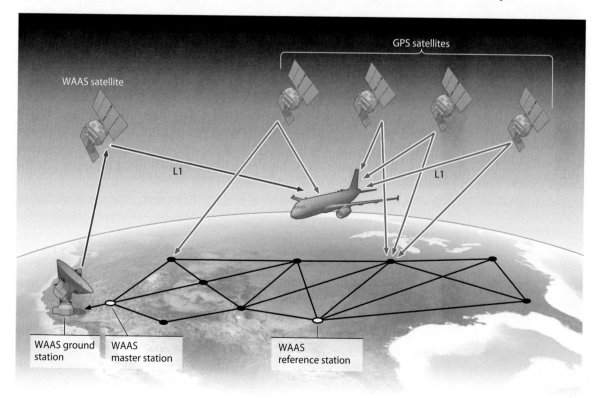

recently been added to Mexico and Canada to greatly expand the range of WAAS benefits). Other systems exist beyond WAAS and DGPS but use similar concepts (some sort of correction is calculated and transmitted to a receiver for it to use in its position determination).

What Other GNSS Are There Beyond GPS?

This chapter has been devoted to the operation, errors, and accuracy issues surrounding GPS, but as mentioned early on, "GPS" refers to the US NAV-STAR GPS. Other countries have developed (or are in the process of developing) their own versions of GPS or GPS augmentation systems, presumably as not to be completely dependent on U.S. technology. There are other full constellations of GPS-style satellites being developed and placed into orbit, with numerous other enhancement systems in operation or in planning around the globe.

GLONASS the former USSR's (now Russia's) GNSS, currently rebuilding to a constellation of satellites.

GLONASS was the name of the USSR's counterpart to GPS and operated in a similar fashion. The GLONASS equivalent to the C/A code operated at horizontal accuracies of about 100 m (or better) with an equivalent of the P code for more accurate use. GLONASS consisted of a full constellation of satellites with an orbital setup similar to GPS. By 2001, however, there were only a few GLONASS satellites in operation, but Russia now has a renewed program and more GLONASS satellites have been launched, to possibly have a full operational constellation in the near future. In 2007, Russia indicated that the civilian signal from GLONASS would be made freely available for use.

Galileo the European Union's GNSS, currently in development.

Galileo is the European Union's version of GPS, which, when completed, is projected to have a constellation of 30 satellites and operate in a similar fashion to GPS. The first of the GIOVE (Galileo In Orbit Validation Element) satellites was launched in 2005, and the program has continued development from then. Galileo promises four different types of signals, including information available to civilians as well as restricted signals. Also in development is China's version of GPS, **Compass**, which is projected to also have a full constellation of satellites. The first Compass satellite was launched in 2007.

Compass China's GNSS, currently in development.

Other Satellite Based Augmentation Systems have been developed across the world to act similar to WAAS but to function in other regions outside of North America. **EGNOS** (European Geostationary Navigation Overlay System) is sponsored by the European Union to provide improved accuracy of GPS throughout Europe. EGNOS functions the same way WAAS does—a series of base stations throughout Europe monitor GPS satellite data, calculate corrections, then transmit this correction to geostationary EGNOS satellites over Europe, and these satellites broadcast this data to Earth. An EGNOS-enabled receiver is necessary to utilize this correction data, and the system is capable of accuracies of about 1.5 meters. Another SBAS operates out of Japan, called **MSAS** (Multifunctional Satellite Augmentation System), which

EGNOS a Satellite Based Augmentation System that covers Europe.

MSAS a Satellite Based Augmentation System that covers Japan and nearby regions.

covers Japan and portions of the Pacific Rim. MSAS operates in a similar way to WAAS and EGNOS, just in a different region of the world and consists of two satellites providing coverage.

Many other GPS enhancement systems exist or are in development throughout the world, including the following:

- ◉ Beidou—an SBAS operating in China
- ◉ GAGAN—an SBAS in development for India
- ◉ IRNSS—an Indian satellite navigation system, consisting of seven satellites, currently in development
- ◉ QZSS—an expanded SBAS based in Japan, currently in development

What Are Some Applications of GPS?

GPS is used in a wide variety of settings. Any application that requires location-based data to be collected firsthand can apply GPS to the problem. The following are just a handful of ways GPS is being used in businesses, jobs, and the public sector.

Emergency Response

When you make a 911 call, the call center is able to fix your location and route emergency services (police, firefighters, or ambulances) to where you are. When you make a 911 call from a cell phone, the Enhanced 911 (E-911) system can determine your position—either using the cell towers to determine where you are, or using a GPS receiver in your phone to find your location.

Farming

Precision farming relies on getting accurate location data about factors such as watering levels, crops, or soils. Having the ability to map this data gives farmers information about where crops are thriving and what areas need attention. GPS serves as a data-collection tool to quickly (and accurately) obtain location information for farmers (that can then be combined with other aspects of geospatial technology, such as the mapping and analysis capabilities of GIS or remotely sensed imagery).

Forensics

Human remains recovery has been greatly assisted through the use of GPS. Rather than having to rely solely on surveying equipment, GPS can provide quick data and measurements, and can provide an additional tool in addition to other field-based measurements. The same concepts have been applied to identifying the locations of artifacts and remains at archeological sites.

Public Utilities

City or county data for items that need to be measured manually (such as culverts, road signs, or manhole locations) can be done quickly and easily with

GPS. When field-based data is necessary to create or update utility data, workers equipped with GPS can collect this type of data quickly and accurately.

Transportation

GPS provides a means of continuously tracking travel, whether on the road, in the water, or in the air. Since GPS receivers can constantly receive signals from the satellites, your travel progress can be monitored to make sure that you stay on track. When your GPS-determined position can be plotted in real-time on a map, you can see just where you've traveled to and measure your progress. Delivery and shipping companies can use this type of data to keep tabs on their fleets of vehicles to see if they deviate from their routes.

Wildlife Management

Similarly to tracking a vehicle, wildlife and endangered species can be tagged with a GPS receiver which monitors their location and transmits it to a source capable of tracking those movements. In this way, migration patterns of animals can be determined and patterns of wildlife movements can be observed.

geocaching using GPS to find locations of hidden objects based on previously obtaining a set of coordinates.

Beyond its uses in the private and public sectors, GPS has become a tool to assist in personal recreation. If you're a runner, GPS can track your position while exercising, along with measuring your speed. If you're a hiker, GPS can keep you from getting lost when you're deep into unfamiliar territory. If you're into fishing, GPS can help locate places supposed to be well stocked with fish. If you're a golfer, you can use GPS to measure your ball's location and the distance to the pin. In fact, there's a whole new form of recreation that's grown in popularity along with GPS, called **geocaching**.

Thinking Critically with Geospatial Technology 4.1

What Happens if GPS Stops Working?

What happens if the Global Positioning System stops working—or at least working at 100% capability? If the satellites stop operating without being replaced, GPS would essentially stop functioning. If the number of satellites would drop below the full operational capacity, then GPS would work some of the time (or could have greatly reduced accuracy). The scenario of an operational failure of GPS is explored in this article from *GPS World*: **http://www.gpsworld.com/gnss-system/news/gps-risk-doomsday-2010-7092**.

If something like this would come to pass, what would be the effect of a lack of GPS? What industries would be impacted without having access to GPS? What would be the effect on navigation of aircraft or other travel? What are the military implications of a world without GPS? How would all of these factors affect your life?

By the same token, is society too reliant on GPS? Since Selective Availability was turned off in 2000, there's been an overwhelming amount of GPS usage in many aspects of everyday life. Is the availability of GPS taken for granted within society? If the system could fail or be negated, what impacts would this have on people reliant on this technology always being available to them?

Hands-on Application 4.2

Things to Do Before You Go Geocaching

Geocaching is a widespread and popular recreational activity today. As long as you have a GPS receiver and some coordinates to look for, you can get started with geocaching. This chapter's lab will have you investigating some specific geocaching resources on the Web, but before you start looking for caches, there are some online utilities to explore first. Open your Web browser and go to **http://www.groundspeak.com**. Groundspeak operates a main geocaching Website as well as providing some tools to use. On the main page are links to several GPS-related sites:

1. Geocaching.com (**http://www.geocaching. com**) – Check around for locations of geocaches near you (we'll use this during the lab).

2. Waymarking.com (**http://www.waymarking. com**) – This Website provides a way to find the coordinates of interesting locations ("waymarks") that have been uploaded by users. Type in your zip code to see what's been labeled as a waymark (and collect its coordinates) near you.

3. Whereigo.com (**http://www.wherigo. com**) – A Website to download a player and a toolset for constructing games (referred to as "cartridges") and other applications for portable GPS devices.

4. Cache In Trash Out (**http://www.geocaching. com/cito**) – An environmental initiative dedicated to cleaning up areas where geocaches are hidden.

In geocaching, certain small objects (referred to as "geocaches") are hidden in an area and the coordinates of the object are listed on the Web (a log is maintained). From there, you can use a GPS receiver to track down those coordinates (which usually involve hiking or walking to a location) and locate the cache. In essence, geocaching is a high-tech outdoor treasure hunt (see *Hands-on Application 4.2: Things to Do Before You Go Geocaching*). This type of activity (using a GPS receiver to track down pre-recorded coordinates to find objects or locations) is also used in educational or classroom activities to teach students the relationship between a set of map coordinates and a real-world location, as well as basic land navigation and GPS usage.

Chapter Wrapup

This chapter provided an introduction to how the Global Positioning System operates as well as how GPS can locate where a receiver is on the ground. A growing application of GPS is navigation systems used in your car, such as brands like Garmin, Magellan, or Tom-Tom. These types of systems plot not only your location on a map using GPS, but also show surrounding roads and nearby locations (such as restaurants or gas stations), locate an address, and then route you from your current location to another destination via the shortest path. The only one of these features that's actually using GPS is finding your current location—the rest are all functions that are covered in Chapter 8. Until then, starting in the next chapter, we'll start focusing on the actual computerized mapping and analysis that underlies those types of in-car systems, all concepts of geographic information systems.

This chapter's lab will start putting some of these GPS concepts to work, even if you don't have a GPS receiver.

Important note: The references for this chapter are part of the online companion for this book and can be found at **http://www.whfreeman.com/ shellito1e.**

Key Terms

GPS (p. 73)

NAVSTAR GPS (p. 74)

GNSS (p. 74)

space segment (p. 74)

constellation (p. 74)

control segment (p. 76)

user segment (p. 77)

channels (p. 77)

single frequency (p. 77)

dual frequency (p. 77)

almanac (p. 77)

ephemeris (p. 77)

C/A code (p. 77)

P code (p. 77)

Y code (p. 77)

pseudorange (p. 78)

trilateration (p. 78)

3D trilateration (p. 80)

Selective Availability (p. 81)

PDOP (p. 81)

multipath effect (p. 82)

DGPS (p. 83)

NDGPS (p. 84)

CORS (p. 84)

SBAS (p. 85)

WAAS (p. 85)

GLONASS (p. 86)

Galileo (p. 86)

Compass (p. 86)

EGNOS (p. 86)

MSAS (p. 86)

geocaching (p. 88)

4.1 Geospatial Lab Application

GPS Applications

This chapter's lab will examine using the Global Positioning System for a variety of activities. The big caveat is that the lab can't assume that you have access to a GPS receiver unit and would be able to use it in your immediate area (and unfortunately, we can't issue you a receiver with this book). It would also be useless to describe exercises involving you running around (for instance) Boston, Massachusetts, collecting GPS data, when you may not reside anywhere close to Boston.

Thus, this lab will use the free Trimble Planning Software for examining GPS satellite positions and other factors used in planning for the use of GPS for data collection. In addition, some sample Web resources are used to examine related GPS concepts for your local area.

For some good introductory information about Trimble Planning Software (whose info helped guide the development of this lab), also check out the article written by Leszek Pawlowicz titled "Determining Local GPS Satellite Geometry Effects On Position Accuracy." This article is available online at http://freegeographytools.com/2007/determining-local-gps-satellite-geometry-effects-on-position-accuracy.

Important note: Though this lab is short, it can be significantly expanded if you have access to GPS receivers. In that case, some of the geocaching exercises described in Section 4.5 of this lab can be implemented. Alternatively, rather than just examining cache locations on the Web (as in Section 4.4), you could use the GPS equipment to hunt for the caches themselves.

Objectives

The goals for you to take away from this lab are:

- Examine satellite visibility charts to determine how many satellites are available in a particular geographic location, as well as when during the day these satellites are available.

- Read a DOP chart to determine what times of day have higher and lower values for DOP for a particular geographic location.

- Explore some Web resources to see where publicly available geocaches or permanent marker sites are that you can track down with a GPS receiver.

Obtaining Software

The current version of Trimble Planning Software (2.9) is available for free download at http://www.trimble.com/planningsoftware_ts.asp.

Important note: Software and online resources sometimes change fast. This lab was designed with the most recently available version of the software at the time of writing. However, if the software or Websites have significantly changed between then and now, an updated version of this lab (using the newest versions) is available online at http://www.whfreeman.com/shellito1e.

Lab Data

There is no data to copy in this lab. All data comes as part of the Trimble Planning Software that is installed with the program or can be downloaded as part of the lab.

Localizing This Lab

Note: This lab uses one location and one date to keep information consistent. Although this lab looks at GPS sky conditions in Orlando, Florida, there are many more geographic locations that can be selected using Trimble Planning Software. Its purpose is to be able to evaluate local conditions for GPS planning—so rather than selecting Orlando, find the closest city to your location and examine that data instead. You can also use today's date rather than the one used in the lab.

Similarly, in Section 4.4 of the lab, you'll be examining Web resources for geocaching locations in Orlando, Florida. Use your own zip code rather than the one for Orlando and see what's near your home or school instead.

4.1 Trimble Planning Software Setup

1. Start the **Trimble Planning Software** program—the default install folder is called Trimble Office—and then look inside the **Utilities** folder where you will find an icon called **Planning**. This will launch the program.

2. The first thing to do is download a copy of the most current almanac file. This can be done from Trimble's Website at http://www.trimble.com/gpsdataresources.shtml and selecting the option for **GPS/GLONASS almanac in Trimble Planning file format**. Download the file (it's a small text file) by right-clicking on the hyperlink (GPS/GLONASS almanac in Trimble Planning file format) and click on **Save Target As . . .** from the dialog box. Click on the **Save** button and the file "almanac.alm" will be downloaded to your computer.

3. In the Planning program, select **Load** from the **Almanac** pull-down menu. In the Almanac Files dialog, navigate to the location where you saved the almanac.alm file, select it, and then click **Open**. A summary of the satellite information contained in the file will appear on the screen and the almanac information will be loaded.

4.2 Local Area GPS Information

The visibility of GPS satellites will change with the time of day and your position on the Earth. The Trimble Planning program will allow you to examine data from stations around the Earth to determine GPS information.

1. From the **File** pull-down menu, select **Station**. You could also select the **Station icon** from the toolbar:

(Source: Courtesy of Trimble)

The Station Editor dialog box will open.

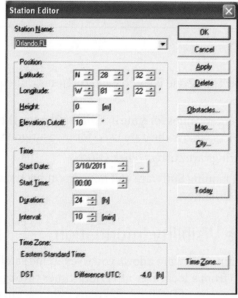

(Source: Courtesy of Trimble)

2. The Station Editor allows you to select any station across the world. For this portion of the lab, we'll be examining GPS satellite visibility near Orlando, Florida.

3. To select an area, click on the **City . . .** button on the right. From the pull-down list, choose **Orlando, FL**.

   Important note: An alternate method for selecting a location is to click on the **Map . . .** button. A world map will appear that allows you to select based on a geographic location—scroll the mouse across the map to see what options are available, and then double-click on the city you want.

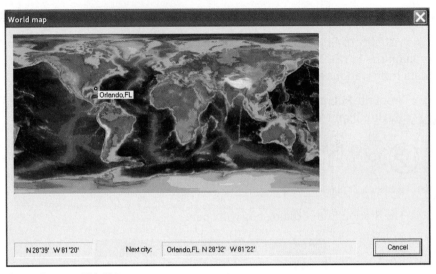

(Source: Courtesy of Trimble)

4. Make sure the Time Zone is set to Eastern Time (United States and Canada).

5. You can also enter values for time and date. Change the date for this lab to **3/10/2011** (as this is the date used for measurements).

6. When the settings are correct, click **OK**.

7. The Trimble Planning Software will now report back satellite information in relation to Orlando.

4.3 Satellite Visibility Information

1. Trimble Planning Software allows you to examine satellite visibility (and other factors) from a local area at various times throughout the day. To choose which group of satellites you're examining, put checkmarks in the Satellite Systems boxes below the toolbar:

(Source: Courtesy of Trimble)

2. For now, only select the **GPS** box.

3. First, we'll examine the number of GPS satellites potentially visible from Orlando at various points of the day. From the **Graphs** pull-down menu, select **Number of Satellites**. You could also choose the **Number of Satellites icon** from the toolbar:

(Source: Courtesy of Trimble)

4. The number of satellites visibility graph will appear, showing the number of visible satellites on the *y*-axis and a 24-hour time frame on the *x*-axis.

Question 4.1 At what times of day are the maximum number of GPS satellites visible? What is this maximum number?

Question 4.2 At what times of day are the minimum number of GPS satellites visible? What is this minimum number?

5. To determine which satellites are visible at specific times, go to the **Graphs** pull-down menu, then choose **Visible Satellites**, then choose **General Visibility**. You could also choose the **Visibility icon** from the toolbar:

(Source: Courtesy of Trimble)

6. The Visibility graph will appear, showing the specific GPS satellite number on the *y*-axis and a 24-hour time frame on the *x*-axis.

Question 4.3 Which satellites are visible from Orlando at noon?

Question 4.4 Which satellites are visible from Orlando at 4:00 PM?

7. To examine the Position Dilution of Precision (PDOP) at a particular time of day, from the **Graphs** pull-down menu, select **DOP**, then select **All Together**. You could also choose the **DOPs icon** from the toolbar:

(Source: Courtesy of Trimble)

8. The graph for Dilution of Precision will appear, showing the values for DOP on the *y*-axis and a 24-hour time frame on the *x*-axis.

Question 4.5 At what time(s) of day will the maximum values for DOP be encountered? What is this maximum value?

Question 4.6 At what time(s) of day will the minimum values for DOP be encountered? What is this minimum value?

9. Take a look at all the graphs and answer Question 4.7.

Question 4.7 Just based on satellite visibility and DOP calculations, what would be some of the peak times to perform GPS field measurements in Orlando? Also, what would be some of the "worst" times?

10. Close all the graphs, then uncheck the **GPS** box in Satellite Systems and place a checkmark in the **WAAS** box instead.

11. Examine the Visibility chart for the available WAAS satellites.

Question 4.8 How many WAAS Satellites does Trimble Planning say are available? At what times are they available? Why is this?

12. Next, examine the Sky Plot (a chart showing the position of the satellites in orbit) for these WAAS satellites. From the **Graphs** pull-down menu, select **Sky Plot**. You could also choose the **Sky Plot icon** from the toolbar:

(Source: Courtesy of Trimble)

Question 4.9 Based on the information from the visibility chart and the sky plot, what satellites are these? (*Hint:* The software says "WAAS," but are they all strictly functional WAAS satellites? You may have to look up some names.)

13. At this point, you can close all windows and exit the Trimble Planning Software.

4.4 GPS and Geocaching Web Resources

With some preliminary planning information available from Trimble Planning, you can begin to head out with a GPS receiver. For some starting points for GPS usage, try the following:

1. Open a Web browser and go to **http://www.geocaching.com**. This Website is home to several hundred thousand geocaches at locations throughout the world. The Website will list the coordinates of geocaches (small hidden items) so that you can use a GPS to track them down.

2. On the main page of the Website, enter **32801** (one of Orlando's many zip codes) in the box that asks for a zip code. Several potential geocaches should appear on the next Web page.

3. You can also search for benchmarks (permanent survey markers set into the ground) by using the Website's **Find A Benchmark** option (at the time of writing, it was listed at the bottom of the Webpage). On the benchmark page, input **32801** for the zip code to search for.

Question 4.10 How many benchmarks can be located within less than one mile of the Orlando 32801 zip code? What are the coordinates of the closest one to the origin of the zip code?

4.5 GPS and Geocaching Applications

As mentioned back in the introduction, there's a lot more that can be done application-wise if you have access to a GPS receiver. For starters, you could find the positions of some nearby caches or benchmarks by visiting http://www.groundspeak.com or http://www.geocaching.com and then tracking them down using the receiver and your land navigation skills. However, there are a number of different ways that geocaching concepts can be adapted in different ways to be incorporated into a classroom setting for lab applications.

The first of these is based on material developed by Dr. Mandy Munro-Stasiuk and published in her article "Introducing Teachers and University Students to GPS Technology and Its Importance in Remote Sensing Through GPS Treasure Hunts" (see the online chapter references for the full citation of the article). Like geocaching, the coordinates of locations have been determined ahead of time by the instructor and given to the participants (such as university students or K–12 teachers attending a workshop) along with a GPS receiver. Having only sets of latitude/longitude or UTM coordinates, the participants break into groups and set out to find the items (such as statues, monuments, building entrances, or other objects around their campus or local area). The students are required to take a photo of the object, and all participants must be present in the photo (this usually results in some highly creative picture taking). The last twist is that the "treasure hunt" is a timed competition with the team that returns first (and having found all the correct items) earning some sort of reward, such as extra credit in the class, or—at the very least—bragging rights for the remainder of the class. In this way, participants learn how to use the GPS receivers, how to tie coordinates to real-world locations, and land navigation skills (to further reinforce concepts such as how northings and eastings work in UTM).

A similar version of this GPS activity is utilized during the OhioView SATELLITES Teacher Institutes (see Chapter 15 for more information on this program). Again, participants break up into groups with GPS receivers, but this time each group is given a set of small trinkets (such as small toys or plastic animal figures) and is instructed to hide each one and register the coordinates of the hiding place with their GPS receiver on a sheet of paper. When the groups reconvene, the papers with the coordinates are switched with another group, and each group now has to find the others' hidden objects. In this way, each group sets up its own "geocaches" and then gets challenged to find another group's caches. Like before, this is a timed lab with a reward awaiting the team of participants that successfully finds all of the hidden items at the coordinates they've been given. This activity helps to reinforce not only GPS usage and land navigation, but also the ties between real-world locations and the coordinates being recorded and read by a GPS receiver. Both of these activities have proven highly successful in linking these concepts to the participants involved.

However you get involved with geocaching, a useful utility for managing geocached data and waypoints from different software packages is the Geocaching Swiss Army Knife, available for free download at http://www.gsak.net. This utility comes with a free trial and helps manage geocaching data.

Closing Time

This lab illustrated some initial planning concepts and some directions in which to take GPS field work (and some caches and benchmarks that are out there waiting for you to find) to help demonstrate some of the chapter's concepts. The two geocaching field exercises described above can also be adapted to classroom use to help expand the computer portion of this lab with some GPS field work as well.

Starting with Chapter 5, you'll begin integrating some new concepts of geographic information systems to your geospatial repertoire.

Working With Digital Spatial Data and GIS

Geographic Information Systems, Modeling the Real World, Vector Data and Raster Data, Attribute Data, Joining Tables, Metadata, Esri, and ArcGIS

Once you've measured and collected your geospatial data, it's time to do something with it. The nature of spatial data lends itself to concepts like analyzing where one location lies in relation to another or the capabilities of one location versus another. When a new library is to be built, it should be placed at the optimal location to serve the greatest number of people without duplicating the usage of other libraries in the area. When a company is deciding where to open a new retail store, they'll need information about the location, the surrounding areas, and populations and demographics of these areas to make a decision. When a family is planning its vacation to Florida, it'll want to know the best routes to travel and where the hotels are located in relation to the theme parks they're planning to visit. All of these concepts involve the nature of spatial data and being able to analyze or compare locations (and the attributes of these locations). Geospatial technology is used to address these ideas and solve these types of problems (and many, many more). Whenever examination, manipulation, or analysis of spatial data is involved (for instance, if you need to tie non-spatial data to a location or if you want to use this information to create a map of your analysis), **Geographic Information Systems (GIS)** are essential to getting these tasks done.

GIS is a powerful tool for analyzing spatial data (the "geographic information" of the title). This ability to explicitly utilize spatial data is what separates GIS from other information systems or methods of analysis. For instance, a spreadsheet would allow you to tabulate housing values and the amount of money paid for a house at a particular address. Even though you're

Geographic Information System (GIS) a computer-based set of hardware and software used to capture, analyze, manipulate, and visualize spatial information.

99

using data that has something to do with a location, there's nothing in the spreadsheet that allows you to examine that data in a spatial context (such as translating the housing values into a map of where these houses are).

The ability to map the real-world locations of the houses, spatially examine trends in housing values, and search for patterns or clusters in the spatial data is what makes GIS unique. Using the coordinate and measurement concepts discussed in the previous chapters allows us to create, map, and analyze digital spatial data in a wide variety of geospatial technology applications (see Figure 5.1 and *Hands-on Application 5.1: Using GIS Online* for an example of GIS data used by a county government).

GIS operates using some kind of computer-based environment—whether it is software running on a desktop, laptop, mobile device, or served off the Web, the "system" being referred to is a computer system. The information being handled by GIS is spatial information—something that has direct ties to a location on the Earth's surface. GIS can also utilize non-spatial data and link it directly to a location. For instance, if you have a set of data points that represent the locations of local car washes (that is, spatial data) and a separate spreadsheet of data of how many cars use that car wash (that is, non-spatial data), you can link the usage data to the car wash location so that the location contains that data. From there, you can do analysis of local population demographics or road conditions about how they relate to the usage statistics of the car wash at that particular location. This wouldn't be possible without GIS being able to link non-spatial data to a location. Lastly, GIS can also perform multiple types of operations related to spatial data—it can be used to analyze, capture, create, manipulate, store, or visualize all manner of spatial information, not just simply design maps (though it can do that too).

FIGURE 5.1 The online GIS utility for Mahoning County, Ohio. (Source: Courtesy Mahoning County GIS Department.)

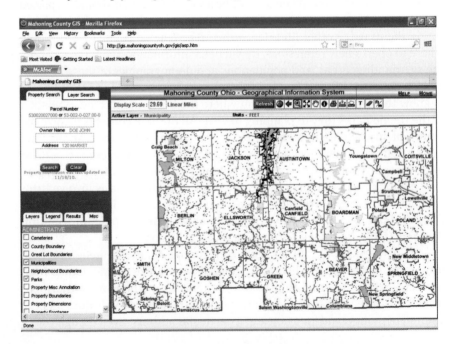

Hands-on Application 5.1

Using GIS Online

There are numerous online GIS resources for viewing and analyzing spatial data. Search on the Internet to see if your county or local government has any online GIS resources or a GIS mapping Website set up. An example of a local GIS is from Mahoning County, Ohio. Open your Web browser and go to **http://gis.mahoningcountyoh.gov** and then select the option for Map Viewer. In the upper left-hand corner of the screen will be the Property Search functions, allowing you to search for a parcel of land by Parcel Number, Owner Name, or Address.

Do a search for parcel ID: 29-042-0-002.00-0 (this is the parcel number for the Boardman Township Park). The map will zoom to the parcel of the park. In the lower left-hand corner of the screen are a series of available layers. Place a checkmark in

the box of the most recent available Ortho Photos (under the Imagery heading). Click on the Refresh button on the toolbar to see the results of adding that layer (that is, high-resolution aerial photos of the park). Select some other layers to examine, and then refresh the map (by pressing the Refresh button) to examine that data. Use the tools on the toolbar (the magnifying glass to zoom in or the hand to pan around the map) to get a closer look at the map and the surroundings of the park.

Check your local state or county government's Websites to see if an online spatial data viewer (or GIS) has been set up for your area. It may be part of your county auditor's Website or a separate GIS department for the state or county.

GIS isn't a development that just popped up over the last couple of years. Large-scale mapping efforts were done using computer-based cartography in the 1950s and 1960s by a number of different government agencies and other labs (including Harvard Laboratory). The term "GIS" first appeared in the same time period of the early 1960s with the implementation of **"CGIS"** (the acronym for the "Canadian Geographic Information System"), which was designed to provide large-scale mapping of land use in Canada. The development of CGIS is attributed to Roger Tomlinson, who has become known as "the father of GIS." GIS developments have continued over the years through the efforts of government agencies and the private sector, until the current time when GIS technology and concepts are extremely widespread (see *Hands-on Application 5.2: ArcGIS Online,* page 102, for another example of using GIS concepts online).

CGIS the Canadian Geographic Information System – a large land inventory system used in Canada and the first time the name "GIS" was utilized for this type of system.

How Does GIS Represent Real-World Items?

GIS has a multitude of applications, ranging from mapping natural gas pipelines to determining which sections of natural land are under the greatest pressure to be developed into some sort of urban land use. No matter the application, the data (such as the locations and lengths of pipelines or the area and extent of land cover) needs to be somehow represented within the GIS. In this sense, GIS provides a means to represent (or model) this data in a computer environment so that you can analyze or manipulate the data. A model indicates some sort of representation (or simplification, or generalization) of

---

Hands-on Application 5.2

ArcGIS Online

There are a lot of GIS applications available online that are used for mapping and analysis of spatial data. For some interactive examples, open your Web browser and go to **http://www.arcgis.com/home** (this is the Website for ArcGIS Online). We'll discuss a lot more about ArcGIS (and its manufacturer, Esri) later in this chapter, but for now, we'll be examining some of the online applications that have been developed with ArcGIS. From the main page, select the option for View the Gallery. Open and examine some of the maps that people have developed and posted online. When opening a map application, you'll be placed into a GIS environment where you can pan around the map, zoom in and out, and use an "identify" tool to get information back about layers on the map. You can also select the option for Make a Map, and a new Web page will open where you can select an area of interest (for instance, your state or county), then add data from ArcGIS Online or select a new basemap to use (including street maps, topographic maps, or high-resolution imagery).

---

real-world information. For instance, if you were tasked to make a map of all of the stop signs in your city, you'd need to be able to represent the locations of all the signs on a map at the scale of the entire city. By laying out a map of the city and placing a dot at each sign location, you would have a good representation of them. The same holds true if you had to come up with a map of all the roads in your town. You could attempt to draw the roads at their correct width, or more likely, just drawing a line along the road's path would suffice for representing the road locations.

Whenever you're deciding how to model real-world phenomena with a GIS, you first have to decide on the nature of the items you're trying to represent. When trying to represent the real world in the GIS, there are two ways of viewing the world. The first is called the **discrete object view**. Under this way of thinking, the world is made up of a series of objects that have a fixed location, or a fixed starting and stopping point, or some sort of fixed boundary. A street starts at one point and ends at another, or there is a definite property boundary between your property and your neighbor's. Around a university campus, there will be numerous objects—buildings, parking lots, stop signs, benches, roads, trees, and sidewalks. By viewing the world in this way, real-world items can be modeled in the GIS as a series of objects (such as a collection of lines to represent roads or a series of points to represent fire hydrant locations).

When adapting the discrete object view of the world to a data model, items in the real world are represented in the GIS by one of three objects (**Figure 5.2**):

discrete object view the conceptualization of the world that all items can be represented with a series of objects.

points zero-dimensional vector objects.

lines one-dimensional vector objects.

polygons two-dimensional vector objects.

- ◉ **Points:** These are zero-dimensional objects, a simple set of coordinate locations.
- ◉ **Lines:** These are one-dimensional objects, created from connecting starting and ending points (and any points in between that give the line its shape).
- ◉ **Polygons:** These are two-dimensional objects that form an area from a set of lines (or having an area defined by a line forming a boundary).

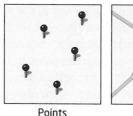 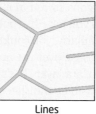 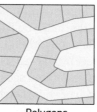

Points   Lines   Polygons

FIGURE 5.2 Basic points, lines, and polygons.

These three items are referred to as **vector objects**, as they make up the basis of the GIS **vector data model**. Basically, the vector data model is a means of representing and handling real-world spatial information as a series of vector objects—items are realized in the GIS as points, lines, or polygons. For instance, locations of fire hydrants, crime incidents, or trailheads can be represented as points; roads, streams, or power conduits are represented as lines; and land parcels, building footprints, and county boundaries are represented as polygons. When dealing with a much larger geographic area (for instance, a map of the whole United States), cities may be represented as points, interstates as lines, and state boundaries as polygons (Figure 5.3).

A lot of GIS data has already been created (such as road networks, parcel boundaries, utilities, etc.), but in order to update or develop your own data, there are a few steps that need to be done. **Digitizing** is a common way to create the points, lines, and polygons of vector data. With digitizing, you are (in essence) "tracing" or "sketching" over the top of areas on a map or other image to model features in that map or image in the GIS. For instance, if you have an aerial photograph of your house being shown on the screen of your GIS software, you could use the mouse to sketch the outline of your house to create a polygon saved in the GIS. If you sketch outlines of four of your neighbors' houses

vector objects points, lines, and polygons that are used to model real-world phenomena using the vector data model.

vector data model a conceptualization of representing spatial data with a series of vector objects (points, line, and polygons).

digitizing the creation of vector objects through sketching or tracing representations from a map or image source.

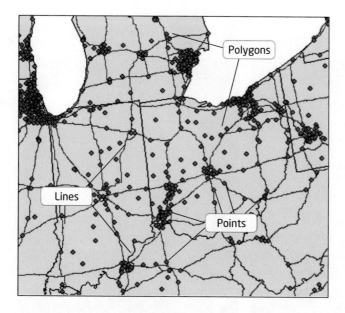

FIGURE 5.3 Points (city locations), lines (interstates), and polygons (state borders) as represented in GIS. (Source: Esri® ArcGIS ArcView graphical user interface Copyright © Esri. All rights reserved.)

as well, you'll have several polygons stored in a single layer in the GIS. When you examine the data, you'll find you have five polygon entries in that data layer (your house and four others). Similarly, you could sketch each house's driveway as a line object by tracing the mouse over the appropriate place in the image, which would result in a second data layer that's comprised of five lines. With digitizing, each of your sketches can be translated into a vector object.

Digitizing is commonly done through a process referred to as "heads-up digitizing" or "on-screen digitizing," which is similar to the sketching just described. In "heads-up" digitizing, a map or other image (usually an aerial photograph or satellite image) is displayed on the screen as a backdrop. Decide if you'll be sketching points, lines, or polygons and create an empty data layer of the appropriate type (to hold the sketches you'll make). From there, you use the mouse to sketch over the objects, using as much detail as you see fit (Figure 5.4). Keep in mind that despite your best efforts, it may likely be impossible or unfeasible to sketch every detail in the image. For instance, when tracing a river, it may not be possible to capture each bend or curve by digitizing straight lines.

When objects are created in GIS, they can be set up with coordinates, but the GIS still needs to have knowledge of how these objects connect to each other and what relation each object has to the others. For instance, when two lines are digitized that represent crossing streets, the GIS needs to have some sort of information that an intersection should be placed where the two lines cross. Similarly, if you have digitized two residential parcels, the GIS would require some information that the two parcels are adjacent to one another. This notion of the GIS being able to understand how objects connect to one another independent of their coordinates is referred to as **topology**. With topology, geometric characteristics aren't changed, no matter how the dataset may be altered (by such things as projecting or transforming the data).

topology how vector objects connect to each other (in terms of their adjacency, connectivity, and containment) independently of the objects' coordinates.

FIGURE 5.4 An example of heads-up digitizing–tracing the boundary of the football field on Youngstown State University's (YSU's) campus to create a polygon object. (Source: Ohio Geographically Referenced Information Program (OGRIP), Ohio Statewide Imagery Program (OSIP), April 2006/Esri® ArcMap ArcInfo graphical user interface Copyright © Esri. All rights reserved.)

Topology establishes adjacency (that is, how one polygon relates to another polygon, in that they share a common boundary), connectivity (that is, how lines can intersect with one another), and containment (that is, how locations are situated inside of a polygon boundary).

Digital Line Graphs (**DLGs**) are examples of vector data applications in GIS. DLGs were created from United States Geologic Survey (USGS) topographic maps (see Chapter 13 for more information about topographic maps), featuring vector datasets representing transportation features (such as streets, highways, and railroads), hydrography features (such as rivers or streams), or boundaries (such as state, county, city, or national forest borders). DLGs are created and freely distributed by the USGS (see Figure 5.5

DLG a Digital Line Graph—the features (such as roads, rivers, or boundaries) digitized from USGS maps.

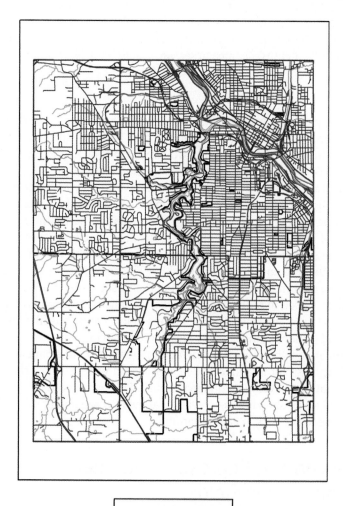

FIGURE 5.5 1 : 24000 Digital Line Graph data (showing roads, railroads, boundaries, and hydrologic features) of Youngstown, Ohio. (Source: USGS/Esri® ArcGIS Layout graphical user interface Copyright © Esri.)

Youngstown DLGs

—— Roads DLG

—— Rails DLG

—— Hydro DLG

—— Boundaries DLG

Hands-on Application 5.3

USGS Digital Line Graphs

Digital Line Graphs are made freely available by the USGS. A good online utility for examining available DLGs is EarthExplorer, a Website designed to facilitate access to a wide range of geospatial data (not just DLGs). To get started, open your Web browser and go to **http://edcsns17.cr.usgs.gov/NewEarthExplorer/**. Using EarthExplorer is a four-step process. First, select the Search Criteria tab and type the name of a location of an address or place name (such as Reno, Nevada). Next, click on Data Sets and place checkmarks in the DLG 1:100k and DLG Large Scale boxes under the Digital Line Graphs option. Next, click Additional Criteria–this is where you could further refine your search, if desired. Last, click Results to see what EarthExplorer turned up. From the available options in the results, you can see the footprint that the DLG would cover. If desired, DLGs can be downloaded to use in GIS (they may need to be imported into a specific format to use, and the USGS requires a user to register and log in with them to download data). Check your local area to see what types of DLGs are available.

for an example of several DLGs in GIS). The DLGs are created by digitizing—a USGS map is used as the source and the map features (like roads or rivers) are turned into digital format using GIS. See *Hands-on Application 5.3: USGS Digital Line Graphs* for more about how to obtain DLGs.

There's also a second way of viewing the world, in which not everything has a fixed boundary or is an object. For instance, things such as temperature, atmospheric pressure, and elevation vary across the Earth and are not best represented in GIS as a set of objects. Instead, they can be thought of as a surface that's made up of a near-infinite set of points. This way of viewing the world is called the **continuous field view**. This view implies that real-world phenomena continuously vary, and rather than a set of objects, a surface filled with values is used to represent things. For instance, elevation gradually varies across a landscape from high elevations to low elevations filled with hills, valleys, and flatlands, and at every location, we can measure the height of the land. Similarly, if you were standing at a ranger station in a park and preparing a rescue plan for stranded hikers, you would want to know the distance from the station to every location in the park. Thus, you could have two layers in your GIS—one showing the elevation at every place and another showing the distance to the park at each location.

continuous field view the conceptualization of the world that all items vary across the Earth's surface as constant fields, and values are available at all locations along the field.

How Can You Represent the Real World as Continuous Fields?

raster data model a conceptualization of representing spatial data with a series of equally spaced and sized grid cells.

When representing these kinds of continuous fields in GIS, the three vector objects (and thus, the vector data model) are often not the best way of representing data. Instead, a different conceptualization called the **raster data model** is usually used. In the raster data model, data is represented

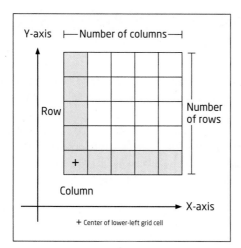

FIGURE 5.6 The grid cells of the raster data model.

using a set of evenly distributed square **grid cells**, with each square cell representing the same area on the Earth's surface (see Figure 5.6 for the layout of the grid cells). For instance, to represent an entire section of the landscape, each grid cell might represent 10 square feet, or 30 square meters, or 1 square kilometer, depending on the grid cell size being used for the model. Regardless of the grid cell size, each cell has the same resolution (that is, the same grid dataset cannot mix cells of 30 square feet together with cells of 5 square feet). Also, each grid cell contains a single value representing the data being modeled. For instance, in an elevation raster, each grid cell could contain a value for the elevation at that area on the ground. Datasets such as land cover, soils, or elevation are all commonly represented using the raster data model.

grid cell a square unit, representing some real-world size, which contains a single value.

The National Land Cover Database (**NLCD**) is an example of a raster data application in GIS. Developed by a consortium of United States agencies (including the USGS and the Environmental Protection Agency), NLCD provides 30-meter raster data covering the entire United States, wherein each grid cell is coded with a value that corresponds to a specific land cover type at that location. NLCD designations include categories such as "Open Water," "Developed, High Intensity," "Deciduous Forest," "Pasture/Hay," and "Woody Wetlands." NLCD provides a means of broad-scale land cover classification at a state, regional, or national scale (Figure 5.7, page 108). NLCD products are available for land cover circa 1992, 2001, and a new dataset circa 2006 available at the time of this writing. A separate dataset is available that details the change in land cover types between the 1992 and 2001 datasets, enabling researchers to not only measure land cover types, but also see how the landscape is changing (such as where agricultural fields or forests are converting over to urban land uses). NLCD data is distributed free of charge via Web download in a format that can be easily read or converted by GIS software. See *Hands-on Application 5.4: The National Land Cover Database (NLCD)* for how to access NLCD data (page 108).

NLCD the National Land Cover Database is a raster-based GIS dataset that maps the land cover types for the entire United States at 30-meter resolution.

Hands-on Application 5.4

The National Land Cover Database (NLCD)

The USGS has an online tool for viewing the 1992 and 2001 NLCD (and some products derived from it), as well as letting you download sections of the dataset for use in GIS. The viewer is available at **http://gisdata.usgs.net/website/MRLC/viewer.htm**. The default view should be the 2001 NLCD data (although 2006 data is also available). From the display options on the right side of the screen, you can select different NLCD data to view. Zoom in to your local area and examine the NCLD—what do the various colors of the grid cells represent (in terms of their land cover)? How are the 1992 and 2001 datasets different? If you want to download the data, the tools in the viewer will allow you to directly download the raster datasets.

FIGURE 5.7 A section of the 2001 NLCD showing land cover in Columbiana County, Ohio, modeled with raster grid cells. (Source: USGS/ Esri® ArcGIS Layout graphical user interface Copyright © Esri.)

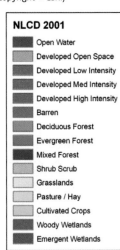

NLCD 2001
- Open Water
- Developed Open Space
- Developed Low Intensity
- Developed Med Intensity
- Developed High Intensity
- Barren
- Deciduous Forest
- Evergreen Forest
- Mixed Forest
- Shrub Scrub
- Grasslands
- Pasture / Hay
- Cultivated Crops
- Woody Wetlands
- Emergent Wetlands

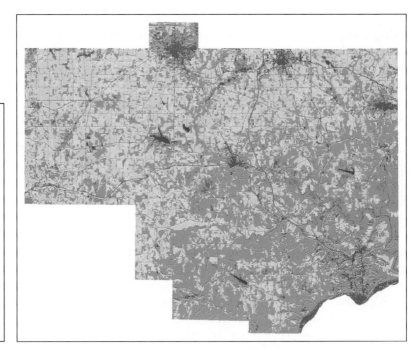

How Is Non-Spatial Data Handled by GIS?

Regardless if you're mapping schools ranked by average SAT scores or the locations of the United States' top tourist destinations, all of this numeric data needs to be somehow represented in the GIS. For instance (going back to the digitizing example), if you have two data layers (one containing polygons representing housing footprints and one containing lines representing the driveways), each would contain multiple objects. The GIS will have information about the spatial properties of the objects in that layer, but what it does not have is any other information about those objects (such as the tax-assessed value of the house, the house's address, the name of the owner, the number of people living in the house, the material the driveway's made of, or how old the driveway is). All of these **attributes** represent other non-spatial information associated with each of the objects. When representing attribute data in GIS, these values can take one of four forms: nominal, ordinal, interval, or ratio data.

attributes the non-spatial data that can be associated with a spatial location.

Nominal data are values that represent some sort of unique identifier. Your Social Security number or telephone number would both be examples of nominal data—both of these values are unique to you. Names or descriptive information that are associated as a location would be nominal data, as they're unique values, the same as a value from a classification scheme. Also, the difference between numerical nominal values is not significant—you can't add your Social Security number to a friend's number and come up with a relative's number—the same way you can't subtract your phone number from one friend's number and come up with another friend's phone number.

nominal data a type of data that is a unique identifier of some kind. If numerical, the differences between numbers are not significant.

Ordinal data is used to represent a ranking system of data. If you have data that is placed in a hierarchy where one item is first, another is second, and another is third, that data is considered ordinal. For instance, if you had a map of the city and were going to identify the locations of the homes of the winners of a local car-racing event, the points on the map would be tagged as to the location of the first-place winner, the second-place winner, etc. Ordinal data deals solely with the rankings themselves, not with the numbers associated with these ranks. For instance, the only values mapped for the car-race winners is their placement in the race, not their winning times or the cars' speed, or any other values. The only thing ordinal data represents is a clear value of what item is first, which is second, and so on. No measurements can be made of how much better the first-place winner's time was than the second-place winner, only that one driver was first and one driver was second.

ordinal data a type of data that refers solely to a ranking of some kind.

Interval data is used when the difference between numbers is significant, but there is no fixed zero point. With interval data, the value of zero is just another number used on the scale. For instance, temperature measured in degrees Celsius would be considered interval data since values can fall below zero, and the value of zero only represents the freezing point of water, not the bottom of the Celsius temperature scale. However, since there is no fixed zero point, we can make differences between values (for instance, if it was 15 degrees yesterday and 30 degrees today, we can say it was 15 degrees warmer), but dividing numbers wouldn't work (since a temperature of 30 degrees is not twice as warm as a temperature of 15 degrees).

interval data a type of numerical data in which the difference between numbers is significant, but there is no fixed non-arbitrary zero point associated with the data.

Ratio data are values with a fixed and non-arbitrary zero point. For instance, a person's age or weight would be considered ratio data since a person cannot be less than zero years in age or weigh less than zero pounds. With ratio data, the values can be meaningfully divided and subtracted. If we want to know how much time separated the car-racing winners, we could subtract the winning driver's time from the second-place driver's time and get the

ratio data a type of numerical data in which the difference between numbers is significant, but there is a fixed non-arbitrary zero point associated with the data.

attribute table a spreadsheet-style form where the rows consist of individual objects and the columns are the attributes associated with those objects.

records the rows of an attribute table.

fields the columns of an attribute table.

FIGURE 5.8 The attribute table associated with a GIS data layer. (Source: Esri® ArcGIS ArcMap ArcView graphical user interface Copyright © Esri.)

necessary data. Similarly, a textbook that costs $100 is twice as expensive as a textbook that costs $50 (as we divide the two numbers). Ratio data is used when comparing multiple sets of numbers and looking for distinctions between them. For instance, when mapping the locations of car wash centers, the data associated with the location is the number of cars that use the service. By comparing the values, we can see how much one car wash outsells the others, or calculate the profit generated by each location.

Each layer in the GIS has an associated **attribute table** that stores additional information about the features making up that layer. The attribute table is like a spreadsheet where each object is stored as a **record** (a row) and the information associated with the records—the attributes—is stored as **fields** (columns). See Figure 5.8 for an example of an attribute table in conjunction with its GIS objects. To use the housing example again, the houses attribute table would consists of five records, each representing a house polygon. New fields could be created, having headings such as "Owner" and "Appraised Value" so that this type of descriptive, non-spatial data could be added for each of the records. These new attributes can be one of the four data types (for instance, "Owner" would be nominal data, while "Appraised Value" would be ratio data). In this way, non-spatial data is associated with a spatial location.

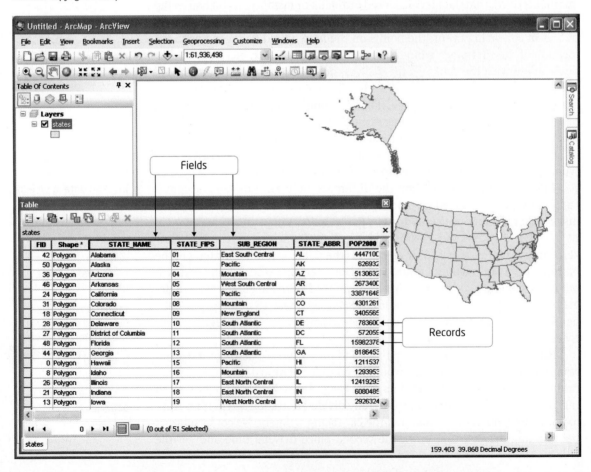

A raster attribute table is handled differently than one for vector objects since a raster dataset is composed of multiple grid cells (for instance, a 30-meter resolution grid of a single county could take millions of cells to model). Rather than having separate records for each grid cell, raster data will often be set up in a table with each record featuring the value for a raster cell and an attribute showing the count of how many cells comprise the dataset. For instance, a county-land-use dataset may contain millions of cells, but only consist of seven values. Thus, the raster would have seven records (one for each value) and another field whose value would be a summation of how many cells have that value. New attributes would be defined by the raster values (for instance, the type of land use that value represents).

Besides creating new fields and populating them with attributes by hand, a **join** is another way GIS allows non-spatial data to be connected to spatial locations. This operation works by linking the information for records in one table to its corresponding records in another table. This is managed by both tables having a field in common (this field is also referred to as a **key**). For instance, Figure 5.9 shows two tables, the first from a shapefile's attribute table that has spatial information about a set of points representing the states of the United States, while the second table has non-spatial Census information about housing stock for those states. However, since both tables have a common field (that is, the state's abbreviation), a join can be performed on the basis of that key.

join a method of linking two (or more) tables together.

key the field that two tables have to have in common with each other in order for the tables to be joined.

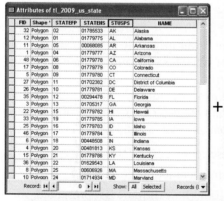

FIGURE 5.9 A joining of two attribute tables, based on their common field (the state's abbreviation attribute). (Source: Esri® ArcGIS ArcExplorer graphical user interface Copyright © Esri.)

After the join, information for the records from the Census information table is now associated with the corresponding records from the states' table. For instance, you could now select the record for Alaska and have access to all of its housing information since the tables have been related to each other. In this way, joining tables is a simple method for linking sets of data together and especially for linking non-spatial spreadsheet data to spatial locations. With all of this other data available (and linked to a spatial location), new maps can be made. For instance, you could now create maps of the United States showing any of the attributes (total number of houses, number of vacation homes, or percentage of the total housing that are vacation homes) linked to the spatial features.

What Other Kind of Information Do You Need to Use GIS Data?

Once you have your spatial and attribute data within the GIS, there's one more important piece of data you're going to need—descriptive information about your data. When creating data or obtaining it from another source, you should have access to a separate file (usually a "readme" text file or an XML document) that fully describes the dataset. Useful things to know would be information about the coordinate system, projection, and datum of the data, when the data was created, what sources were used to create it, how accurate the data is, and what each one of the attributes represents. All of this kind of information (and more) is going to be critical to understanding the data to properly use it in GIS. For instance, if you have to reproject the data, you're going to have to know what projection it was originally in, or if you're going to be mapping certain attributes, then you'll have to know what each one of those strangely named fields in the attribute table actually represents. In GIS, this "data about your data" is referred to as **metadata**.

metadata descriptive information about geospatial data.

The Federal Geographic Data Committee (FGDC) has developed some standards for the content and format that metadata should have. These include:

◉ Identification of the dataset (a description of the data)

◉ Data quality (information about the accuracy of the data)

◉ Spatial data organization information (how the data is represented, such as vector or raster)

◉ Spatial reference information (the coordinate system, datum, and projection used)

◉ Entity and attribute information (what each of the attributes means)

◉ Distribution information (how you can obtain this data)

◉ Metadata reference information (how current the metadata is)

◉ Citation information (if you want to cite this data, how it should be done)

◉ Time period information (what date does the data represent)

◉ Contact information (whom to get in touch with for further information)

Thinking Critically with Geospatial Technology 5.1

What Happens When You Don't Have Metadata?

Say you're trying to develop some GIS resources for your local community, which includes needing data about the area's sewer system. You get the data in a format that your GIS software will read, but it doesn't come with any metadata—meaning you have no information on how the sewer data was created, the accuracy of the data, what the various attributes indicate, the date it was created, or the datum, coordinate system, or projection the data is in. How useful is this dataset for your GIS work? When metadata for a GIS layer is unavailable (or incomplete), what sorts of issues can occur when trying to work with that data?

What Kinds of GIS Are Available?

There are many different types of GIS software packages available, everything from free lightweight "data viewers," to numerous Open Source programs, to expensive commercially available software packages. There are numerous companies producing a wide variety of GIS software. According to a 2007 GIS Salary Survey conducted by the Urban and Regional Information Systems Association (URISA), GIS products from the Environmental Systems Research Institute (**Esri**) were found to be among the most popular and widespread in their usage. See *Hands-on Application 5.5: Esri TV* for more information about Esri and their current GIS innovations and applications.

Esri the Environmental Systems Research Institute, a key developer and leader of Geographic Information Systems products.

Located in Redlands, California, Esri was founded by Jack and Laura Dangermond in 1969. By 1982, Esri released the first of its "Arc" products, the initial version of **Arc/Info**, a key GIS software package. Though extremely powerful in its analytical capabilities, Arc/Info was also a product of the DOS command-line era of computing. As such, there were literally hundreds of different Arc/Info commands, each with numerous variations in their usage, optional commands, and syntax. Today, new iterations of this version of Arc/Info are no longer being made, but the command-line software is available and referred to as "Workstation Arc/Info."

Arc/Info an Esri GIS product that required command-line entry for tools and capabilities.

Esri would later release **ArcView**, a Windows-based tool for viewing and examining spatial data. ArcView featured things common to a graphical user interface, such as pull-down menus, icons, and windows, and as

ArcView an Esri GIS product developed with a Windows interface.

Hands-on Application 5.5

Esri TV

Esri is constantly coming up with new developments, software innovations, and GIS applications. As a way of distributing this type of news, Esri runs their own channel on YouTube, called Esri TV. To check out their online videos, open your Web browser and go to **http://www.youtube.com/user/Esritv**. What kinds of applications or information about Esri software are being distributed in video form?

such, was much simpler to use than the command-line interface. ArcView went through several iterations to add a greater variety of functions and expansions to it to increase its analytical capabilities and was very commonly used by the late 1990s. The final version of ArcView was version 3.3, which is still available.

Esri released its newest software package, **ArcGIS** (version 8, which built off the numbering of the previous Arc/Info versions) in stages between 1999 and 2001. ArcGIS was a blend of the analytical capabilities of Arc/Info with a Windows graphical interface (similar to ArcView) in one package. As of this writing, the most recent iteration of the software is ArcGIS Desktop 10. When purchased, ArcGIS Desktop is available in three different varieties (which recycle the previous names of some products): ArcView, ArcEditor, and ArcInfo. These can best be thought of as different levels of the same software package—all of them are considered ArcGIS Desktop, but the ArcView version has the smallest number of functions, the ArcEditor version has the mid-range of functions, and the ArcInfo version has the largest number of functions.

No matter which of the three versions of ArcGIS you're using, the main component is referred to as **ArcMap**, which is used for viewing and analyzing data (see Figure 5.10). Within ArcMap, GIS datasets can be added and treated as different map layers. ArcGIS offers multiple sets of tools for analysis, data manipulation, or making a map from your data (things you'll do in the labs for Chapters 6, 7, and 8). Previous versions of ArcGIS also featured ArcCatalog, a

ArcGIS Esri's current primary GIS software package.

ArcMap the component of ArcGIS used for viewing and analyzing data.

FIGURE 5.10 The basic layout of Esri's ArcMap. (Source: Esri® ArcGIS ArcMap ArcView graphical user interface Copyright © Esri.)

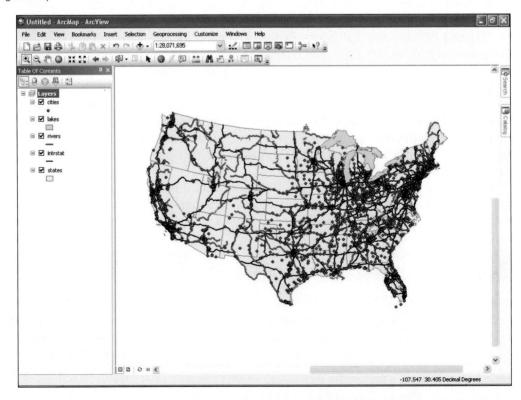

separate utility for managing available GIS data. With ArcGIS 10, the **Catalog** functions are now contained as a feature within ArcMap itself.

While it is the leading software product, ArcGIS is expensive, so Esri has several other software options available for utilizing its tools. These include ArcExplorer Java Edition for Educators (**AEJEE**), a program that allows a user to view GIS data, pan and zoom around data, do simple analysis like queries and buffers, and create maps of data (which you'll use in the labs for this chapter, as well as the labs in Chapters 6, 7, and 8). AEJEE is an easy-to-use GIS software program that works like a simplified version of ArcMap. AEJEE and ArcGIS Desktop share a number of similarities in their layout, setup, menus and icons, and functionality. For instance, you add data and work with it in AEJEE in very similar ways as you would in ArcMap—likewise, the Catalog functions of AEJEE are a simplified version of ArcGIS's Catalog. Learning GIS basics with AEJEE should give you a good background and (due to its similar setup and structure) reduce the learning curve if you use ArcGIS Desktop itself. See Figure 5.11 for the layout of AEJEE (and note the similarities to ArcGIS' ArcMap design from Figure 5.10). The labs for this section of the book are designed so that you can do them with either ArcGIS 10 or AEJEE.

Another free Esri product is **ArcGIS Explorer**, a free downloadable tool that Esri touts as being "GIS For Everyone" (see Figure 5.12, page 116), which should not be confused with AEJEE, despite the very similar names.

Catalog the component of ArcGIS used for managing data (which contains the functionality of the previous ArcCatalog).

AEJEE the ArcExplorer Java Edition for Educators—a free GIS program produced by Esri that can display data, create maps, and perform basic analysis.

ArcGIS Explorer a virtual Earth tool that can be downloaded for free from Esri.

FIGURE 5.11 The basic layout of Esri's AEJEE program. (Source: Esri® ArcGIS ArcExplorer graphical user interface Copyright © Esri.)

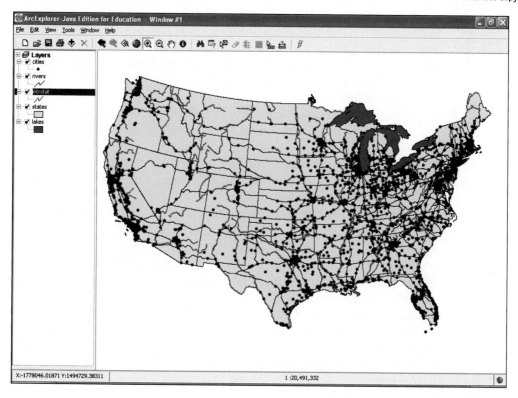

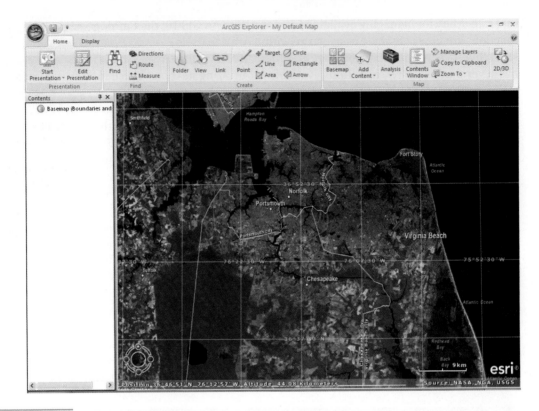

FIGURE 5.12 The Hampton Roads area as seen in Esri's ArcGIS Explorer software program. (Source: NASA/USGS/NGA/Esri® ArcGIS ArcExplorer graphical user interface Copyright © Esri.)

coverage a data layer represented by a group of files in a workspace consisting of a directory structure filled with files and also associated files in an INFO directory.

ArcGIS Explorer has multiple functions for handling GIS data and other data made available for use via the Internet (you'll work with some of ArcGIS Explorer's capabilities later in Chapter 15). See *Hands-on Application 5.6: Esri ArcNews, ArcUser, and ArcWatch* for more information about Esri software and applications.

With the proliferation of numerous Esri products on the market and in use today, it's no surprise that a lot of GIS data that is available is in an Esri data format. Raster data is commonly available in (or can be easily imported into) Esri Grid format, which can be directly read by Esri software. Vector data can usually be found in one of the following three different varieties: coverage, shapefile, and geodatabase.

Coverage is the original file format for Arc/Info. The coverage itself consists of multiple files inside of a directory structure. The folder that a coverage is stored in is referred to as a workspace and a workspace may contain multiple coverages. A workspace will also contain a separate folder called the INFO directory, which contains extra files needed by each coverage in the workspace. Because of this structure of holding files, Esri created a method to take all the necessary files that make up one coverage and export them to a single file for portability. This exported file (which is called an "interchange file" that ends with the file extension .e00) can then be moved to a different location and reimported to rebuild the coverage structure on another folder or computer.

Hands-on Application 5.6

Esri ArcNews, ArcUser, and ArcWatch

Esri data and software products are used in numerous applications across the globe. To keep informed regarding new Esri updates, software, and real-world uses of GIS, Esri publishes a pair of print publications (and makes articles and content available free online) as well as an e-magazine. Check the following sites for their publications:

1. ArcNews: **http://www.Esri.com/news/arcnews/arcnews.html**

2. ArcUser: **http://www.Esri.com/news/arcuser/index.html**

3. ArcWatch: **http://www.Esri.com/news/arcwatch/index.html**

Look for some articles related to fields of interest to you. For instance, see how Esri software is used in environmental monitoring, national security, urban planning, or law enforcement—and see what new types of developments are currently underway in the GIS field.

Shapefile is the original file format for ArcView. A shapefile can hold only one type of vector object—thus there are point shapefiles, line shapefiles, and polygon shapefiles. Despite the name, a shapefile actually consists of multiple files that have to be present together to properly represent the data. The files all start with the same prefix (that is, if the shapefile is called "roads," then all the files that it consists of will be called "roads") but will have different extensions, including .shp, .shx. .dbf, and others. When copying a shapefile, you need to move all files with the same prefix to their new location for the shapefile to function properly.

> **shapefile** a series of files (with extensions such as .shp, .shx, and .dbf) that make up one vector data layer.

Geodatabase is a new file format established for ArcGIS. A geodatabase consists of a single file that contains all of the spatial information for a dataset. Its object-oriented structure is set up such that multiple datasets can be stored in a single geodatabase, with each one being its own feature class. For example, a geodatabase called "Yellowstone" could store the park boundaries as a polygon feature class, the roads running through the park as line feature classes, and the hiking trails as another line feature class.

> **geodatabase** a single file that can contain multiple datasets, each as its own feature class.

Esri is not the only company making GIS software—there are several other software packages and GIS companies providing a wide range of products used throughout the industry. For instance, there are many Open Source GIS products available for use and download via the Internet. An example is GRASS (Geographic Resources Analysis Support System), a long-running free (and open source) GIS software that can handle vector, raster, and image data with many different types of analysis. Some notable commercial GIS vendors and their software products (and this list is not comprehensive) are:

- Geomedia (from Intergraph)
- IDRISI Taiga (from Clark Labs)
- Manifold System (from Manifold)

⊙   MapInfo Professional (from Pitney Bowes Business Insight)

⊙   Maptitude (from Caliper Corporation)

Chapter Wrapup

No matter the provider, the software, or the data representation format, GIS has a wide variety of uses, ranging from public utility mapping to law enforcement analysis, to fire tracking, to landscape planning. The next chapter will examine a number of different types of GIS analysis and address how to take this spatial data and start doing things with it.

The lab for this chapter provides an introduction to GIS and some basic handling of spatial data. Two versions of the lab have been designed—the first one, *Geospatial Lab Application 5.1: GIS Introduction: AEJEE Version,* uses the free AEJEE software package, and the second one, *Geospatial Lab Application 5.2: GIS Introduction: ArcGIS Version,* uses ArcGIS 10. Both labs (5.1 and 5.2) use the same data and concepts, but implement them differently, depending on which software package you have available to you.

Important note: the references for this chapter are part of the online companion for this book and can be found at **http://www.whfreeman. com/shellito1e.**

Key Terms

Geographic Information System (p. 99)
CGIS (p. 101)
discrete object view (p. 102)
points (p. 102)
lines (p. 102)
polygons (p. 102)
vector objects (p. 103)
vector data model (p. 103)
digitizing (p. 103)
topology (p. 104)
DLG (p. 105)
continuous field view (p. 106)
raster data model (p. 106)
grid cell (p. 107)
NLCD (p. 107)
attributes (p. 109)
nominal data (p. 109)
ordinal data (p. 109)

interval data (p. 109)
ratio data (p. 109)
attribute table (p. 110)
records (p. 110)
fields (p. 110)
join (p. 111)
key (p. 111)
metadata (p. 112)
Esri (p. 113)
Arc/Info (p. 113)
ArcView (p. 113)
ArcGIS (p. 114)
ArcMap (p. 114)
Catalog (p. 115)
AEJEE (p. 115)
ArcGIS Explorer (p. 115)
coverage (p. 116)
shapefile (p. 117)
geodatabase (p. 117)

GIS Introduction: AEJEE Version

This chapter's lab will introduce you to some of the basic features of GIS. You will be using one of Esri's GIS programs to navigate a GIS environment and begin working with spatial data. The labs in Chapters 6, 7, and 8 will utilize several more GIS features; the aim of this chapter's lab is to familiarize you with the basic functions of the software. This lab uses the free ArcExplorer Java Edition for Educators (AEJEE).

Objectives

The goals for you to take away from this lab are:

- Familiarize yourself with the AEJEE software environment, including basic navigation and tool use with both the map and Catalog.
- Examine characteristics of spatial data, such as their coordinate system, datum, and projection information.
- Familiarize yourself with adding data and manipulating data layer properties, including changing the symbology and appearance of Esri data.
- Familiarize yourself with data attribute tables in AEJEE.
- Make measurements between objects in AEJEE.

Obtaining Software

The current version of AEJEE (2.3.2) is available for free download at http://edcommunity.esri.com/software/aejee.

Important note: Software and online resources sometimes change fast. This lab was designed with the most recently available version of the software at the time of writing. However, if the software or Websites have significantly changed between then and now, an updated version of this lab (using the newest versions) is available online at http://www.whfreeman.com/shellito1e.

Lab Data

There is no data to copy in this lab. All data comes as part of the AEJEE sample data that gets installed with the software.

Localizing This Lab

The dataset used in this lab is Esri sample data for the entire United States. However, starting in Section 5.6, the lab focuses on Ohio and the locations of some of its cities. With the sample data covering the state boundaries and city locations for the whole United States, it's easy enough to select your city (or ones nearby) and perform the same measurements and analysis using those cities more local to you than ones in northeast Ohio.

5.1 An Introduction to ArcExplorer Java Edition for Educators (AEJEE)

1. Start **AEJEE** (the default install folder is called AEJEE). AEJEE will open in its initial mode. The left-hand column (where the word Layers is) is referred to as the Table of Contents (TOC)—it's where you will have a list of available data layers. The blank screen that takes up most of the interface is the View, where data will be displayed.

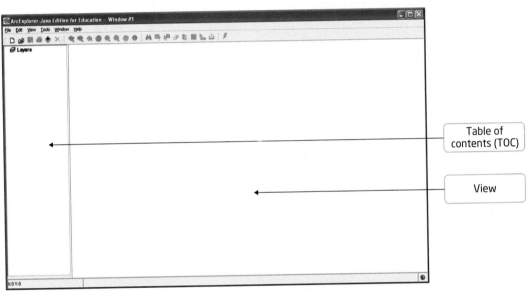

Table of contents (TOC)

View

(Source: Esri® ArcGIS ArcExplorer graphical user interface Copyright © Esri.)

Important note: Before you begin adding and viewing data, the first thing to do is examine the data you have available to you. When AEJEE installs, it also gives you a sample set of data to work with. To examine this data (or any other GIS data you'll want to work with), you'll use AEJEE's Catalog—a utility designed to allow you to organize and manage GIS data.

2. From the **Tools** pull-down menu, select **Catalog**. The Catalog dialog box will open in a new window.

3. The Catalog can be used to manage your data as well as to view a preview of it. In the **Catalog** tree (the section going down the right-hand side of the screen), navigate to the **C:\drive**, open the **ESRI** folder, and then open the **AEJEE** folder. Next, open the **Data** folder and expand the **usa** folder. You'll see a list of all the shapefiles available for use in the sample data, as well as a couple of JPEG graphics.

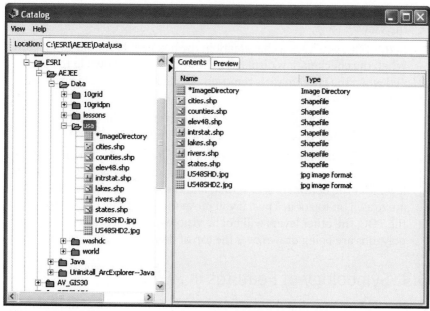

(Source: Esri® ArcGIS ArcExplorer graphical user interface Copyright © Esri.)

4. In the Catalog tree, select the **states.shp** file. In the other section of the Catalog, information about the dataset will be given. Click on the **Preview** tab to view what the dataset looks like.

5. Do the same for the rest of the shapefiles, switching between the Contents tab and the Preview tab.

Question 5.1 How many polygons are in the lakes dataset (remember, each feature is a separate polygon)? How many lines are in the interstates dataset?

5.2 Adding Data to AEJEE and Working with the TOC

1. Back in the main window of AEJEE, click on the yellow and black **"plus"** button to start adding the data you've previewed in the Catalog onto the map.

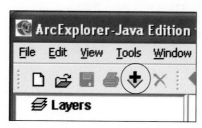

2. In the Content Chooser dialog box, navigate to the **Data** folder, open the **usa** folder, and then select the **states.shp** shapefile. Click **OK**.

3. You'll now see the states.shp in the TOC, and its content is displayed in the View.

4. In the TOC, you'll see a checkmark in the box next to the states shapefile. When the checkmark is displayed, the layer will be shown in the View, and when the checkmark is removed (by clicking on it), the layer will not be displayed.

5. Now add two more layers, cities (a point layer), and intrstat (a line layer). All three of your layers will now be in the TOC.

6. You can manipulate the "drawing order" of items in the TOC by grabbing a layer with the mouse and moving it up or down in the TOC. Whatever layer is at the bottom is drawn first, and the layers above it in the TOC are drawn on top of it. Thus, if you move the states layer to the top of the TOC, the other layers will not be visible in the View since the states' polygons are being drawn over the top of them.

5.3 Symbology of Features in AEJEE

1. You'll also notice that the symbology generated for each of the three objects is simple—points, lines, and polygons have been assigned a random color. You can significantly alter the appearance of the objects in the View to customize your maps.

2. Right-click on the **states** layer and select **Properties**.

3. In the Properties dialog box, you can alter the appearance of the states by changing their Style and Color, and then elements of how the polygons are outlined.

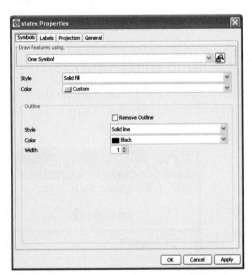

(Source: Esri® ArcGIS ArcExplorer graphical user interface
Copyright © Esri.)

4. Change the states dataset to something more appealing to you, click **Apply** to make the changes, and then click **OK** to close the dialog box.

5. Do the same for the cities and interstates shapefiles. Note that you can alter the points for the cities into shapes like stars, triangles, or crosses, and change their size as well. Several different styles are available for the lines of the interstates file as well.

5.4 Obtaining Projection Information

1. Chapter 2 discussed the importance of projections, coordinate systems, and datums. All of the information associated with these concepts can be accessed using AEJEE.

2. Right-click on the **states** shapefile and select **Properties**. Click on the **Projection** tab in the dialog box. Information about units, projections, coordinate systems, and datums can all be accessed for each layer.

3. Click **OK** to exit the states Properties dialog box.

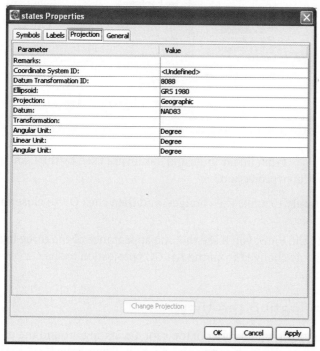

(Source: Esri® ArcGIS ArcExplorer graphical user interface Copyright © Esri.)

Question 5.2 What units of measurement are being used in the interstates dataset? What type of coordinate system (UTM, SPCS, etc.) is being used?

Question 5.3 What datum and projection are being used for the cities dataset?

4. AEJEE also allows you to change the projected appearance of the data in the View from one projected system to another. From the main AEJEE window, choose the **Tools** pull-down menu and select **Projection**. Numerous projection options are available to use. For this lab, select the projection options as follows:

   a. **Regional Projections**

   b. **Lambert Conformal Conic**

   c. **North America**

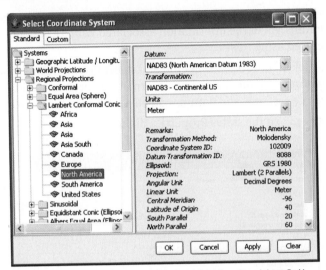

(Source: Esri® ArcGIS ArcExplorer graphical user interface Copyright © Esri.)

Question 5.4 What linear units and datum are used by the Lambert Conformal Conic projection?

5. Click **Apply** to make the changes, and then click **OK** to close the dialog box.

6. Back in the View, you'll see that the appearance of the three layers has completely changed from the flat GCS projection to the curved LCC projection.

5.5 Navigating the View

1. AEJEE provides a number of tools for navigating around the data layers in the View. Clicking on the **globe** icon will zoom the View to the extent of all the layers. It's good to use if you've zoomed to far in or out in the View or need to restart.

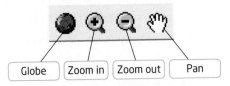

Globe   Zoom in   Zoom out   Pan

2. The magnifying glass icons allow you to zoom in (the **plus** icon) or out (the **minus** icon). You can zoom by clicking in the window or clicking and dragging a box around the area you want to zoom into.

3. The **hand** icon is the Pan tool that allows you to "grab" the View by clicking on it, and dragging the map around the screen for navigation.

4. Use the **Zoom** and **Pan** tools to center the View on Ohio so that you're able to see the entirety of the state and its cities and roads.

5.6 Interactively Obtaining Information

1. Even with the View centered on Ohio, there are an awful lot of point symbols there, representing the cities. We're going to want to identify and use only a couple of them. AEJEE has a searching tool that allows you to locate specific objects—on the toolbar, select the **Find** tool (the binoculars icon) as follows:

2. In the **Find** dialog box, select the **Features** tab.

3. In the **Value** field type: **Akron**.

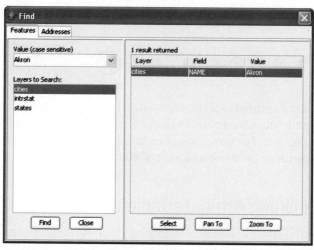

(Source: Esri® ArcGIS ArcExplorer graphical user interface Copyright © Esri.)

4. Choose **Cities** for the Layers to Search.

5. Click on the **Find** button. The returned results will be shown on the right of the dialog box.

6. Drag the dialog box out of the way, so you can see both it and the View at the same time.

7. Click on the **Select** button in the dialog box. You'll see that the point symbol for the city of Akron has changed to a yellow color. This means it has been "selected" by AEJEE—any actions performed on the cities layer will affect the selected items, not all of the cities.

However, all that Find does is locate and select an object. To obtain information about it, you can use the **Identify** tool (the icon on the toolbar is the white "i" in a purple circle).

8. Select the **Identify** tool, and then click on **Akron** (zooming in as necessary). A new dialog box will open, listing all of the attribute information associated with Akron (it's actually giving you all of the field attribute information that goes along with that particular record).

9. Return to **Find** and look for the city called Youngstown, and then use the **Identify** tool on the selected city.

Question 5.5 According to AEJEE, what is the year 2000 population of Youngstown (carefully examine the attributes returned from Identify)?

10. Another method of interactively selecting features is to use the **Select Features** tool. To select features from the cities layer, click on **Cities** in the TOC. Next, on the toolbar, the icon is a cursor with a white and cyan polygon to its right.

11. The **Select Features** tool allows you to select several objects at once by defining the shape you want to use to select them with: rectangle, circle, line, or polygon. For now, choose rectangle, and select the four cities to the immediate northwest and west of Youngstown.

5.7 Examining Attribute Tables

1. Each of the layers has an accompanying attribute table. To open the attribute table, click on the **Attribute Table** icon on the toolbar (it's the one that looks like a three-column chart). You can also open a layer's attribute table by right-clicking on the **layer** in the TOC, and then selecting **Attribute Table**.

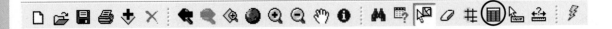

Important note: At the bottom of the cities' attribute table, you'll see that four records have been selected (the four cities near Youngstown). That's all well and good, but you'll also see that there are 3557 records in the entire attribute table (and thus, there are 3557 point objects in the cities data layer). To find those four records would mean a lot of digging through the data. However, AEJEE has some helpful sorting features, including one that will sort the table and move the selected records to the top.

2. Scroll across the attribute table to find the field called ST (which has the two-letter abbreviations for each state). It's between the CLASS and STFIPS fields and may require stretching its boundaries out. Right-click on the **name** of the ST field and from the new list of options, choose **Sort Selected Data To The Top**. The selected records will be highlighted in cyan and moved to the top records in the table.

Question 5.6 Without using Identify, which four cities did you select?

Question 5.7 How many total women (the attribute table lists this statistic as "Females") live in these four cities combined?

3. Close the **attribute table** when you're done.
4. To clear the selected features, select the **eraser** icon on the toolbar (the Clear All Selection tool).

5.8 Labeling Features

Rather than dealing with several points on a map and trying to remember which city is which, it would be easier to simply label each city so its name appears in the View. AEJEE gives you the ability to do this by creating a label for each record and allowing you to choose the field used to create the label.

1. Right-click on the **cities** layer in the TOC and select **Properties**.
2. Click on the **Labels** tab.
3. Under **Label Features Using**, choose **NAME**.
4. For now, accept all the defaults and click **Apply**. Back in the View, you'll see that labels of all the city names have been created and placed near each point.
5. Change around any of the options—font, color, text size, effects (plus options like bold or italics), and even the rotation angle to set up the labels however you feel is best for examining the map data.

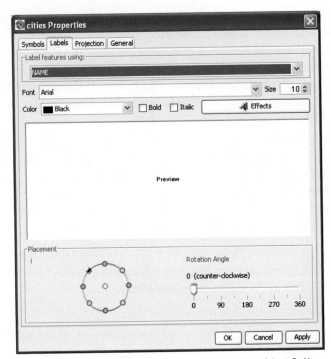

(Source: Esri® ArcGIS ArcExplorer graphical user interface Copyright © Esri.)

5.9 Measurements on a Map

With all the cities now labeled, it's easier to keep track of all of them. With this in mind, your next task is to make a series of measurements between points to see what the Euclidian (straight line) distance is between cities in northeast Ohio.

1. Zoom in tightly so that your View contains the cities of Youngstown and Warren and the other cities between them.

2. Select the **Measurement** tool from the toolbar—it's the one that resembles a ruler with a question mark positioned overtop.

3. From the available options that appear when you select the Measurement tool, choose **Miles**.

4. You'll see that the cursor has turned into a crosshairs. Place the crosshairs on the point representing Youngstown and left-click the mouse. Hold down the left mouse button and drag the crosshairs north and west to the point representing Girard. The distance measurement will appear in a box in the upper left section of the screen. The value for segment is the distance measured on the map, but the value for geodesic refers to the "surface" distance measured on a sphere.

5. Continue another line from Girard to Niles.

Question 5.8 What is the (geodesic) distance from Girard to Niles? What is the total (geodesic) distance from Youngstown to Niles (via Girard)?

6. You can clear all of the lines of measurement by again clicking on the **Measurement** tool on the toolbar, and then selecting **Clear Measure Totals**.

Question 5.9 What is the (geodesic) distance from Boardman to Warren, and then from Warren to Niles?

Question 5.10 What is the (geodesic) distance from Austintown, Ohio, to New Castle, Pennsylvania, and then from New Castle to Hermitage, Pennsylvania?

5.10 Saving Your Work (and Working on It Later)

When using AEJEE, you can save your work at any time and return to it. When work is saved in AEJEE, an Extended ArcXML file is written to disk. Later, you can re-open this file to pick up your work where you left off.

1. Saving to an ArcXML file is done by clicking on the **Save** icon on the toolbar (the floppy disk icon) or by choosing **Save** from the **File** pull-down menu.

2. Files can be re-opened by choosing the **Open** icon on the toolbar (the folder icon) or by choosing **Open** from the **File** pull-down menu. If the Catalog dialog box is still open, close it by selecting the **red X** in its window.

3. Exit AEJEE by selecting **Exit** from the **File** pull-down menu.

Closing Time

This lab was pretty basic, but served to introduce you to how AEJEE operates and how GIS data can be manipulated and examined. You'll be using either AEJEE or ArcGIS 10 in the next three labs, so the goal of this lab was to get the fundamentals of the software down. The lab in Chapter 6 takes this GIS data and starts to do spatial analysis with it, while the lab in Chapter 7 will have you start making print-quality maps from the data. The lab in Chapter 8 will involve some further GIS analysis involving road networks.

GIS Introduction: ArcGIS Version

This chapter's lab will introduce you to some of the basic features of GIS. You will be using one of Esri's GIS programs to navigate a GIS environment and begin working with spatial data. The labs in Chapters 6, 7, and 8 will utilize several more GIS features; the aim of this chapter's lab is to familiarize you with the basic functions of the software. The previous *Geospatial Lab Application 5.1: GIS Introduction: AEJEE Version* asked you to use the free ArcExplorer Java Edition for Educators (AEJEE), however, this lab provides the same activities but asks you to use ArcGIS 10.

Objectives

The goals for you to take away from this lab are:

- Familiarize yourself with the ArcGIS software environment, including basic navigation and tool use with both ArcMap and Catalog.
- Examine characteristics of spatial data, such as their coordinate system, datum, and projection information.
- Familiarize yourself with adding data and manipulating data layer properties, including changing the symbology and appearance of Esri data.
- Familiarize yourself with data attribute tables in ArcGIS.
- Make measurements between objects in ArcGIS.

Obtaining Software

The current version of ArcGIS (10) is not freely available for use. However, instructors affiliated with schools that have a campus-wide software license may request a 1-year student version of the software online at **http://www. esri.com/industries/apps/education/offers/promo/index.cfm.**

Important note: Software and online resources sometimes change fast. This lab was designed with the most recently available version of the software at the time of writing. However, if the software or Websites have significantly changed between then and now, an updated version of this lab (using the newest versions) is available online at **http://www.whfreeman.com/ shellito1e.**

Lab Data

There is no data to copy in this lab. All data comes as part of the AEJEE sample data that gets installed with the software.

Important note: In order to keep the data and results similar, both the AEJEE and ArcGIS 10 versions of the lab use the same set-up sample data that comes with AEJEE. Thus, if you're using ArcGIS 10 for the lab, please download and install AEJEE in order to use its sample data (see page 119).

Localizing This Lab

The dataset used in this lab is Esri sample data for the entire United States. However, starting in Section 5.6, the lab focuses on Ohio and the locations of some Ohio cities. With the sample data covering the state boundaries and city locations for the whole United States, it's easy enough to select your city (or ones nearby) and perform the same measurements and analysis using those cities more local to you than ones in northeast Ohio.

5.1 An Introduction to ArcMap

1. Start **ArcMap** (the default install folder is called ArcGIS). ArcMap will begin by asking you questions about "maps. " In ArcMap, a "map document" is the means of saving your work—since you're just starting, click on **Cancel** in the **Getting Started** dialog box. The left-hand column (where the word Layers is) is referred to as the Table of Contents (TOC)—it's where you will have a list of available data layers. The blank screen that takes up most of the interface is the View, where data will be displayed.

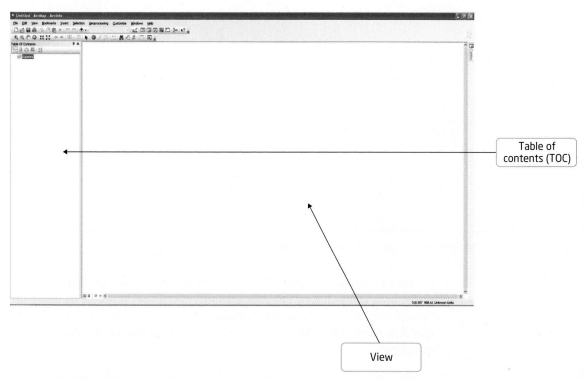

Table of contents (TOC)

View

(Source: Esri® ArcGIS ArcMap ArcInfo graphical user interface Copyright © Esri.)

Before you begin adding and viewing data, the first thing to do is examine the data you have available to you. To examine data layers, you can use the Catalog window in ArcMap—a utility designed to allow you to organize and manage GIS data.

Important note: ArcGIS 10 also contains a separate application called ArcCatalog that can be used to perform the same data management tasks. However, ArcGIS 10 incorporates the same functionality into ArcMap's Catalog window, so you'll be using that in this Lab, rather than a separate program.

2. By default, the Catalog window is "pinned" to the right-hand side of ArcMap's screen as a tab. Clicking on the **tab** will expand the **Catalog** window. You can also open the Catalog window by clicking on its **icon** on the main toolbar.

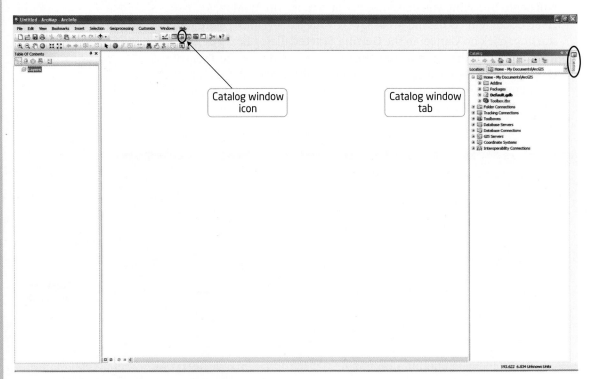

(Source: Esri® ArcGIS ArcMap ArcInfo graphical user interface Copyright © Esri.)

Important note: Various folders (representing network locations, hard drives on a computer, or external USB drives) can be accessed by using the Connect to Folder button. For instance, the default install location for the Esri data used in this lab should be stored in a folder on the C:\drive of your computer.

3. Press the **Connect to Folder** button on the Catalog's toolbar.

(Source: Esri® ArcGIS ArcMap ArcInfo graphical user interface Copyright © Esri.)

4. In the **Connect to Folder** dialog box, choose your computer's **C:\drive** (or the drive or folder where the Esri data is stored) from the available options and click **OK**.

5. Back in the Catalog, whichever folder you chose (such as C:\) will be available by expanding the Folder Connections option.

Important note: This lab assumes the Esri sample data is stored on the C:\ESRI\AEJEE\Data\path. Navigate to that folder (or its equivalent on your computer) and open the **usa** folder. Several layers will be available, including cities, lakes, and states).

6. The Catalog can be used to preview each of your available data layers as well as manage them. To see a preview of the layers you'll be using in this lab, do the following:

   a. Right-click on one of the layers in the Catalog (for instance, **states**).

   b. Select the option for **Item Description**.

   c. A new window will open (called "Item Description - STATES"). Click on the **Preview** tab in this window to see what the dataset looks like. Tools will be available in the window to zoom in and out (the magnifying glasses), pan around the data (the hand), or return to the full extent of the dataset (the globe).

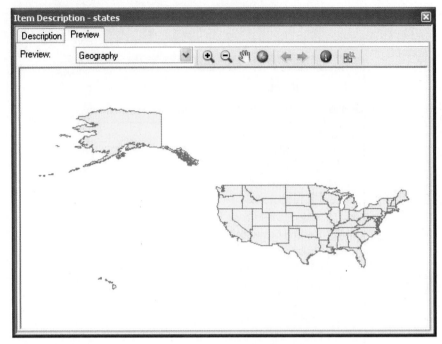

(Source: Esri® ArcGIS ArcMap ArcInfo graphical user interface Copyright © Esri.)

Preview each of the shapefiles (selecting a layer's **Item Description** will place it into the new window). Note that once the Item Description window is open (and the Preview of the Geography option selected), you can click once on the name of the file in the Catalog and it will display a preview of it in the Item Description window.

Question 5.1 What objects are each of the following datasets consisting of: cities, rivers, and counties?

5.2 Adding Data to ArcMap and Working with the TOC

1. Back in the main window of ArcMap, click on the yellow and black "**plus**" button to start adding the data you've previewed in the Catalog onto the map.

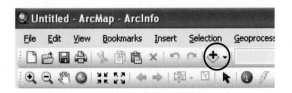

2. In the **Add Data** dialog box, navigate to the **ESRI** folder and open the **AEJEE** folder, the **Data** folder, the **usa** folder, and select the **states.shp** shapefile. Click **Add** (or double-click on the name of the shapefile).

3. You'll now see the states. shp in the TOC and its content is displayed in the View.

4. In the TOC, you'll see a checkmark in the box next to the states shapefile. When the checkmark is displayed, the layer will be shown in the View, and when the checkmark is removed (by clicking on it) the layer will not be displayed.

5. Now add two more layers, **cities** (a point layer), and **intrstat** (a line layer). All three of your layers will now be in the TOC.

6. You can manipulate the "drawing order" of items in the TOC by grabbing a layer with the mouse and moving it up or down in the TOC. Whatever layer is at the bottom is drawn first, and the layers above it in the TOC are drawn on top of it. Thus, if you move the states layer to the top of the TOC, the other layers will not be visible in the View since the states' polygons are being drawn over top of them.

5.3 Symbology of Features in ArcMap

1. You'll also notice that the symbology generated for each of the three objects is simple—points, lines, and polygons assigned a random color. You can significantly alter the appearance of the objects in the View to customize your maps.

2. Right-click on the **states** layer and select **Properties**.

3. Select the tab for **Symbology**.

4. Press the **large colored rectangle** in the **Symbol** box—in the new Symbol Selector dialog box, you can alter the appearance of the states by changing their color (choose one of the defaults or select from the Fill Color options), as well as the outline color and thickness of the outline width itself.

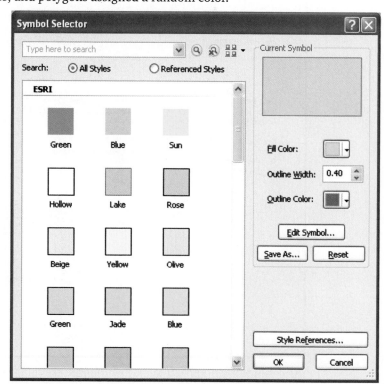

(Source: Esri® ArcGIS ArcMap ArcInfo graphical user interface Copyright © Esri.)

5. Change the states dataset to something more appealing to you, then click **OK**. In the **Layer Properties** dialog box, click **Apply** to make the changes, then **OK** to close the dialog. Do the same for the cities and roads shapefiles. Note that you can alter the points for the cities into shapes like stars, triangles, or crosses, and change their size as well. Several different styles are available for the lines of the roads file as well.

5.4 Obtaining Projection Information

1. Chapter 2 discussed the importance of projections, coordinate systems, and datums, and all of the information associated with these concepts can be accessed using ArcMap.

2. Right-click on the **states** shapefile and select **Properties**. Click on the **Source** tab in the dialog box.

3. Information about units, projections, coordinate systems, and datums can all be accessed for each layer.

4. Click **OK** to exit the states **Layer Properties** dialog box.

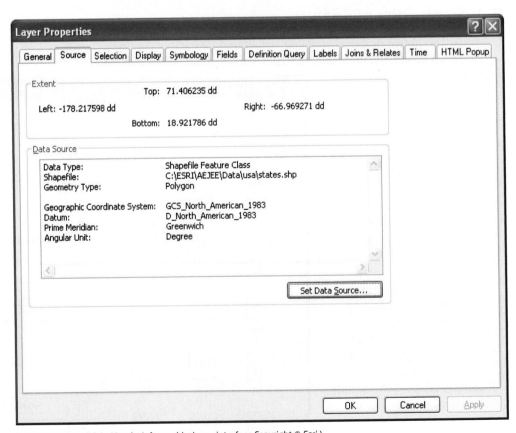

(Source: Esri® ArcGIS ArcMap ArcInfo graphical user interface Copyright © Esri.)

Question 5.2 What units of measurement are being used in the interstates dataset? What type of coordinate system (UTM, SPCS, etc.) is being used?

Question 5.3 What datum and projection are being used for the cities dataset?

Important note: ArcMap also allows you to change the projected appearance of the data in the View from one projected system to another. In ArcMap, all of the layers that are part of the View are held within a Data Frame—the default name for a Data Frame is "Layers," which you can see at the top of the TOC. By altering the properties of the Data Frame, you can change the appearance of how layers appear in the View. Note that this does not alter the actual projection of the layers themselves, simply how you're seeing them displayed.

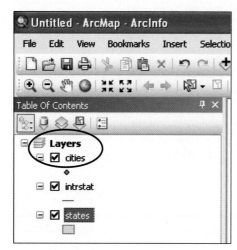

(Source: Esri® ArcGIS ArcMap ArcInfo graphical user interface Copyright © Esri.)

5. Right-click on the **Data Frame** in the TOC (the "Layers" icon) and select **Properties**.

6. In the **Data Frame Properties** dialog box, select the **Coordinate System** tab.

7. From the available options, select **Predefined**, then **Projected Coordinate Systems**, then **Continental**, and then **North America**. Lastly, choose **North America Lambert Conformal Conic**.

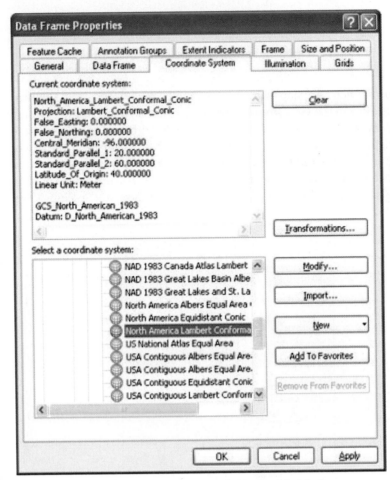

(Source: Esri® ArcGIS ArcMap ArcInfo graphical user interface Copyright © Esri.)

Question 5.4 What linear units and datum are used by the Lambert Conformal Conic projection?

8. Click **Apply** to make the changes, and then click **OK** to close the dialog box.

9. Back in the View, you'll see that the appearance of the three layers has completely changed from the flat GCS projection to the curved LCC projection. You could select other projections for the Data Frame if you wish. Try out some other appearances—when you're done, return to the Lambert Conformal Conic projection for the remainder of the Lab.

5.5 Navigating the View

1. ArcMap provides a number of tools for navigating around the data layers in the View. By default, these items are on a separate toolbar docked under the pull-down menus. ArcMap has a variety of separate toolbars to access—the one with the main navigation tools is simply called "Tools. " If it is not present, select the **Customize** pull-down menu, select **Toolbars**, and then select **Tools** from the available choices.

   Important note: The magnifying glass icons allow you to zoom in (the plus icon) or out (the minus icon). You can zoom by clicking in the window or clicking and dragging a box around the area you want to zoom into.

2. The **hand** icon is the Pan tool that allows you to "grab" the View by clicking on it, and dragging the map around the screen for navigation.

3. The **globe** icon will zoom the View to the extent of all the layers. It's good to use if you've zoomed to far in or out in the View or need to restart.

4. The four pointing arrows allow you to zoom in (arrows pointing inward) or zoom out (arrows pointing outward) from the center of the View.

5. The **blue arrows** allow you to step back or forward from the last set of zooms you've made.

6. Use the **Zoom** and **Pan** tools to center the View on Ohio so that you're able to see the entirety of the state and its cities and roads.

5.6 Interactively Obtaining Information

1. Even with the View centered on Ohio, there are an awful lot of point symbols there, representing the cities. We're going to want to identify and use only a couple of them. ArcMap has a searching tool that allows you to locate specific objects—on the toolbar, select the **Find tool** (the binoculars icon):

2. In the **Find** dialog box, select the **Features** tab.

   In the **Find** field type: **Akron**.

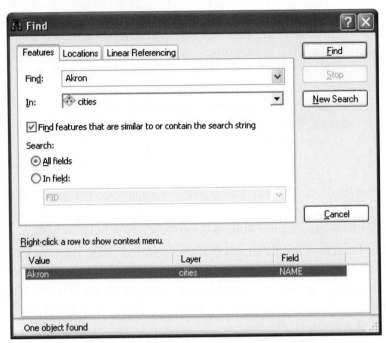

(Source: Esri® ArcGIS ArcMap ArcInfo graphical user interface Copyright © Esri.)

3. Choose **Cities** for the Layers to look In.

4. Select the option for **All fields** for the Search.

5. Click the **Find** button. The returned results will be shown in the bottom of the dialog box.

6. Drag the dialog box out of the way so you can see both it and the View at the same time.

7. Double-click on the result for **Akron** in the bottom of the dialog box.

8. You'll see that the point symbol for the city of Akron briefly highlights and changes color.

9. Right-click on the same results in the Find dialog box and choose **Select** from the available options. You'll see that the point representing Akron has changed color to a light blue. This means it has been "selected" by ArcMap—any actions performed on the cities layer will affect the selected items, not all of the cities.

10. However, all that Find does is locate and select an object. To obtain information about it, you can use the **Identify tool** (the icon on the toolbar is the white "i" in a purple circle).

11. Select the **Identify tool**, and then click on Akron (zooming in as necessary). A new dialog box will open, listing all of the attribute information associated with Akron (it's actually giving you all of the field attribute information that goes along with that particular record).

12. In the **Find** dialog box, right-click on the results and choose **Unselect** to remove Akron from selection (this will prevent it from being used in the analysis for the next sections).

13. Return to **Find** and look for the city called Youngstown, and then use the **Identify tool** on the selected city.

Question 5.5 According to ArcMap, what is the year 2000 population of Youngstown (carefully examine the attributes returned from Identify)?

14. Another method of interactively selecting features is to use the **Select Features** tool. To select features from the cities layer, click on **cities** in the TOC. Next, on the toolbar, the icon is a **cursor** with a white and cyan polygon to its right.

15. The **Select Features tool** allows you to select several objects at once by defining the shape you want to use to select them with: rectangle, polygon, lasso, circle, or line. For now, choose **rectangle** or **polygon**, and select the four cities to the immediate northwest and west of Youngstown (note that this will also select features from other layers, although we will just be working with cities in this lab).

5.7 Examining Attribute Tables

1. Each of the layers has an accompanying attribute table. To open a layer's attribute table, right-click on the layer in the TOC, then select **Open Attribute Table**. Open the attribute table for cities.

2. At the bottom of the cities' attribute table you'll see that four records have been selected (the four cities nearby Youngstown to its west and northwest). That's all well and good, but you'll also see that there's 3557 records in the entire attribute table (and thus, there are 3557 point objects in the cities data layer). To find those four records would mean

a lot of digging through the data. However, ArcMap has some helpful sorting features, including one that will sort the table and only show the selected records.

3. At the bottom of the attribute table, select the icon for **Show Selected Records** (it's the short stack icon in light blue).

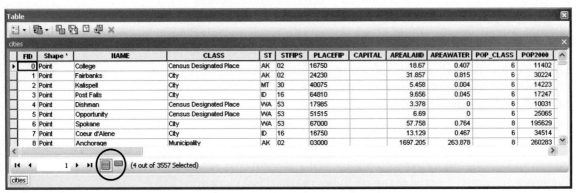

FID	Shape	NAME	CLASS	ST	STFIPS	PLACEFIP	CAPITAL	AREALAND	AREAWATER	POP_CLASS	POP2000
0	Point	College	Census Designated Place	AK	02	16750		18.67	0.407	6	11402
1	Point	Fairbanks	City	AK	02	24230		31.857	0.815	6	30224
2	Point	Kalispell	City	MT	30	40075		5.458	0.004	6	14223
3	Point	Post Falls	City	ID	16	64810		9.656	0.045	6	17247
4	Point	Dishman	Census Designated Place	WA	53	17985		3.378	0	6	10031
5	Point	Opportunity	Census Designated Place	WA	53	51515		6.69	0	6	25065
6	Point	Spokane	City	WA	53	67000		57.758	0.764	8	195629
7	Point	Coeur d'Alene	City	ID	16	16750		13.129	0.467	6	34514
8	Point	Anchorage	Municipality	AK	02	03000		1697.205	263.878	8	260283

(4 out of 3557 Selected)

(Source: Esri® ArcGIS ArcMap ArcInfo graphical user interface Copyright © Esri.)

The four selected records will be highlighted in cyan and be the only records shown in the table.

Question 5.6 Without using Identify, which four cities did you select?

Question 5.7 How many total women (the attribute table lists this statistic as "Females") live in these four cities combined?

4. Close the **attribute table**. To clear the selected features, select the **clear icon** on the toolbar (the Clear Selected Features) tool.

5.8 Labeling Features

1. Rather than dealing with several points on a map and trying to remember which city is which, it would be easier to simply label each city so its name appears in the View. ArcMap gives you the ability to do this by creating a label for each record and allowing you to choose the field used to create the label.

2. Right-click on the **cities** layer in the TOC and select **Label Features**. Default labels for the names of the cities will appear next to the points.

3. To change the appearance of the labels, right-click on the **cities** layer and select **Properties**. In the **Properties** dialog box, choose the **Labels** tab.

4. Change around any of the options—font, color, text size, (plus options like bold, italics, or underlining), and even the placement of the labels however you feel is best for examining the map data.

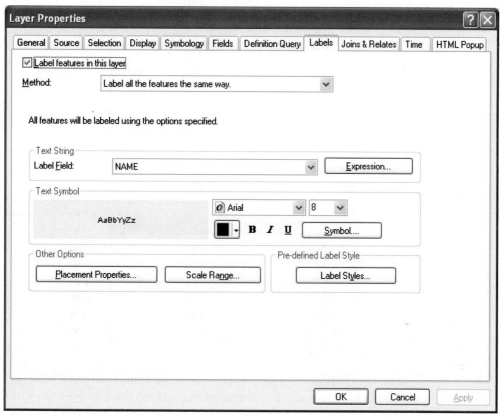

5.9 Measurements on a Map

1. With all the cities now labeled, it's easier to keep track of all of them. With this in mind, your next task is to make a series of measurements between points to see what the Euclidian (straight line) distance is between cities in northeast Ohio.

2. Zoom in tightly so that your View contains the cities of Youngstown and Warren and the other cities between them.

3. Select the **Measure tool** from the toolbar—it's the one that resembles a ruler with two blue arrows over it.

4. In the **Measure** dialog box, select the **Choose Units** pull-down menu, then select **Distance**, and then select **Miles** (this will return all measurements to you in miles rather than the default of meters).

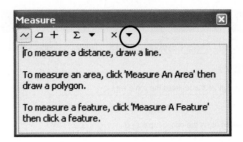

5. Select the **Measure Line** tool from the toolbar. You'll see that the cursor has turned into a crosshairs framed by a ruler.

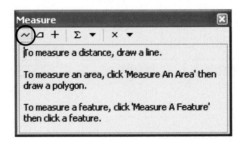

6. Place the crosshairs on the point representing Youngstown and left-click the mouse (you'll also see a circle appear around Youngstown). Drag the crosshairs north and west to the point representing Girard (once you've reached the point representing Girard, a circle will appear around that point) and left-click the mouse again. The distance measurement will appear in a box in the Measure dialog box. The value for Segment is the planar distance measured on the map for one leg of the measure, while Length is the value for all segments being measured.

7. Continue another line from Girard to Niles.

Question 5.8 What is the (planar) distance from Girard to Niles? What is the total (planar) distance from Youngstown to Niles (via Girard)?

8. Double-clicking the mouse will stop the measurement tool and remove the lines from the screen. To clear the results shown in the measurement dialog box, select the **Clear and Reset Results** button in the Measure dialog.

9. There are a few more measurements to make to complete this chapter's questions.

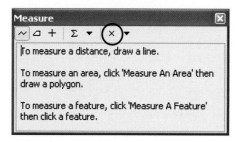

Question 5.9 What is the (planar) distance from Boardman to Warren, and then from Warren to Niles?

Question 5.10 What is the (planar) distance from Austintown, Ohio, to New Castle, Pennsylvania, and then from New Castle to Hermitage, Pennsylvania?

5.10 Saving Your Work (and Working on It Later)

When using ArcMap, you can save your work at any time and return to it. When work is saved in ArcMap, a map document file is written to disk. Later, you can re-open this file to pick up your work where you left off.

1. Saving to an ArcMap document file is done by clicking on the **Save** icon on the toolbar (the floppy disk icon) or by choosing **Save** from the **File** pull-down menu.

2. Files can be re-opened by choosing the **Open** icon on the toolbar (the folder icon) or by choosing **Open** from the **File** pull-down menu.

3. Exit ArcMap by selecting **Exit** from the **File** pull-down menu. There's no need to save any data in this lab.

Closing Time

This lab was pretty basic but served to introduce you to how ArcGIS operates and how GIS data can be manipulated and examined. You'll be using either AEJEE or ArcGIS 10 in the next three chapter labs, so the goal of this lab was to get the fundamentals of the software down. The lab in Chapter 6 takes this GIS data and starts to do spatial analysis with it, while the lab in Chapter 7 will have you start making print-quality maps from the data. Chapter 8's lab will involve some further GIS analysis involving road networks.

Using GIS for Spatial Analysis

Database Query and Selection, Buffers, Overlay Operations, Geoprocessing Concepts, and Modeling with GIS

The previous chapter described the basic components of GIS and how it operated, so now it's time to start doing some things with that GIS data. Anytime you'll be examining the spatial character of data, or how objects relate to one another across distances, you're performing **spatial analysis**. A very early (pre-computer) example of this is Dr. John Snow's attempt to solve the mystery of several cholera-related deaths in London in 1854. Snow's analysis was able to identify an infected pump, which was the source of the outbreak. At some point, Snow mapped the locations of the cholera deaths in relation to the water pumps (**Figure 6.1**). This type of spatial analysis

spatial analysis examining the characteristics or features of spatial data, or how features spatially relate to each other.

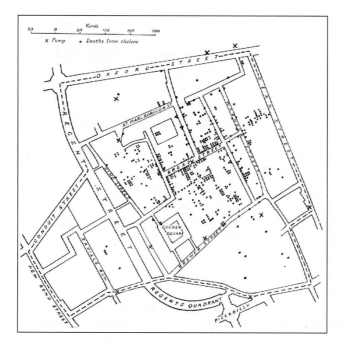

FIGURE 6.1 A version of John Snow's map showing locations of cholera deaths and well pumps.

(relating locations of cholera deaths to the positions of water pumps) was innovative for the mid-nineteenth century, but commonplace today—with GIS, these types of spatial analysis problems can be easily addressed.

Other common forms of spatial analysis questions are some of the following:

◉ How many objects are within a certain distance of a particular location? All manner of these types of questions can be answered, including marketing topics such as, "How many fast food restaurants are within one mile of an interstate exit?"; real estate topics regarding how many rental properties are within two blocks of a house that is for sale; or planning topics to determine the acreage of wetland threatened with removal by a new proposed development plan.

◉ How do you choose the most suitable location based on a set of criteria? Many different types of questions related to this topic can be answered, such as those involving where to locate new commercial developments, new real estate options, or the best places for animal habitats.

This chapter examines how GIS can be used with spatial data to answer questions like these in the context of spatial analysis. Keep in mind the two ways that GIS is used to model real-world data from the previous chapter—vector and raster. Due to the nature of some of these properties, some types of spatial analysis are best handled with vector data and some with raster data.

How Can Data Be Retrieved From a GIS for Analysis?

One type of analysis would be finding which areas or locations meet particular criteria—for instance, being able to take a map of U.S. Congressional Districts and identify which districts had a Democratic representative and which had a Republican representative. Alternatively, you might have a map of all residential parcels and would want to see which houses have been sold in the last month. In both of these examples, you would perform a database **query** to select only those records from the attribute table (see Chapter 5) that reflect the qualities of the objects you want to work with.

In other cases, before beginning the analysis, you may only want to utilize a subset of your GIS data. For instance, rather than dealing with all Congressional Districts in the entire United States, you might want to examine only those in Ohio. Or in the housing example, if you have a dataset of all residential parcels in a county (which is likely thousands of parcels), you may only want to do analysis related to one (or a handful) of the parcels, rather than examining all of them. You'd want to go into the dataset and select only those parcels required for your study. In these cases, the query would again only select the records you want to deal with.

In GIS, queries are composed in the Structured Query Language (**SQL**) format, like a mathematical function. SQL is a specific format that is used for

query the conditions used to retrieve data from a database.

SQL the Structured Query Language—a formal setup for building queries.

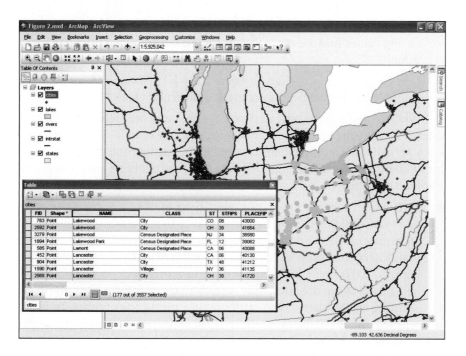

FIGURE 6.2 A database query for Ohio cities returning multiple records (selected records and their corresponding cities are in cyan). (Source: Esri® ArcGIS ArcMap ArcView graphical user interface Copyright © Esri.)

querying a layer or database to find what attributes meet certain conditions. For example, a layer made up of points representing cities in Ohio also has a number of fields representing its attributes, such as city name, population, and average income. If you just wanted to do analysis with one of these records (the city called "Boardman"), then a query could be built to find all records where the field called CITY_NAME contained the characters of 'Boardman' as: CITY_NAME = 'Boardman'. This should return one record for you to use in your analysis. Another example would be to select all cities in Ohio with a population greater than or equal to 50,000 persons. The attribute table has a field called POP2010 (representing the 2010 city population) that contains this information. Thus, a query of POP2010 >= 50000 could be built, and all records whose population field contains a value of greater than or equal to the number 50,000 would be selected for your use (see Figure 6.2 for an example of a query in GIS).

A query will use one of the following **relational operators**:

⊙   Equal (=): This is used when you want to find all values that match the query. For instance, querying for CITY_NAME = 'Poland' will locate all records that exactly match the characters in the string that comprise the word 'Poland'. If the record was called 'Poland Township' it would not be located since the character string looks for an exact match.

⊙   Not Equal (<>): This is used when you want to find all the records that do not match a particular value. For instance, querying for CITY_NAME <> 'Poland' will return all the records that do not have a character string of the word 'Poland' (which would probably be all of the other cities with names in the database).

relational operator
one of the six connectors (=, < >, <, >, >=, or =<) used to build a query.

compound query a query that contains more than one operator.

Boolean operator one of the four connectors (AND, OR, NOT, XOR) used in building a compound query.

AND the Boolean operation that corresponds with an Intersection operation.

Intersection the operation wherein the chosen features are those that meet both criteria in the query.

OR the Boolean operator that corresponds with Union operation.

Union the operation wherein the chosen features are all that meet the first criteria as well as all that meet the second criteria in the query.

NOT the Boolean operator that corresponds with a Negation operation.

Negation the operation wherein the chosen features are those that meet all of the first criteria and none of the second criteria (including where the two criteria overlap) in the query.

XOR the Boolean operator that corresponds with an Exclusive Or operation.

◉   Greater Than (>) or Greater Than Or Equal To (>=): This is used for selecting values that are more than (or more than or equal to) a particular value.

◉   Less Than (<) or Less Than Or Equal To (=<): This is used for selecting values below (or below and equal to) a particular value.

While a simple query only uses one operator and one field, a **compound query** enables you to make selections using multiple criteria. There are different ways of linking multiple criteria together for creating one of these types of compound queries, each using a different option of querying (referred to as a **Boolean operator**):

◉   **AND**: For instance, you want to select cities in Ohio that have a population of over 50,000 (a variable called POP2010 in the attribute table) and an average household income (a variable called AVERAGEHI) of more than $30,000. A compound query can combine these two requests by saying: POP2010>= 50000 AND AVERAGEHI > 30000. Because "AND" is being used as the operator, only records that match both the criteria get returned by the query. If an attribute only meets one criterion, it is not selected. An "AND" query is referred to as an **Intersection** since it returns what the two items have in common.

◉   **OR**: A different query could be built by saying: POP2010>= 50000 OR AVERAGEHI > 30000. This would return records of cities with a population of greater than or equal to 50,000, or cities with an average household income of greater than $30,000, or cities that meet both. When an OR operator is being used, records that meet one or both criteria are selected. An "OR" query is referred to as a **Union** since it returns all the elements of both operations.

◉   **NOT**: A third query could be built around the concept that you want all of the data related to one criteria, but to exclude what relates to the second criteria. For instance, if you want to select cities with a high population, but not those with a higher average household income, you could build a query like: POP2010>= 50000 NOT AVERAGEHI > 30000 to find those particular records. A "NOT" query is referred to as a **Negation** since it returns all the elements of one dataset, but not what they have in common with the other.

◉   **XOR**: A final type of query can be built using the idea that you want all of the data from each layer, except for what they have in common. For instance, if you wanted to find those cities with a high population or those that had high incomes, but not both types of cities, your query would be: POP1990>= 50000 XOR AVERAGEHI > 30000. XOR works as the opposite of AND since it returns all data, except for the intersection of the datasets. An "XOR" query is referred to as an **Exclusive Or**, since it acts like a Union but leaves out the Intersection data.

Compound queries can use multiple operations to select multiple records by adhering to an order of operations, such as: (POP2010>=50000 AND

Hands-on Application 6.1

Working with Queries in GIS

Durham, North Carolina, has an online GIS Spatial Data Explorer utility that allows the user to do a variety of spatial analyses with the city's data, including performing queries. To work with this tool, open your Web browser and go to **http://gisweb. durhamnc.gov/gomaps/map/index.cfm**. In the upper left-hand corner are several tools (such as zooming and panning) to get oriented to the GIS data. Included in the tools is a "hammer" icon, which allows you to build queries of the data. Press the "hammer" icon and a new set of options will appear on the right, allowing you to construct simple queries of the GIS. First, select an available layer (for instance, select 2000 Census Block Groups). Second, select an attribute to query (for instance, choose "Housing Units"). Third, select an operator (for instance, choose the greater than ">" symbol). Press the "add to query" button and you'll see the SQL statement appear. To finish the query, type a value (for instance, type 1000), and press the "Send the Query" button. All records meeting your query will be retrieved (in this case, all Census block groups with more than 1000 housing units in them). The selected block groups will be highlighted on the map, while the records (and the attributes that go along with them) will be displayed in a spreadsheet at the bottom of the screen. Investigate some of the other available data layers and the types of queries that can be built with them.

AVERAGEHI > 30000) OR CITY_NAME = 'Boardman'. This would select cities that meet both the population and the income requirement and would select cities named 'Boardman' as well. The result of a query will be a subset of the records that are chosen, and then those records can be used for analysis, rather than the whole data layer. Selected records can also be exported to their own dataset. For instance, the results of the previous query could be extracted to compose a new GIS layer, made up of only those records (and their corresponding spatial objects) that met the query's criteria. Once you have your selected records or new feature layers, you can start to work with them in GIS (see *Hands-on Application 6.1: Working with Queries in GIS* for an example of doing queries with online GIS utility).

Exclusive Or the operation wherein the chosen features are all of those that meet the first criteria as well as all of those that meet the second criteria, except for the features that have both criteria in common in the query.

How Can You Perform Basic Spatial Analysis in GIS?

One you have your selected data, it's time to start doing something with it. A simple type of analysis is the construction of a **buffer** around the elements of a data layer. A buffer refers to an area of proximity set up around one or more objects. For instance, if you wanted to know how many rental properties were within a half-mile of a selected house, you would create a half-mile buffer around the object representing the house and then determine how many rental properties were within the buffer. Buffers can be created for multiple points, lines, or polygons (**Figure 6.3** on page 152). Buffers are a simple tool to utilize, but they can provide useful information when creating areas of analysis for various events.

buffer a polygon of spatial proximity built around a feature.

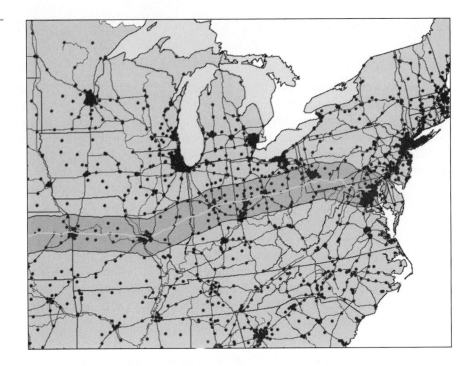

dissolve the ability of the GIS to combine polygons with the same features together.

GIS can also perform a **dissolve** operation, where boundaries between adjacent polygons (that have the same properties) are removed, merging the polygons into a single, larger shape. Dissolve is useful when you don't need to examine each polygon, only regions with similar properties. For example, you have a land-use map of a county consisting of polygons representing each individual parcel and how it's being used (commercial, residential, agricultural, etc.). Rather than having to examine each individual parcel, you could dissolve the boundaries between parcels, creating regions of "residential" land use or "commercial" land use for ease of analysis. Similarly, if you have overlapping buffers generated from features, dissolve can be used to combine buffer zones together. See Figure 6.4 for an example of dissolve in action.

Buffers and querying are basic types of spatial analysis, but they only involve a single layer of data (such as creating a buffer around roads or selecting all records that represent a highway). If you were going to do spatial analysis along the lines of John Snow's hunt for a cholera source, you'd have to be able to do more than just select a subset of the population who died or just selecting the locations of wells within the city borders—you'd have to know something about how these two things interacted with each other over a distance. For instance, you would want to know something about how many cholera deaths were found within a certain distance of each well. Thus, you wouldn't just want to create a buffer around one layer (the wells), but you'd want to know something about how many features from a different layer (the death locations) were in that buffer. Many types of GIS analysis involve combining the features from multiple layers together to allow examination

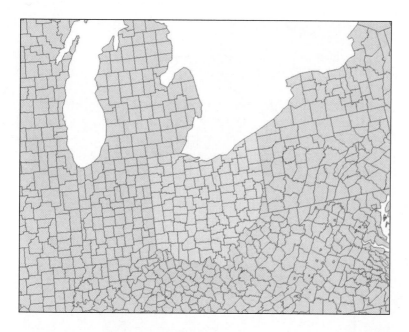

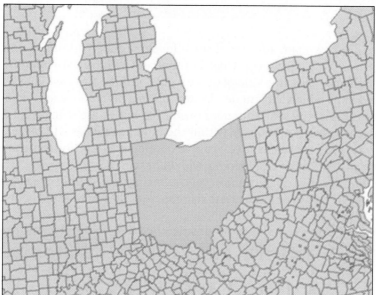

FIGURE 6.4 Before and after the dissolve operation—Ohio county boundaries are dissolved on the basis of their state's FIPS code. (Source: Esri® ArcGIS ArcMap ArcView graphical user interface Copyright © Esri.)

of several characteristics at once. In GIS, when one layer has some sort of action performed to it and the result is a new layer, this process is referred to as **geoprocessing**. There are numerous types of geoprocessing methods and they are frequently performed in a series to solve a spatial analysis question.

GIS gives you the ability to not just perform a regular SQL-style query, but to also perform a **spatial query**—to select records or objects from one layer based upon their spatial relationships with other layers (rather than using attributes). With GIS, queries can be based on spatial concepts—to determine

geoprocessing the term that describes when an action is taken to a dataset that results in a new dataset being created.

spatial query selecting records or objects from one layer based upon their spatial relationships with other layers (rather than using attributes).

FIGURE 6.5 Effect of selecting point features (cities) by using a buffer (fifty miles around I-70). (Source: Esri® ArcGIS ArcExplorer graphical user interface Copyright © Esri.)

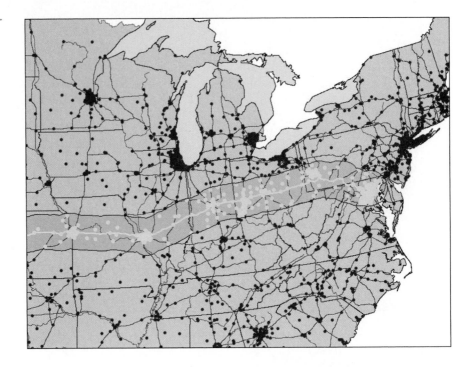

things like how many registered sex offenders reside within a certain distance away from a school or which hospitals are within a particular distance of your home. The same kind of query would be done if a researcher wanted to determine which cholera deaths are within a buffer zone around a well. In these cases, the spatial dimensions of the buffer are being used as the selection criteria—all objects within the buffer are what will be selected. See Figure 6.5 for an example of using a buffer as a selection tool as well as *Hands-on Application 6.2: Working with Buffers in GIS*.

Other analyses can be done beyond selecting with a buffer or selecting by location. When two or more layers share some of the same spatial boundaries

Hands-on Application 6.2

Working with Buffers in GIS

Honolulu has an online GIS application that looks at parcels and zoning for the area. On this Website, you can select a location, generate a buffer around it, and all parcels touched by the buffer can be selected (and their parcel records displayed). To use this, open your Web browser and go to **http://gis. hicentral.com/fastmaps/parcelzoning**—then zoom in to a developed or residential area (if you're familiar with the area, you can also search by address).

To generate a buffer, select the Tools icon, and then choose the option for Circle-Select Parcels. At this point, you can specify a buffer distance (the radius of the circle) and click on a place on the map. A buffer will be generated around your placemark and the parcels selected by the buffer will be highlighted on the map and their records returned. Try examining some areas with various buffer sizes to see how buffers can be used as a selection tool.

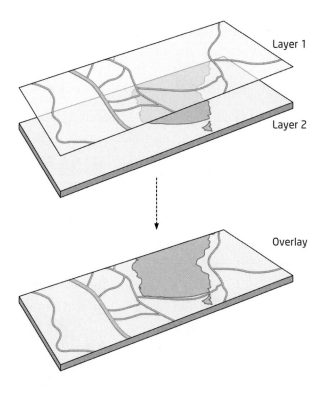

FIGURE 6.6 An example of overlaying two layers.

(but have different properties) and are combined together, this is referred to as an **overlay** operation in the GIS. For instance, if one layer contains information about the locations of property boundaries and a second layer contains information about water resources, combining these two layers in an overlay can help determine which resources are located on whose property. An overlay is a frequently used GIS technique for combining multiple layers together for performing spatial analysis (Figure 6.6).

There are numerous ways that polygon layers can be combined together through an overlay in GIS. Some overlay methods (Figure 6.7 on page 156) are:

- **Intersect:** In this operation, only the features that both layers have in common are retained in a new layer. This type of operation is commonly used when you want to determine an area that meets two criteria—for instance, a location needs to be found within a buffer and also on an agricultural land use—you would intersect the buffer layer and the agricultural layer together to find where, spatially, they have in common.

- **Identity:** In this operation, all of the features of an input layer are retained, and all the features of a second layer that intersect with them are also retained. For example, you may want to examine all of the nearby floodplain and also what portions of your property are on it.

- **Symmetrical difference:** In this operation, all of the features of both layers are retained—except for the areas that they have in common. This

overlay the combining of two or more layers in the GIS.

intersect a type of GIS overlay that retains the features that are common to two layers.

identity a type of GIS overlay that retains all features from the first layer along with the features it has in common with a second layer.

symmetrical difference a type of GIS overlay that retains all features from both layers except for the features that they have in common.

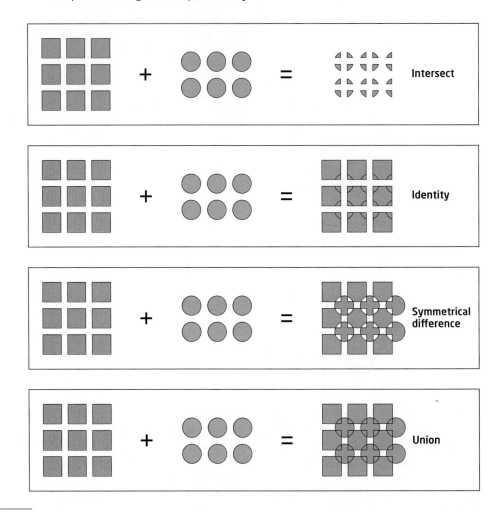

operation could show you, for example, all potential nesting areas along
with the local plan for development, except where they overlap (so those
areas can be taken out of the analysis).

⦿ **Union (overlay)**: In this operation, all of the features from both layers
are combined together into a new layer. This operation is often to merge
features from two layers (for instance, in order to evaluate all possible options for a development site, you may want to overlay the parcel layer and
the water resources layer together).

Another way of performing spatial overlay is by using raster data. With
raster cells, each cell has a single value and two grid layers can be overlaid in
a variety of ways. The first of these is by using a simple mathematical operator
(sometimes referred to as **Map Algebra**), such as addition or multiplication.
To simplify things, we're only going to assign our grids values of 0 or 1, where
a value of 1 means that the desired criteria are met and a value of 0 is used

union (overlay) a
type of GIS overlay that
combines all features
from both layers.

Map Algebra
combining datasets
together using
simple mathematical
operators.

when the desired criteria are not met. Thus, grids could be designed using this system with values of 0 or 1 to reflect whether a grid cell meets or does not meet particular criteria.

Let's start with a simplified example—say you're looking to find a plot of land to build a vacation home on, and your two key criteria are that it be close to more forested terrain for hiking and be close to a river or lake for water recreation. What you could do is get a grid of the local land cover and assign all grid cells that have forests on the landscape a value of 1 and all non-forested areas a value of 0. The same holds true for a grid of water resources—any cells near a body of water get a value of 1 and all other cells get a 0. The "forests" grid has values of 1 where there are forests and values of 0 for non-forested regions, while the "water" grid has values of 1 where a body of water is present and values of 0 for non-river areas (see Figure 6.8 for a simplified version of setting this up).

Now you want to overlay your two grids and find what areas meet your two criteria to build your vacation home, but there are two ways of combining these grids. Figure 6.9 on page 158 shows how the grids can be added and multiplied together with varying results. By multiplying grids containing values of 0 and 1 together, only values of 0 or 1 can be generated in the resulting overlay grid. In this output, values of 1 indicate where (spatially) both criteria (forests and rivers) can be found, while values of 0 indicate where both cannot be found. In the multiplication example, a value of 0 could have one of the criteria, or none, but we're only concerned with areas that meet both criteria, or else we're not interested. Thus, the results either match what we're looking for or they don't.

In the addition example, values of 0, 1, or 2 are generated from the overlay. In this case, values of 0 contain no forests or rivers, values of 1 contain one or the other, and values of 2 contain both. By using addition to overlay the grids, a gradient of choices is presented—the ideal location would be at locations with a value of 2, but values of 1 represent possible second-choice alternatives, and values of 0 represent unacceptable locations.

This type of overlay analysis is referred to as **Site Suitability**, as it seeks to determine which locations are the "best" (or most "suitable") for meeting certain criteria. Site Suitability analysis can be used for a variety of topics,

Site Suitability the determination of the "useful" or "non-useful" locations based on a set of criteria.

FIGURE 6.8 A hypothetical landscape consisting of forests, rivers, and non-forested or dry areas.

=

FIGURE 6.9 Raster
overlay examples for
multiplication and
addition.

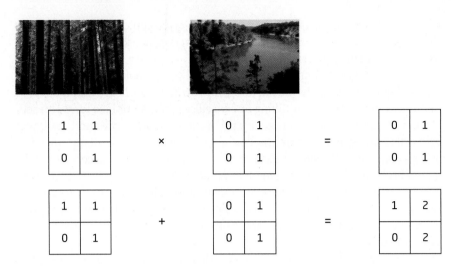

including identifying ideal animal habitats, prime locations for urban developments, or prime crop locations. When using the multiplication operation,
you will be left with sites classified as "suitable" (values of 1) or "non-suitable"
(values of 0). However, the addition operation leaves other options that may
be viable in case the most suitable sites are unavailable, in essence providing
a ranking of suitable sites. For instance, the best place to build the vacation
home is in the location with values of 2 since it meets both criteria, but the
cells with a value of 1 provide less suitable locations (as they only meet one
of the criteria).

How Can Multiple Types of Spatial Analysis Operations Be Performed in GIS?

Often there are many criteria that play into the choice of site suitability or determining what might be the "best" or "worst" locations for something, and often
it's not as cut and dried as "the location contains water" (values of 1) or "the
location does not contain water" (values of 0). Think about the vacation home
example—locations were classified as either "close to water" or "not close to
water." When choosing a vacation home location, there could be locations right
on the water, somewhat removed from it but still close enough to recreation,
and some that were much further away. Similarly, terrain is so changeable, it's
difficult to assign it a characteristic of "mountainous" or "not mountainous."
Instead, a larger range of values could be assigned—perhaps on a scale of
1 through 5, with 5 being the best terrain for hiking or the closest to the water
and 1 being furthest away from the water or the least desirable terrain.

In order to set something like this up, a distance calculation could be
performed. GIS can calculate the distance from one point to all other points
on a map or calculate the distance a location is from all other features. For
example, when examining a location for building a house in a rural locale,

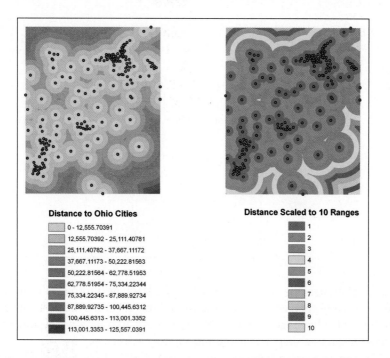

Distance to Ohio Cities

- 0 - 12,555.70391
- 12,555.70392 - 25,111.40781
- 25,111.40782 - 37,667.11172
- 37,667.11173 - 50,222.81563
- 50,222.81564 - 62,778.51953
- 62,778.51954 - 75,334.22344
- 75,334.22345 - 87,889.92734
- 87,889.92735 - 100,445.6312
- 100,445.6313 - 113,001.3352
- 113,001.3353 - 125,557.0391

Distance Scaled to 10 Ranges

- 1
- 2
- 3
- 4
- 5
- 6
- 7
- 8
- 9
- 10

FIGURE 6.10 An example of a raster distance surface calculation, where each grid cell contains a value representing the distance from a point (cities in Ohio) and the same grid sliced into 10 ranges. (Source: Esri® ArcGIS ArcMap ArcView graphical user interface Copyright © Esri.)

you may want to know the distance that spot is from a landfill or waste-water treatment plant. Alternatively, you may want to know the distance a vacation home is away from a lake or river. A way of conceptualizing distance (as is done in ArcGIS) is to think of it has a continuous surface (see Chapter 5), where each spot on that surface is has a value for the distance it is away from a feature. ArcGIS makes these measurements by calculating a distance surface of raster grid cells, with each cell containing a value of the distance from the features in a layer. Figure 6.10 illustrates this by mapping the location of a set of cities in Ohio and calculating the distance that each cell is away from a city. Next, the distance from the cities is sliced into ten equal ranges (that is, all of the areas closest to a city are given a value of 1, the next closest are given a value of 2, and so forth).

For this example, we'll come up with a new set of grids classified along these lines and throw in a third option as well—distance away from roads (using the assumption that a vacation home would be best placed away from busy roads and the accompanying automobile noise). In all of these, the highest values represent the "best" options (such as the furthest places away from busy roads), and the lowest values represent the "worst" options (such as right next to a busy road). Figure 6.11 on page 160 shows the new grids and how they are overlaid together through the addition operation to create a new output grid. Rather than choices of 0, 1, and 2, the output grid contains a wide range of options, where the highest number reflects the most suitable site based on the criteria and the lowest number reflects the least suitable sites. This new grid represents a very simple **Suitability Index**, where a range of values are generated, indicating a ranking of the viability of locations (these would be ordinal values as discussed back in Chapter 5). In the example in Figure 6.11, the values of 13 represent the

Suitability Index a system whereby locations are "ranked" according to how well they fit a set of criteria.

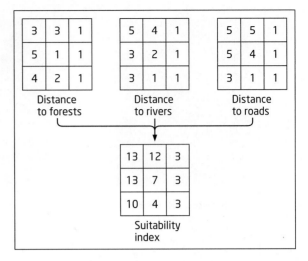

FIGURE 6.11 A raster overlay example for combining three criteria together to form a simple Suitability Index.

most suitable sites, the values of 12 and 10 are possible alternative venues, the value of 7 is a likely mediocre site, and the values of 4 and 3 are probably unacceptable. The advantage of the Suitability Index is to offer a wider gradient of choices for suitable sites instead of some of the previous options.

GIS can process numerous large spatial data layers together at once. Thinking about the vacation home example, there's many additional factors that play into choosing the site of a second home—access to recreational amenities, the price of the land, the availability of transportation networks, and how far the site is away from large urban areas. All of these factors (and more) come into play when identifying the optimal location for a vacation home. Similarly, if you work in the retail industry and are assessing which place is the best location for a new store, there are a lot of criteria that will influence your decision—how close the potential sites are to your existing store, how close that area is to similar stores, the population of the area, the average household income for each neighborhood, and the rental price of the property. GIS can handle all of these spatial factors by combining them together (see **Figure 6.12** for an example of combining several data layers together).

GIS Model a representation of the factors used for explaining the processes that underlie an event or for predicting results.

A **GIS Model** refers to a way of combining these spatial dimensions or characteristics together in an attempt to describe or explain a process or predict results. For instance, trying to predict where future urban developments will occur based on variables such as a locations' proximity to existing urban areas, the price of the land parcels, or the availability of transportation corridors could be performed using GIS, with the end result being a map of the places most likely to convert to an urban land use. There are many different types of GIS Models with different features and complexities to them, and it's far beyond the scope of this book to catalog and examine all of them.

GIS Models are often used to examine various criteria together to determine where, spatially, some phenomenon is occurring or could potentially occur. These are often used in decision making processes or to answer questions (like the previous example of where to build a vacation home). An example

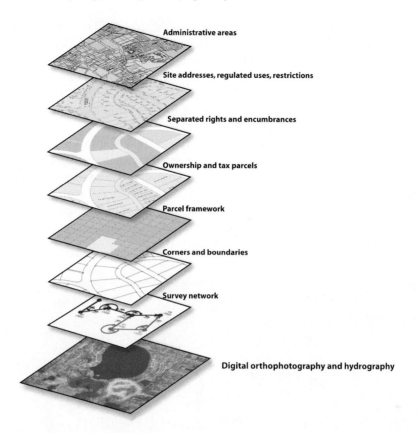

Administrative areas

Site addresses, regulated uses, restrictions

Separated rights and encumbrances

Ownership and tax parcels

Parcel framework

Corners and boundaries

Survey network

Digital orthophotography and hydrography

of this would be to try and determine which parcels of land surrounding Civil War battlefields are under the greatest pressure to convert to an urban land use. Battlefield land is often considered a historic monument to the men who died during the Civil War, and while the core areas of several sites are under the protection of the U. S. National Park Service, there are many other sites that have been developed over or are in danger of having residential or commercial properties built on them, forever losing the "hallowed ground" of history.

GIS can be used to combine a number of factors together that influence where new developments could be occurring (such as proximity to big cities, existing urban areas, or water bodies) similar to Site Suitability to try and find the areas under the greatest development pressures. This type of information could be used by preservationists when attempting to purchase lands to keep them undeveloped.

The factors that contribute to a phenomenon can be constructed as layers within GIS, weighted (to emphasize which layers have a greater effect than others) and combined together. The end result is a ranking of areas, to determine their suitability. An example of this process is referred to as Multi-Criteria Evaluation (**MCE**), which takes layers that have been standardized and initially ranked against each other, then weights and combines them together. The MCE process produces a map of various "pressures" at each location (along the lines of the suitability index map) to determine which places have the greatest

MCE Multi-Criteria Evaluation—the use of several factors (weighted and combined) to determine the suitability of a site.

influence placed on them (based on the combination of all the input layers). This acts like a more complex form of the basic site suitability and can be used in decision making processes to help something like determining which pieces of battlefield land have the greatest development pressures on them to convert to an urban land use. Other problems, such as locating suitable animal habitats, areas of high flood potential, or places for retail sites can be approached using similar types of techniques.

A different example of a GIS Model is the Land Transformation Model (LTM), a tool used to combine various weighted raster layers together to determine where land has changed to an urban land use, and then also to predict urban land use change into the future. See Figure 6.13 for an example of the output of the LTM's predictive capability for areas in Michigan out to the year 2040. Also see *Hands-on Application 6.3: The Land Transformation Model* for further examination of the LTM as a GIS Model.

FIGURE 6.13 The use of the Land Transformation Model in examining and predicting areas that will change to an urban land use in the future. (Source: HEMA Laboratory Purdue University/South East Michigan Council of Governments/Michigan Department of Natural Resources/Esri® ArcGIS ArcMap ArcView graphical user interface Copyright © Esri.)

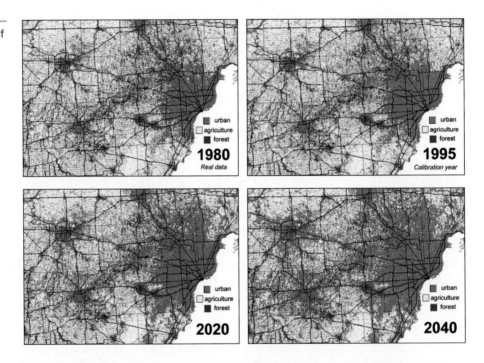

Hands-on Application 6.3

The Land Transformation Model

The Land Transformation Model is headquartered at Purdue University. Open your Web browser and go to **http://ltm.agriculture.purdue.edu/default_ltm. htm** to visit the Website for LTM. Check it out for further background on an example of a GIS Model.

The actual model itself (and sample datasets) can be downloaded as well, if you're feeling adventurous, as well as a selection of graphics and papers related to the model.

Thinking Critically with Geospatial Technology 6.1

What Are Potential Societal or Policy Impacts of GIS Models?

The output of GIS Models is often used to aid in decision-making processes and could have a significant impact. For instance, trying to determine which sites would make the most suitable locations for a new casino development would greatly affect the social, economic, and land use conditions of an area. If the end result of a model is determining what areas of a residential area are under the greatest potential to get flooded, this can have a strong impact on an area's real-estate market. What kinds of impacts can these types of models have on local, state, or federal policy?

Also, consider, what if the results of the model are incorrect? If they predict one area is the optimal location for a new shopping center—and it turns out not to be—how could these results affect this area and future decision making? Keep in mind that you could have a great model or design, but have inaccurate data for one (or more) of the layers. If that's the case, how can having this kind of poor-quality data in a model affect the output (and what kinds of impacts could the results of models using inaccurate data have on policy decisions)?

Chapter Wrapup

Spatial analysis is really at the core of working with geospatial data and GIS. While there are many more types of spatial analysis that can be performed with GIS, this chapter simply introduces some basic concepts of how it can be performed (and some applications of it). Next chapter, we're going to look at taking GIS data (for instance, the results of an analysis or a model) and making a map from the data. This chapter's labs will return you to working with GIS software (either AEJEE or ArcGIS), to start in on more functions, like querying and buffers.

Important note: The references for this chapter are part of the online companion for this book and can be found at http://www.whfreeman.com/shellito1e.

Key Terms

spatial analysis (p. 147)
query (p. 148)
SQL (p.148)
relational operator (p. 149)
compound query (p. 150)
Boolean operator (p. 150)
AND (p. 150)
Intersection (p. 150)
OR (p. 150)
Union (p. 150)
NOT (p. 150)
Negation (p. 150)
XOR (p. 150)
Exclusive Or (p. 151)

buffer (p. 151)
dissolve (p. 152)
geoprocessing (p. 153)
spatial query (p. 153)
overlay (p. 155)
intersect (p. 155)
identity (p. 155)
symmetrical difference (p. 155)
union (overlay) (p. 156)
Map Algebra (p. 156)
Site Suitability (p. 157)
Suitability Index (p. 159)
GIS Model (p. 160)
MCE (p. 161)

GIS Spatial Analysis: AEJEE Version

This chapter's lab builds on the basic concepts of GIS from Chapter 5 by applying several analytical concepts to the GIS data that you used in that chapter's labs. This lab also expands on Chapter 5 by introducing several more GIS tools and features and how they are used.

Similar to the two Chapter 5 geospatial lab applications, two versions of this lab are also provided. The first version (*6.1 Geospatial Lab Application: GIS Spatial Analysis: AEJEE Version*) uses the free ArcExplorer Java Edition for Educators (AEJEE). The second version (*6.2 Geospatial Lab Application: GIS Spatial Analysis: ArcGIS Version*) provides the same activities for use with ArcGIS 10. Although ArcGIS contains many more functions than AEJEE, these two programs share several useful spatial analysis tools.

Objectives

The goals for you to take away from this lab are:

◉ Constructing database queries in AEJEE.

◉ Creating and using buffers around points, lines, and polygons.

◉ Using different selection criteria (from queries) to perform simple spatial analysis.

Obtaining Software

The current version of AEJEE (2.3.2) is available for free download at **http://edcommunity.esri.com/software/aejee**.

Important note: Software and online resources sometimes change fast. This lab was designed with the most recently available version of the software at the time of writing. However, if the software or Websites have significantly changed between then and now, an updated version of this lab (using the newest versions) is available online at **http://www.whfreeman.com/shellito1e**.

Lab Data

There is no data to copy in this lab. All data comes as part of the AEJEE sample data that gets installed with the software.

Localizing This Lab

The dataset used in this lab is Esri sample data for the entire United States. However, starting in Section 6.2, the lab focuses on North Dakota and South

Dakota and the locations of some cities and roads in the area, and then switches to the Great Lakes region in Section 6.6. With the sample data covering the state boundaries and city locations for the whole United States, it's easy enough to select your city (or ones nearby), as well as major roads or lakes and perform the same measurements and analysis using those places more local to you than ones in these areas.

6.1 Some Initial Pre-Analysis Steps

1. Start **AEJEE**.

2. From the **USA** folder, add the following shapefiles: **states, cities, lakes, rivers**, and **intrstat**.

3. Change the color schemes and symbology of each layer to make them distinctive from each other for when you'll be doing analysis. (See *5.1 Geospatial Lab Application: GIS Introduction: AEJEE Version* for how to do this.)

4. Position the layers in the Table of Contents (TOC) so that all layers are visible (that is, the states layer is at the bottom of the TOC and the other layers are arranged on top of it).

5. When the layers are set how you'd like them, zoom in to South Dakota, so the state (and the nearby data) fills up the screen.

6.2 Database Queries

A database query is a standard feature in AEJEE. We'll start off with a simple query (using only one expression).

1. To build a query in AEJEE, select the layer you want to query (in this case, cities) by clicking on its name in the TOC. Then, select the **Query Builder icon** from the toolbar:

The Query Builder dialog box will appear.

2. To build a query, first select the Field to use. In this case, choose "ST" (which contains the state name abbreviations).

3. Next, select an operator to use. In this case, select the "=" operator.

4. Lastly, select the appropriate values. In this case, choose "SD" (which stands for South Dakota). The query will appear in the dialog box (ST = 'SD'). This will select all records whose ST field are equal to the characters 'SD'.

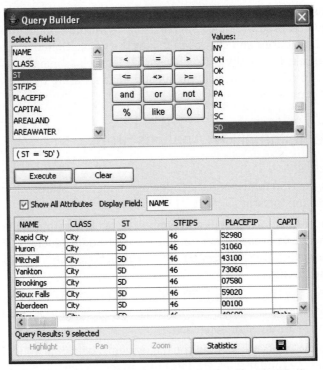

(Source: Esri® ArcGIS ArcExplorer graphical user interface Copyright © Esri.)

5. Click **Execute** to run the query. The returned records will be shown at
    the bottom of the dialog box.

6. This particular query returned nine records (listed as "Query Results" at
    the bottom of the dialog box).

7. Take a look at the View again and you'll see that those nine cities in
    South Dakota have turned yellow (and have been selected).

8. Next, we'll build a compound query to find all of the cities in South
    Dakota with a Year 2000 population of greater than 50,000 persons.
    Bring up the **Query Builder** again, this time creating a query looking for
    ST = 'SD'(it may still be there from the last time the query was created).

9. Now click the button for the "**and**" operator, then finish the second
    half of the compound query by selecting **POP2000** as the Field (click
    No if asked about displaying unique values), then select the **operator**,
    and lastly, type in the number **50000** in the query box. Your final query
    should look something like this:

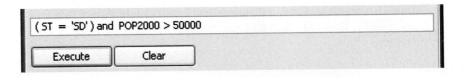

10. Click **Execute** to query the database.

Question 6.1 How many cities in South Dakota have a population greater than 50,000 persons? What are they?

6.3 Creating and Using Buffers around Points

1. AEJEE allows you to create buffers around objects for examination of spatial proximity, along with the ability to use the spatial dimensions of the buffer as a means of selecting other objects.

2. To create a buffer, select the layer whose selected features you wish to buffer from the TOC (in this case, select the **cities** layer).

3. Click on the **buffer icon** on the toolbar:

4. The Buffer dialog box will appear:

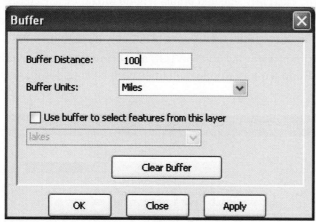

(Source: Esri® ArcGIS ArcExplorer graphical user interface Copyright © Esri.)

5. We'll start with a simple buffer around the cities.

6. For Buffer Distance, type in **100**.

7. For Buffer Units, select **Miles**.

8. Click **Apply** to create the buffer and then click **OK** to close the dialog box.

9. The circular 100-mile buffers will be created around the selected cities.

10. The buffer can also be used as a selection tool (like the various shapes for interactively selecting objects). Return to the **Buffer** dialog box and use the same settings again (100 miles around the selected cities), but this time, put a checkmark in the "**Use buffer to select features from this layer**" box, and choose **cities** as the layer to select from. Click **Apply**

and then **OK.** The buffers will be created again, but this time, the cities that fall within the buffer will be selected as well.

11. Open the attribute table and sort the **Name** field so that the selected records will be placed at the top (see *5.1 Geospatial Lab Application: GIS Introduction: AEJEE Version* on page 119 for how to do this).

Question 6.2 How many cities are within a 100-mile radius of cities found in Question 6.1? (Be very careful to note what actually got selected in the "select with a buffer" procedure when answering this question.)

6.4 Creating and Using Buffers around Lines

1. To create some new buffers, you'll first have to clear (unselect) the currently selected features. Click on the **Clear All Selection** icon on the toolbar:

2. All of the currently selected points will no longer be selected.

3. For this section, you'll be creating buffers around line features and examining the spatial relations of other objects to these lines.

4. Select the **intrstat** layer in the TOC and bring up the **Query Builder** again. Build a query to select all lines with a route number equal to I94.

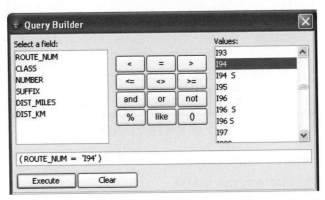

(Source: Esri® ArcGIS ArcExplorer graphical user interface Copyright © Esri.)

5. The field that corresponds with route number is **ROUTE_NUM.**

6. Click **Yes** when AEJEE asks you to display all the values.

7. Select "=" for the operator, then from the values table, choose **I94** (note that AEJEE doesn't necessarily sort all values in numerical order).

8. Click **Execute** when the query is built. Zoom out so that you can see the length of I-94 (a major artery running east to west through the northern United States) on the screen.

9. Next, bring up the **Buffer tool** and create a 20-mile buffer around I-94, using the buffer to select cities that fall within the buffer.

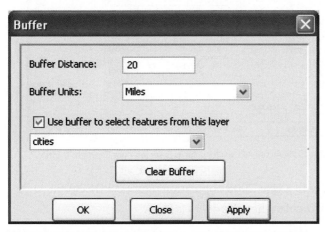

(Source: Esri® ArcGIS ArcExplorer graphical user interface Copyright © Esri.)

10. When the buffer is set up correctly, click **Apply**, and then click **OK** to close the dialog box.

11. Select the **cities** layer from the TOC, then open the attribute table for the cities and sort the results to the top.

Question 6.3 How many cities are within 20 miles of I-94?

12. Clear the selected features and redo the analysis, this time with a 10-mile buffer.

Question 6.4 How many cities are within 10 miles of I-94?

13. Clear the selected features and redo the analysis one last time, now with a 5-mile buffer.

Question 6.5 How many cities are within 5 miles of I-94?

14. Clear all of the currently selected features and buffers.

6.5 Creating and Using Buffers around Polygons

1. The next question you'll answer is how many cities are within states that have a high population. Unfortunately, the states layer only has total population values, not any information about numbers of cities or populated places. This type of question can be answered spatially using the query and buffer tools you've been using so far in this lab.

2. Right-click on the **states** layer in the TOC and bring up the **Query Builder**.

3. Construct a query that will select all states with a year 2005 population of more than 2,000,000 persons.

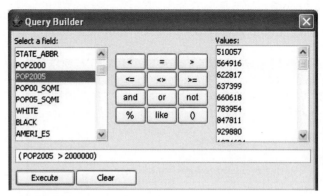

(Source: Esri® ArcGIS ArcExplorer graphical user interface Copyright © Esri.)

4. The field with a year 2005 population figure is **POP2005**.

5. Click on **Execute** when the query is ready.

Question 6.6 How many states had a year 2005 population of greater than 2,000,000 persons?

6. Bring up the **Buffer tool** to create a 1-foot buffer around the selected states. Also, use the buffer to select cities that fall within the buffer zone.

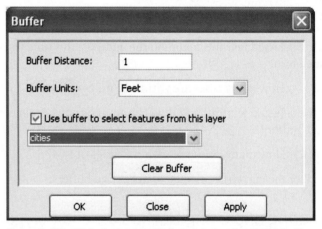

(Source: Esri® ArcGIS ArcExplorer graphical user interface Copyright © Esri.)

7. The reasoning here is that by creating a buffer of a minimum size, it will select all cities within the polygons and only anything else that would be within 1 foot of the buffer's edges (which should be no other cities outside of the selected states).

8. Click **Apply** when the settings are correct and then click **OK** to close the dialog box.

9. Answer Question 6.7 (this may involve examining and sorting the cities' attribute table).

Question 6.7 How many cities are contained within the United States with a population of more than 2,000,000 persons?

10. Clear all of the currently selected features and buffers.

6.6 More Spatial Analysis

1. The next analysis that you'll perform involves analysis of the features in relation to the five Great Lakes, so zoom in on the View so that you can see all five Great Lakes. The first thing you'll have to do is select the Great Lakes to work with.

2. Select the **lakes** shapefile in the TOC and bring up the **Query Builder**. You'll see that the two available fields that are available for query are the Name and Area of the lake records. You'll want to build a query that looks for the name of each of the five Great Lakes as follows: (NAME = 'Lake Erie' or NAME = 'Lake Huron' or NAME = 'Lake Michigan' or NAME = 'Lake Ontario' or NAME = 'Lake Superior').

(Source: Esri® ArcGIS ArcExplorer graphical user interface Copyright © Esri.)

3. Answer Question 6.8, and then click **Execute**. All five Great Lakes will now be selected. Close the **Query Builder**.

Question 6.8 Why is the "or" operator used in this expression rather than the "and" operator? What would the same query give you back is you used "and" rather than "or" in each instance?

4. Next, select the **buffer tool** and find out how many rivers are within 5 miles of the Great Lakes. Note that even if part of the river is within the 5-mile buffer, the entire river system will be selected.

Question 6.9 Which rivers does AEJEE consider within 5 miles of the Great Lakes? Which river system are they part of?

5. Lastly, create a new buffer to find out how many cities are within 5 miles of the Great Lakes.

Question 6.10 How many cities are within 5 miles of the Great Lakes?

6. Exit AEJEE by selecting **Exit** from the **File** pull-down menu. There's no need to save any data in this lab.

Closing Time

Now that you have the basics of working with GIS data and performing some initial spatial analysis down, the lab in Chapter 7 will describe how to take spatial data and create a print-quality map using AEJEE or ArcGIS 10.

Geospatial Lab Application

GIS Spatial Analysis: ArcGIS Version

This chapter's lab builds on the basic concepts of GIS from Chapter 5 by apply-ing several analytical concepts to the GIS data you used in *5.2 Geospatial Lab Application: GIS Introduction: ArcGIS Version*. This lab also expands on Chapter 5 by introducing several more GIS tools and features and how they are used.

The previous *6.1 Geospatial Lab Application, GIS Spatial Analysis: AEJEE Version* uses the free ArcExplorer Java Edition for Educators (AEJEE); how-ever, this lab provides the same activities for use with ArcGIS 10.

Although ArcGIS contains many more functions than AEJEE, the two pro-grams share several useful spatial analysis tools.

Objectives

The goals for you to take away from this lab are:

⦿ Constructing database queries in ArcGIS.

⦿ Using different selection criteria (from queries and selecting by location) to perform simple spatial analysis.

Obtaining Software

The current version of ArcGIS (10) is not freely available for use. However, instructors affiliated with schools that have a campus-wide software license may request a 1-year student version of the software online at http://www. esri.com/industries/apps/education/offers/promo/index.cfm.

Important note: Software and online resources sometimes change fast. This lab was designed with the most recently available version of the software at the time of writing. However, if the software or Websites have significantly changed between then and now, an updated version of this lab (using the new-est versions) is available online at http://www.whfreeman.com/shellito1e.

Lab Data

There is no data to copy in this lab. All data comes as part of the AEJEE sample data that gets installed with the software.

Important note: In order to keep the data and results similar, both the AEJEE and ArcGIS 10 portions of the lab use the same sample data that comes with AEJEE. Thus, if you're using ArcGIS 10 for the lab, please download and install AEJEE in order to use its sample data.

Localizing This Lab

The dataset used in this lab is Esri sample data for the entire United States. However, starting in Section 6.2, the lab focuses on North Dakota and South

Dakota and the locations of some cities and roads in the area, and then switches to the Great Lakes region in Section 6.6. With the sample data covering the state boundaries and city locations for the whole United States, it's easy enough to select your city (or ones nearby), as well as major roads or lakes and perform the same measurements and analysis using those places more local to you than ones in these areas.

6.1 Some Initial Pre-Analysis Steps

1. Start **ArcMap**.
2. From the **USA** folder, add the following shapefiles: **states**, **cities**, **lakes**, **rivers**, and **intrstat**.
3. Change the color schemes and symbology of each layer to make them distinctive from each other for when you'll be doing analysis. (See Lab 5.2 on page 130 for how to do this.)
4. Position the layers in the Table of Contents (TOC) so that all layers are visible (that is, the states layer is at the bottom of the TOC and the other layers are arranged on top of it).
5. When the layers are set how you'd like them, zoom in to South Dakota, so the state (and the nearby data) fills up the screen.

6.2 Database Queries

1. A database query is a standard feature in ArcMap. We'll start off with a simple query (using only one expression). To build a query in ArcMap, choose **Select by Attributes** from the **Selection** pull-down menu.
2. The Select by Attributes dialog box will appear.
3. You'll be building a simple query to find which points (representing cities) are in South Dakota (SD).
4. To build a query, first select the Layer to use. In this case, choose **cities** (since it's the layer that we will be querying).
5. Next, select the Field to query. In this case, choose **"ST"** by double-clicking it with the mouse.
6. Next, select an operator to use. In this case, select the = operator, clicking on the symbol with the mouse.
7. Lastly, select the appropriate values. Press the **Get Unique Values** button to get a list of all the possible values associated with the "ST" Field. In this case, choose '**SD**'.
8. The query will appear in the dialog box ("ST" = 'SD'). This will select all records whose ST field is equal to the characters 'SD'.
9. Click **OK** to run the query.

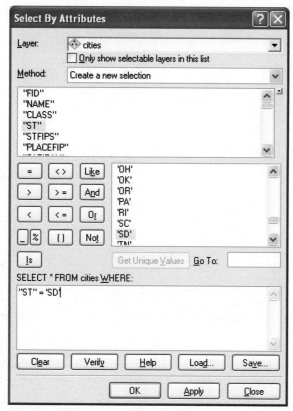

(Source: Esri® ArcGIS ArcMap ArcInfo graphical user interface
Copyright © Esri.)

10. In the View, you'll see several points selected as a result of the query.

11. Open the cities layer's attribute table and only show the selected records (see Lab 5.2 on page 130 for how to do this).

12. This particular query returned nine records (listed as the nine records shown selected in the attribute table).

13. Next, we'll build a compound query to find all cities in South Dakota with a population of greater than 50,000 persons. Bring up the **Select by Attributes** dialog box again, this time creating a query looking for ("ST" = 'SD') (it may still be there from the last time the query was created—make sure to put in the parentheses).

14. Now click the button for the **"And"** operator, then finish the second half of the compound query by selecting the () symbol to place the new query in parentheses.

15. Then choose **POP2000** as the Field (don't display its unique values), then select the greater than (>) operator, and lastly, type the number

50000 in the box. Your final query should look something like this:

(Source: Esri® ArcGIS ArcMap ArcInfo graphical user interface
Copyright © Esri.)

16. Click **OK** to query the database.

Question 6.1 How many cities in South Dakota have a population of greater than 50,000 persons? What are they?

6.3 Selecting By Location with Points

1. ArcMap allows you to select features based on their proximity to other features. To do so, choose **Select By Location** from the **Selection** pull-down menu. This will allow you to select features that are nearby your selected cities.

2. In **Select By Location**, you will specify one of more layers as the Target layer (the layer you will be selecting features from) and a second layer as the Source (the layer whose spatial dimensions you will be using to perform the selection). There are several different ways these two layers can spatially intersect with each other—however, in this case, you'll be finding which of the features from the Target layer (the cities) are within the boundaries of the Source layer (the selected cities).

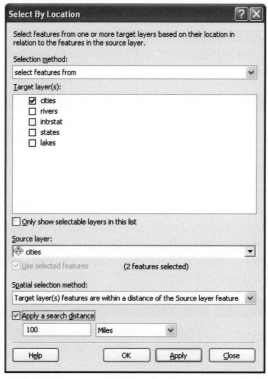

(Source: Esri® ArcGIS ArcMap ArcInfo graphical user
interface Copyright © Esri.)

3. For the Selection method, choose **select features from**.

4. Choose **cities** for your Target layer.

5. Choose **cities** as your Source layer. You will notice that the two
features you selected back in Question 6.1 remain selected—thus,
you will be finding other cities based on their relation to only
these two.

6. Choose **Target layer(s) features are within a distance of the Source
layer feature** as the Spatial selection method.

7. Click **Apply a search distance** and use **100 miles**.

8. Leave the other options unused, and then click **OK**.

9. Open the **cities** attribute table and examine the selected cities (see the
lab in Chapter 5 for how to do this).

Question 6.2 How many cities are within a 100-mile radius of the
cities you selected in Question 6.1? (Be very careful to note what actually
got selected in the Select By Location procedure when answering this
question.)

6.4 Selecting By Location with Lines

1. To create some new selections, you should first clear (unselect) the currently selected features. Click on the **Clear Selected Features icon** on the toolbar:

All of the currently selected points will no longer be selected.

2. For this section, you'll be using Select By Location with line features and examining the spatial relations of other objects to these lines.

3. Bring up the **Select By Attributes** dialog box again, and this time build a simple query to select all lines in the roads layer with a route number equal to I-94 (a major cross-country east-west road that runs through the northern United States).

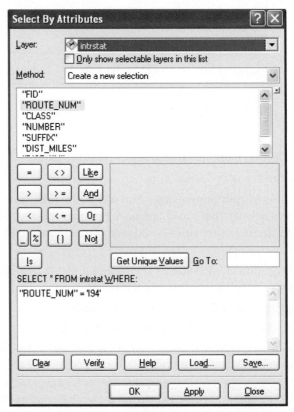

(Source: Esri® ArcGIS ArcMap ArcInfo graphical user interface Copyright © Esri.)

4. Use **intrstat** as the Layer to select from.

5. Use **Create a new selection** as your selection method.

6. Use "ROUTE_NUM" as the Field to select from.

7. Choose = as your operator.

8. Press the **Get Unique Values** button and choose 'I94' as the value to use.

9. Your query should read "ROUTE_NUM" = 'I94'

10. Click **OK** when the query is built. Zoom out so that you can see the length of I-94 (a major artery running east to west through the northern United States) on the screen.

11. Next, use **Select By Location** to find all cities that lie within 20 miles of I-94.

12. Open the **cities** attribute table and examine the selected records.

Question 6.3 How many cities are within 20 miles of I-94?

13. Clear the selected results and redo the analysis as follows:

   a. Use **Select By Location** to find all cities that lie within 10 miles of I-94.

   b. Open the **cities** attribute table and examine the selected records.

Question 6.4 How many cities are within 10 miles of I-94?

14. Clear the selected results and redo the analysis one last time as follows:

   a. Use **Select By Location** to find all cities that lie within 5 miles of I-94.

   b. Open the **cities** attribute table and examine the selected records.

Question 6.5 How many cities are within 5 miles of I-94?

15. Clear all of the currently selected features.

6.5 Selecting By Location with Polygons

1. The next question you'll answer is how many cities are within states that have a high population. Unfortunately, the states layer only has total population values, not any information about numbers of cities or populated places. This type of question can be answered spatially using the select by attribute and select by location tools you've been using so far in this lab.

2. Bring up the **Select By Attributes** dialog box again.

3. Construct a query that will select all states with a year 2005 population (the Field is called POP2005) of greater than 2,000,000 persons.

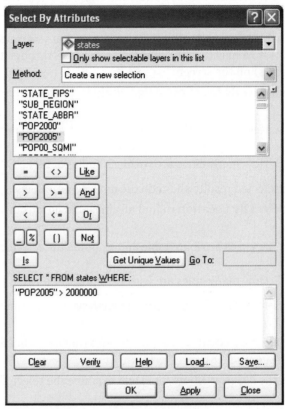

(Source: Esri® ArcGIS ArcExplorer graphical user interface
Copyright © Esri.)

4. Click **OK** when the query is ready.

Question 6.6   How many states had a year 2005 population of greater than 2,000,000 persons?

5. Next, bring up the **Select By Location tool** to select cities that fall within one of these selected states (that have a population of more than 2,000,000 persons).

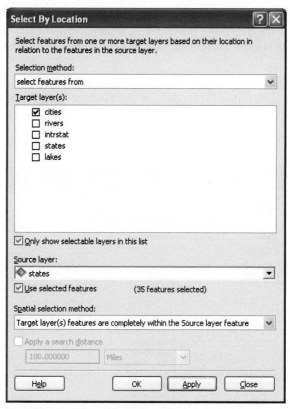

(Source: Esri® ArcGIS ArcMap ArcInfo graphical user interface
Copyright © Esri.)

6. You'll want to select features from the cities layer (the Target) that are completely within the Source layer features (the States).Click **Apply** when the settings are correct and **OK** to close the dialog box.

7. Answer Question 6.7 (this will involve examining the selected cities in the attribute table).

Question 6.7 How many cities are contained within the United States with a population of more than 2,000,000 persons?

8. Clear all of the currently selected features.

6.6 More Spatial Analysis with the Selection Tools

1. The next analysis that you'll perform involves analysis of the features in relation to the five Great Lakes, so zoom in on the View so that you can see all five Great Lakes. The first thing you'll have to do is select the Great Lakes to work with.

2. Bring up the **Select By Attributes** dialog box. Choose **lakes** as the Layer to select from. You'll see that three fields are available to you – FID, AREA, and NAME. Select "NAME" and press the **Get Unique Values** button.

3. You'll want to build a query that looks for the name of each of the five Great Lakes as follows: (NAME = 'Lake Erie' OR NAME = 'Lake Huron' OR NAME = 'Lake Michigan' OR NAME = 'Lake Ontario' OR NAME = 'Lake Superior').

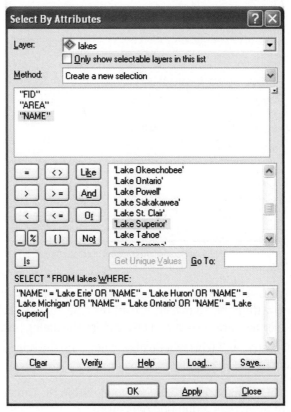

(Source: Esri® ArcGIS ArcMap ArcInfo graphical user interface
Copyright © Esri.)

4. Answer Question 6.8, and then click **OK**. All five Great Lakes will now be selected. Close the **Select By Attributes** dialog box.

Question 6.8 Why is the "OR" operator used in this expression rather than the "AND" operator? What would the same query give you back is you used "AND" rather than "OR" in each instance?

5. Use the **Select By Location** tool to find out how many rivers are within 5 miles of the Great Lakes. Note that even if part of the river is within the 5-mile distance, the entire river system will be selected.

Question 6.9 Which rivers are within 5 miles of the Great Lakes? Which river system are they part of?

6. Lastly, use **Select By Location** to find out how many cities are within 5 miles of the Great Lakes.

Question 6.10 How many cities are within 5 miles of the Great Lakes?

7. Exit ArcMap by selecting **Exit** from the **File** pull-down menu. There's no need to save any data in this lab.

Closing Time

Now that you have the basics of working with GIS data and performing some initial spatial analysis down, the lab in Chapter 7 will describe how to take spatial data and create a print-quality map using AEJEE or ArcGIS 10.

Using GIS to Make a Map

Scale, Map Elements, Map Layouts, Type, Thematic Maps, Data Classification Methods, Color Choices, and Digital Map Distribution Formats

One thing that makes spatial data unique is that it can be mapped. Once you've used GIS to create the boundaries of a Civil War battlefield and analyze its proximity to new housing and commercial developments, the next step is to make a map of the results of your work. However, making a good map with GIS involves more than taking the final product of the analysis, throwing on a title and legend, clicking on "print," and then calling it a day. There are numerous design elements and considerations that should be taken into account when laying out a map. In fact, there's a whole art and science of mapmaking (called **cartography**), involving things like color selection, the positioning of items on the map, or what kind of message the final map product is conveying to the reader.

A **map** is a representation of spatial data that is designed to convey information to its reader. Making digital maps with GIS offers a wide variety of options for creating an end-product, but there are a few considerations to take into account before you start the mapping process. For example, say there's a nearby municipal wooded area that gets used by the local populace for walking trails, biking, picnicking, or walking the dogs. The only map of note is an old hand-drawn one you can pick up at a kiosk at the park entrance and the local government wants to update it. As you're the local geospatial technology expert, they come to you to do the job. Before you break out your GIS tools, there are some basic questions to ask that are going to define the map you design.

First, what's the purpose of the map and who will be using it? If the park map you've been asked to design is intended for the park's visitors to use, then some of the important features the map should contain are locations of park entrances, parking, comfort stations, and designated picnic areas. It should be at an appropriate scale, the trails should be accurately mapped, and trailhead

cartography the art and science of creating and designing maps.

map a representation of geographic data.

Thinking Critically with Geospatial Technology 7.1

Why is Map Design Important?

Why is the design of a map's appearance so important? After all, you could argue that a poorly put together map of New Zealand is still (at the end of the day) a map of New Zealand. However, a well-designed map is going to be more useful than a poorly designed one, but why? Why are things like the choice of colors, lettering, and placement of items on the map critical to the usefulness of a map? Consider what kind of effect a poorly designed map can have and what sort of impact it can make for some of the following concepts—a map of a local county fair, a real estate map of a new subdivision, a promotional map showing the location of a small business, or a map of a proposed new urban development readied for promotional purposes. Also, how can the design of a map influence how the map's information can be perceived by its reader?

locations and points of interest throughout the park will likely be highlighted. If the map is to be used for zoning purposes, things like exact boundaries, road systems, and parcel information will likely be some of the most important factors.

Second, is the information on the map being easily conveyed to the map reader? For instance, if you're designing a trail map, would all trails be named or marked, or would that cause too much congestion and clutter on the map? Are the colors and symbols appropriate for the assumed novice user of the map? For instance, the trail map would likely not need an inset map of the surrounding area to put the park into a larger context, nor should the map reader be left guessing as to what symbols really mean on the map. Third, is the map well designed and laid out properly? A good map should be well balanced, in regard to the placement of the various map elements, for best ease of use by the map reader. This chapter will examine several of these cartographic design and data display functions, and by the time we reach the lab, you'll be designing a professional-looking map of your own using GIS.

How Does the Scale of the Data Affect the Map (and Vice Versa)?

geographic scale
the real-world size or extent of an area.

map scale a metric used to determine the relationship between measurements made on a map and their real-world equivalents.

A basic map item would be information about the scale of the map, and there are a couple different ways of thinking about scale. First is the **geographic scale** of something—things that take up a large area on the ground (or have large boundaries) would be considered large geographic scale. For instance, studying a global phenomenon is a much larger geographic scale than studying something at a city level, which would be a much smaller geographic scale. Something different is **map scale**, a value representing that x number of units of measurement on the map equals y number of units in the real world. This relationship between the real world and the map can be expressed as a representative fraction (**RF**). An example of an RF would be a map scale of 1:24000—a measure of one unit on the map would be equal to 24,000 units in the real world. For instance, measuring one inch on the map would be the same as 24,000 inches in the real world or one foot on the map is equal to 24,000 feet in the real world (and so on).

Maps are considered **large-scale maps** or **small-scale maps** depending on that representative fraction. Large-scale maps show a smaller geographic area and have a larger RF value. For instance, a 1:4000-scale map would be considered a large-scale map—due to the larger scale, it would show a smaller area. The largest-scale map you could make would be 1:1—where one inch on the map was equal to one inch of measurement in the real world (that is, the map would be the same size as the ground you were actually mapping—a map of a classroom would be the same size as the classroom itself). Conversely, a small-scale map would have a smaller RF value (such as 1:250,000) and show a much larger geographic area.

For instance, on a very small-scale map (such as one that shows the entire United States), cities would be represented by points, and likely only major cities will be shown. On a slightly larger-scale map (one that shows all of the state of New York), more cities are likely to be shown as points, along with other major features (additional roads can be shown as lines, for example). On a larger-scale map (one that shows only Manhattan), the map scale allows for more detail to be shown—points will now show the locations of important features and many more roads will be shown with lines. On an even larger-scale map (one that shows only a section of lower Manhattan) buildings may now be shown as polygon shapes (to show the outline or footprint of the buildings) instead of points, and additional smaller roads may also be shown with lines.

The choice of scale will influence how much information the map will be able to convey and what symbols and features can be used in creating the map in GIS. Figure 7.1 on page 188 shows a comparison between how a feature (in this case, Salt Lake City International Airport) is represented on large-scale and small-scale maps. The actual sizes of the maps greatly vary, but you can see that more detail and definition of features is available on the larger-scale map than on the smaller-scale one (also see *Hands-on Application 7.1: Powers of 10 – A Demonstration of Scale,* page 188, for a cool example of visualizing different scales).

The same holds true for mapping of data—for instance, the smaller-scale map of all of the state of New York could not possibly show point locations of all of the buildings in Manhattan. However, as the map scale grows larger, different types of information can be conveyed. For instance, in Figure 7.1, the large-scale map can convey much more detail concerning the dimensions of the airport runways, while the smaller-scale map has to represent the airport as a set of simplified lines. If you were digitizing the lines of the airport runways, you'd end up with two very different datasets (one more detailed, one very generalized).

This relationship between the scale of a map and data that can be derived from a map can be a critical issue when dealing with geospatial data. For

RF representative fraction—a value indicating how many units of measurement in the real world are equivalent to how many of the same units of measurement on a map.

large-scale map a map with a higher value for its representative fraction. Such maps will usually show a small amount of geographic area.

small-scale map a map with a lower value for its representative fraction. Such maps will usually show a large amount of geographic area.

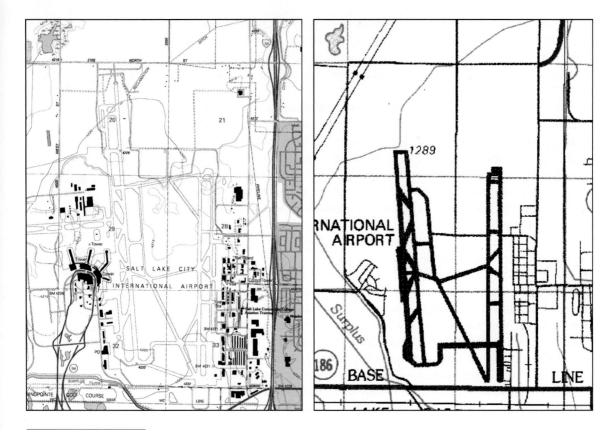

FIGURE 7.1 Salt Lake City International Airport—a comparison of its representation on a 1:24,000 scale map (left) and a 1:100,000 scale map (right).
(Source: Utah State Geographic Information Database (SGID).)

instance, say you're doing field mapping and GIS work of a university campus. At this small geographic scale, you're going to require detailed information that fits your scale of analysis. If the hydrologic and transportation data you're working with was derived from 1:250,000-scale maps, it's likely going to be way too coarse to use. Data generated from smaller-scale maps is probably going to be incompatible with the small geographic scale you're working

Hands-on Application 7.1

Powers of 10 - A Demonstration of Scale

Though it's not a map, an excellent demonstration of scale and how new items appear as the scale changes is available at **http://micro.magnet.fsu.edu/primer/java/scienceopticsu/powersof10**.

This Website (which requires Java on a computer for it to properly run) shows a view of Earth starting from 10 million light years away. Then the scale

changes to a view of 1 million light years away, and then 100,000 light years away (a factor of ten each time). The scale continues changing until it reaches Earth—and then continues all the way down to sub-atomic particles. Let it play through, and then use the manual controls to step through the different scales for a better view of the process.

at. For instance, digitizing features on a small-scale map (like a 1:250,000 scale) are going to be much more generalized than data derived from larger-scale maps (like a 1:24,000 scale) or things such as aerial photos at a larger scale (for example, 1:6000). See Chapter 9 for more information on using aerial photos for analysis.

What Are Some Design Elements Included in Maps?

There are several elements that should show up on a good map. For instance, someone reading a map should be able to quickly figure out what scale the map is. A map element might simply be text of the RF (such as 1:24,000, 1:100,000, or whatever the map scale would be). A graphical representation of equivalent distances shown on a map would be a **scale bar** (Figure 7.2). The scale bar provides a means of measuring the map scale itself, except using a measurement of *x* distance on the scale bar is equal to *y* units of distance in the real world.

A second map element is a **north arrow**, a graphical device used to orient the direction of the map. However the map is oriented, the north arrow is used to point toward the direction that is due north. A north arrow may sometimes be drawn as a compass rose, which shows all cardinal directions on the map. North arrows can be as simple as an arrow with the letter "n" attached, or as complex as a work of art. See Figure 7.3 on page 190 for examples of north arrows used in map design.

Another item is the map's **legend**—a guide to what the various colors and symbols on the map represent. A good legend should be a key to the symbology

scale bar a graphical device used on a map to represent map scale.

north arrow a graphical device on a map used to show the orientation of the map.

legend a graphical device used on a map as an explanation of what the various map symbols and color represent.

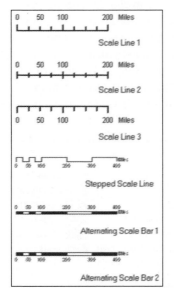

FIGURE 7.2 Examples of various scale bars. (Source: Esri® ArcGIS ArcExplorer graphical user interface Copyright © Esri.)

FIGURE 7.3 Examples of various north arrows. (Source: Esri® ArcGIS ArcMap graphical user interface Copyright © Esri.)

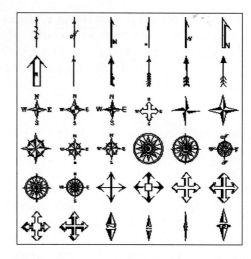

of the map (see Figure 7.4 for an example of a map legend). Since the legend is the part of the map where you can explain things to the map's reader, you may want to call it something other than "legend." Instead, consider calling the legend something more useful like "park trail guide" or "county population change."

type the lettering used on a map.

label text placed on a map to identify features.

fonts various styles of lettering used on maps.

The choice of **type** used on the map is also important, as it's used for such things as the title, the date of the map's creation, the name of the map's creator, the origin of the data sources used to make the map, and the lettering attached to the map's features. GIS programs will usually allow you to **label** map features (adding things like the names of rivers, roads, or cities to the map) using an automatic placement tool or allowing you to interactively select, move, and place map labels.

When selecting the type to use for various map elements, a variety of different **fonts** (or lettering styles) are available. GIS programs (like word-processing programs such as Microsoft Word) will often have a large number of fonts to select from, everything from common fonts (such as "Arial" or "Times New Roman") to much flashier fonts (such as "Mistral" or "Papyrus"). See Figure 7.5 for some examples of different versions of type fonts available in ArcExplorer Java Edition for Educators. Often, a map will contain only two different fonts, carefully selected to fit well with each other instead of

FIGURE 7.4 Examples of various legend items. (Source: Esri® ArcGIS ArcExplorer graphical user interface Copyright © Esri.)

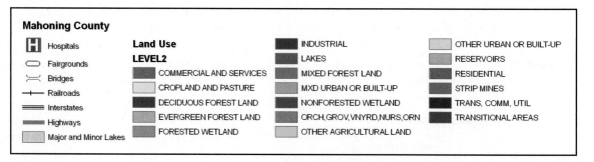

Mahoning County

Hospitals		
Fairgrounds		
Bridges		
Railroads		
Interstates		
Highways		
Major and Minor Lakes		

Land Use
LEVEL2

COMMERCIAL AND SERVICES
CROPLAND AND PASTURE
DECIDUOUS FOREST LAND
EVERGREEN FOREST LAND
FORESTED WETLAND

INDUSTRIAL
LAKES
MIXED FOREST LAND
MXD URBAN OR BUILT-UP
NONFORESTED WETLAND
ORCH,GROV,VNYRD,NURS,ORN
OTHER AGRICULTURAL LAND

OTHER URBAN OR BUILT-UP
RESERVOIRS
RESIDENTIAL
STRIP MINES
TRANS, COMM, UTIL
TRANSITIONAL AREAS

Map Source: Esri Data	Map Source: Esri Data
Map Source: Esri Data	**Map Source: Esri DATA**
Map Source: Esri Data	**Map Source: Esri Data**
Map Source: Esri Data	Map Source: Esri DATA
Map Source: Esri Data	Map Source: Esri Data

FIGURE 7.5 Examples of various fonts available for type in map layouts.

clashing. Using too many fonts or crafting a map's type out of several of the more elaborate fonts is likely to make a map difficult to read and reduce its usefulness. See *Hands-on Application 7.2: Typebrewer Online* for a tool you can use as an aid in selecting appropriate type for a map.

In GIS, a map is put together by assembling all of the elements together in a **layout**. A good way to think of a layout is a digital version of a blank piece of paper that you will then arrange the various map elements together in a cartographic design to create a final map. Sometimes, the software (such as ArcGIS) will include several **map templates**, which provide pre-created designs for your use. Using a template will take your GIS data and place things like the title, legend, and north arrow at pre-determined locations and sizes on the layout. Templates are useful for creating a quick printable map layout, but GIS will also allow you to design a map the way you want it in regard to size, type, and placement of elements.

layout the assemblage and placement of various map elements used in constructing a map.

map template a premade arrangement of items in a map layout.

When designing a layout, it's important to not simply slap a bunch of elements on the map and decide that the map's ready. Balancing the placement of items is important, so as not to overload the map with too much information, but also to provide a useful, readable map. Good map balance provides a means of filling in the "empty" spaces on the map, so it doesn't have all its information crammed into the top or bottom, leaving other parts of the map blank. Items should be of uniform size (instead of, say, making the north arrow gigantic just to fill in some empty space) and placed in proportion to one another (see Figure 7.6 on page 192 for an example of trying to balance out the placement, size, and proportion of items in a map layout with a template).

Hands-on Application 7.2

Typebrewer Online

Typebrewer is a very neat (and free) online utility that allows you to examine different font styles and combinations used on maps. With Typebrewer, you can interactively change font size, density, or appearance. Open your Web browser and go to **http://www.typebrewer.org** to get started (you'll need to have Adobe Flash Player installed on your computer for it to work properly). With Typebrewer, you can select from several different pre-set styles of type on a map, then alter aspects of the type (such as its size, density, and tracking) to see the effect of those changes on the map. The aim of Typebrewer is to examine different forms of map type so you can apply those aspects to your own maps. Try looking at several different formats of the type, then decide on what the most appropriate one would be and why.

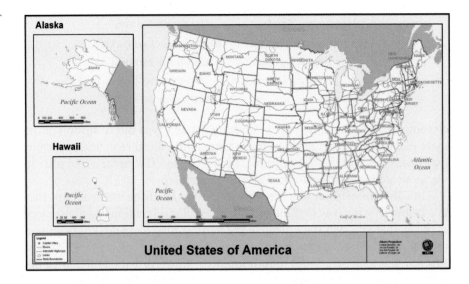

How Is Data Displayed on a GIS Map?

reference map a map that serves to show the location of features, rather than thematic information.

There are several different types of maps that can be made using GIS tools. The purpose of a **reference map** (like what you would see in an atlas or a road map) is to give location information or highlight different features. A map of park trails, a Manhattan restaurant guide map, a map of the casinos on the Las Vegas Strip, or a zoning map of your neighborhood are all examples of reference maps. Topographic maps (that also show landforms) are another example of reference maps (and we'll deal with them in Chapter 13).

thematic map a map that displays a particular theme or feature.

Another main type of map is the **thematic map**, which is geared toward conveying one particular theme to the reader. Thematic maps might be used to show things such as the increase in U.S. population for each state or U.S. Presidential election results by county (see *Hands-on Application 7.3: Census Bureau Thematic Maps* for examples of different types of thematic maps online). In GIS, a layer's attributes are used for displaying information on the map. For instance, to create a map of the 2008 U.S. Presidential election

Hands-on Application 7.3

Census Bureau Thematic Maps

The U.S. Census Bureau makes a number of thematic maps of the United States available for viewing online, using different census demographic information (such as states' population, income levels, housing costs, and retail sales levels). Open your Web browser and go to **http://factfinder.census.gov/jsp/saff/SAFFInfo.jsp?_pageId=thematicmaps** to check them out. Use the tools on the Website to create quick thematic maps using pre-defined census data (such as population, housing, or retail sales values). You can choose to create maps by divisions including state, county, or region, as well as zoom in and out of the maps and identify features.

UNITED STATES PRESIDENTIAL ELECTION 2008
RESULTS BY COUNTY
NOVEMBER 6, 2008

FIGURE 7.7 A thematic
map showing the
results of the 2008 U.S.
Presidential election by
county (the vote in blue
counties going for Barack
Obama and the vote in red
counties going for John
McCain). (Source: Library of
Congress Geography and Map
Division.)

results by county, each polygon representing that county would have an attribute designating whether Barack Obama or John McCain had a higher
number of votes for that county (colored red for counties won by McCain and
blue for counties won by Obama). This attribute would then be the data being
displayed on the map (**Figure 7.7**). Other thematic maps may use **graduated
symbols** for display. In this case, points (or other symbols) are plotted of differing sizes to represent the thematic factors.

Of course, many thematic maps don't rely on a simple two-choice attribute
such as the election map (in which each county will be marked Obama or Mc-
Cain—or else contain no data). Attributes such as the percentage of colleges
and universities (per state) that are using this textbook in a class (or some
other values) are mapped using a type of thematic map called a **choropleth
map**. In order to best display this information on a map, the data will have
to be classified or divided into a few categories. GIS gives several options for
data classification of multiple values—each method classifies data differently and can thus result in some very different results being displayed on the
maps. (See **Figure 7.8** on page 194 for four different maps of the percentage
of the total number of houses per state that are considered seasonal homes or
vacation homes by census, each created using a different data classification
technique.) Note that each map has the data broken into the same number of
classes (four) for comparison purposes.

The first of these data classification methods is called **Natural Breaks**
(also referred to as the Jenks Optimization). Like the name implies, this

graduated symbols
the use of different
sized symbology
to convey thematic
information on a map.

choropleth map
a type of thematic
map in which data is
displayed according to
one of several different
classifications.

data classification
various methods used
for grouping together
(and displaying) values
on a choropleth map.

Natural Breaks a data
classification method
that selects class break
levels by searching
for spaces in the data
values.

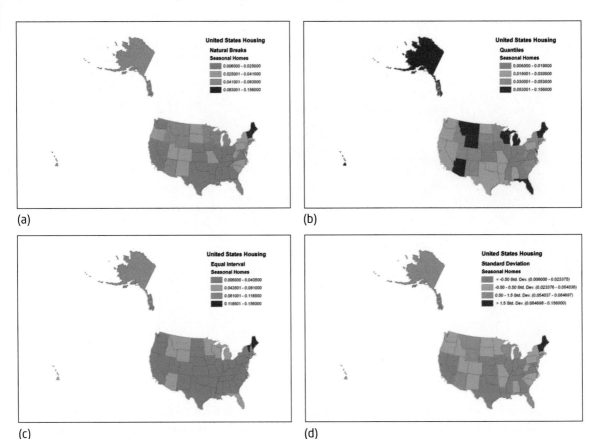

(a)   (b)

(c)   (d)

FIGURE 7.8 Four choropleth map examples created using the same data (the year 2000 percentage of the total number of houses that are considered seasonal or vacation homes) but different data classification methods as follows: (a) Natural Breaks, (b) Quantiles, (c) Equal Interval, and (d) Standard Deviation. (Source: Esri® ArcGIS ArcMap graphical user interface Copyright © Esri. Data: US Census Bureau.)

Quantile a data classification method that attempts to place an equal number of data values in each class.

method takes all of the values being mapped and looks at how they're grouped together. The spaces in between the data values are used to make different classes (or ranges of data) that are displayed. This is shown in Figure 7.8a—states with the lowest percentages of seasonal homes values (such as Nebraska, Oklahoma, and Texas) end up in one class, and states with the highest percentages of seasonal homes (such as Maine, Vermont, and New Hampshire) end up together in another class.

The map in Figure 7.8b shows the results of using the **Quantile** method of data classification. This method takes the total number of data values to be mapped and splits them up into a number of classes. It tries to distribute values so that each range has a similar number of values in it. For instance, with 51 states being mapped (plus the District of Columbia as a 51st area), each of the four ranges will have about 13 counties worth of data being shown in each class. Since the break points between the ranges are based on the total number of items being mapped (that is, how many states ending up in each range), rather than the actual data values being mapped, the Quantile method causes a relatively even distribution of values on the map.

The third method (shown in the map in Figure 7.8c) uses **Equal Intervals** for data classification. It works like it sounds—it creates a number of equally sized ranges and then splits the data values into these ranges. The sizes of each range are based on the total span of values to be mapped. For instance, in the seasonal home maps, the data is divided into four classes, and the range of values goes from the state with the lowest seasonal home percentage (0.6% of the total housing stock in Illinois) to the state with the highest seasonal home percentage (15.6% in Maine). Equal Interval takes the complete span of data values (there is 15% separating the lowest and highest values) and divides it by the number of classes (in this case, four), and that value (in this case, 3.75%) is used to compute the breaking point between classes. So the first class represents states that have a seasonal home value of 3.75% more than the class' lowest end (for instance, the first class would have values between 0.6% and 4.35%). Note that this method simply classifies data based on the range of all values (including the highest and lowest) but does not take into account clusters of data or how the data is distributed. As such, only a few states end up in the upper class because their percentages of seasonal homes were greater than three-fourths of the total span of values.

Equal Interval a data classification method that selects class break levels by taking the total span of values (from highest to lowest) and dividing by the number of desired classes.

The final method is presented in the map in Figure 7.8d, the **Standard Deviation** method. A standard deviation is the average distance that a single data value is away from the mean (the average) of all data values. The breakpoints for each range are based on these statistical values. For instance, the GIS would calculate the average of all United States seasonal home values (3.9%) and the standard deviation for them (3.1%). So when it comes to the percentage of the total housing stock that is seasonal, each state's percentage is an average of 3.1% away from the average state's percentage. These values for the mean and standard deviation of the values are used to set up the breakpoints. For instance, the breakpoint of the first range is of all states whose seasonal home values are less than half a standard deviation value lower than the mean—those states with a seasonal home percentage of less than the mean minus 0.5 times the standard deviation (1.55%), or 2.34%. The fourth range consists of those counties with a value greater than 1.5 times the standard deviation away from the mean. The other ranges are similarly defined by the mean and standard deviation values of the housing data.

Standard Deviation a data classification method that computes class break values by using the mean of the data values and the average distance a value is away from the mean.

Like Figure 7.8 shows, the same data can produce some very different looking choropleth maps depending on which method is used to classify the data, with differing messages from the maps. For instance, the map in Figure 7.8d (Standard Deviation) shows that most states have roughly an average (or below average) percentages of seasonal homes, while the other maps show various distinctions between which states are classified as a higher or lower percentage of homes that are seasonal. Thus, the same data can result in different maps, depending on the classification method chosen. When selecting a method, having information about the nature of the data itself (that is, if it is evenly distributed, skewed toward low or high numbers, or all very similar values) will aid in ending up with the best kind of mapping.

Also keep in mind, while different data classification methods can affect the outcome of the map, the type of data values being mapped can greatly affect the outcome of the choropleth map. An issue involved with choropleth mapping is displaying the values of data that can be counted (for instance, the total population values per state or the total number of housing sales per county) when the sizes of the areas being mapped are very different. If you were mapping the number of vacation homes of each state, a very large state like California is probably going to have a larger number of homes (12,214,549 total homes with 239,062 of them being seasonal, according to the 2000 Census) than a smaller state like Rhode Island (439,837 homes total with 13,002 of them seasonal). Thus, if you're making a choropleth map of the number of vacation homes in each state, California will show many more vacation homes than Rhode Island, just because it has a lot more houses (due to a larger area and population).

However, a better measure to map would be the percentage of the total number of houses that are vacation homes—in this way, the big difference in housing counts between California and Rhode Island isn't a factor in the map. Instead, you're mapping a phenomenon that can be comparably measured between the two states. When you map the percentages of the total housing that are considered vacation homes, California's seasonal homes only make up about 2% of the total houses, while Rhode Island has about 3% seasonal homes. To make a choropleth map of count data, the data should first be **normalized**, or have all count values brought onto the same level. For instance, dividing the number of seasonal homes by the total number of houses would be a way of normalizing the data. See *Hands-on Application 7.4: Interactive Thematic Mapping Online* for an online tool that is used for creating thematic maps.

normalized altering count data values so that they are at the same level of representing the data (such as using them as a percentage).

Hands-on Application 7.4

Interactive Thematic Mapping Online

It's now time to start making your own thematic maps. Open your Web browser and go to **http://thematicmapping.org/engine**. This is the Website of the Thematic Mapping Engine, which allows you to create numerous kinds of thematic maps of data from a variety of topics. First, select an indicator (start with something simple like Population) and choose a year (like 2010). Choose Choropleth for the technique and pick a set of colors. For the Classification method, choose Quantiles. Lastly, click on the Preview button to see a thematic map of world population displayed in an interactive Google Earth interface in your Web browser (alternatively, you can select the Download option to actually download the thematic map as a new layer in Google Earth). After the Quantile map is set up, try the same settings but with the Equal Interval classification method to check out what kind of map is generated.

Try some of the other indicators (such as CO_2 emissions, GDP per capita, Infant mortality rate, or Mobile phone subscribers) over a range of years. These thematic maps can be created using graduated symbols, choropleth maps, or even 3D-style prism maps (we'll use the Thematic Mapping Engine again in Chapter 14 with these types of maps). You can specify which data classification method to use and finally display the mapped results using Google Earth or the online preview.

What Kinds of Colors Are Best to Use with GIS Maps?

The choice of color for a map is also important—first to create something pleasing for the map reader, but also for graphic design and use in a digital format. For instance, the choropleth maps in Figure 7.8 used a range of values of dark green for low percentages and bright blue for high percentages, but there are a lot of ways those classifications could have been colored. There are different color schemes for use in GIS maps—one of these is **RGB**, a setup used for design on computer screens. RGB utilizes the colors red, green, and blue as the primary colors, mixing them together as needed to create other hues. Another is **CMYK**, used in graphic design, which uses the colors of cyan, magenta, yellow, and black (the "K" of the title) for creating hues. Chapter 10 will deal further with generating different colors from an initial set of primary colors in the context of examining digital imagery.

With a reference map, different colors are used to represent the various objects and symbols. On a choropleth map (like those shown in Figure 7.8), a range of shades or colors is needed to show the lowest classes of data through the highest classes. In this case, a **color ramp** is used to show a set of colors to represent the classes of values. A color ramp is a selected set of colors that will show changes of a color scheme—for instance, using a single hue of a color in a color ramp might run the range of lightest to darkest blue or light green to very dark green. Other color ramps (like those shown in Figure 7.8) incorporate multiple hues ranging from green through blue. Several different examples of color ramps are shown in Figure 7.9.

Part of why color choice becomes important with a map is the wide range of visualization media that are available for viewing it. Colors of map items may appear one way on a computer monitor, another way when they're projected onto a screen with an LCD projector, and also appear differently yet again when printed by an inkjet or laserjet printer. Just because you like the look of the shades of blue and green on the GIS map on your monitor, it doesn't

RGB a color scheme based on using the three primary colors red, green, and blue.

CMYK a color scheme based on using the colors cyan, magenta, yellow, and black.

color ramp a range of colors that are applied to the thematic data on a map.

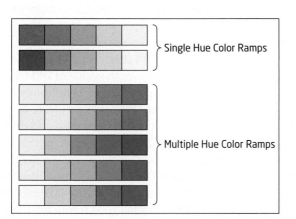

Single Hue Color Ramps

Multiple Hue Color Ramps

FIGURE 7.9 Examples of various color ramps.

Hands-on Application 7.5

ColorBrewer Online

To further investigate the impact of color choices on a map, check out ColorBrewer 2.0, which is available online at **http://colorbrewer2.org**. ColorBrewer is a great tool for color choice and color ramp selection for a variety of different map data and formats. Color-Brewer 2.0 sets up a pre-determined choropleth map while allowing you to examine different color selections, see how different color choices on choropleth maps can be distinguished from one another, or examine the best choices for printing, photocopies, and color-blind safe schemes. Try out several of the options to determine (based on the pre-set choropleth maps) what the best color scheme would be. You can use ColorBrewer in a similar way as Typebrewer—to determine the optimum setup of colors and then apply those settings to your own maps.

mean that the same colors will translate the same way to a printed page or a photocopy of the printout. It might even have a "washed out" appearance when the map is projected onto a big screen. Check out *Hands-on Application 7.5: ColorBrewer Online* to investigate the effect that different color choices will have on a final map.

How Can GIS Maps Be Exported and Distributed?

Once a map has been designed and formatted the way you want it, it's time to share and distribute the results. Rather than just printing a copy of the map, there are several digital formats that maps can be quickly and easily converted to for ease of distribution. A simple way is to export the map as a graphical raster file—this saves a "snapshot" of the map as a digital graphic that can be viewed like a picture, either as a file or an image placed on a Website. There are many different formats for map export, and two common ones are **JPEG** (Joint Photographic Experts Group) and **TIFF** (Tagged Image File Format). Images saved in JPEG format can experience some data loss due to the file compression involved with JPEGs. Consequently, JPEG images usually have smaller file sizes for viewing or downloading. Images saved in TIFF format have a much larger file size but are a good choice for clearer graphics.

JPEG the Joint Photographic Experts Group image, or graphic file format.

TIFF the Tagged Image File Format used for graphics or images.

DPI Dots Per Inch—a measure of how coarse (lower values) or sharp (higher values) an image or map resolution will be when exported to a graphical format.

The clarity of an image is a function of what **DPI** (Dots Per Inch) setting is used when the map is exported to a graphic. The lower the value of DPI (such as a value of 72), the less clarity the resultant image will have. Very low values of DPI will result in the exported image being very blocky or pixilated. Maps exported with higher values of DPI (such as 300) will be very crisp and clear. However, the higher the DPI value, the larger the file size will be, which becomes important when distributing map data online, as the larger the file, the more time will be needed to transfer or display the images. For professional-print-quality maps (such as the maps and graphics in books), the TIFF file format is used at a higher DPI value (such as 300) to create good resolution maps.

A **GeoPDF** is another option for exporting and distributing maps. A Geo-PDF allows the user to export a GIS map in the commonly used PDF file format, which can be opened by the free Adobe Reader software using a special free plug-in. A GeoPDF differs from a regular PDF in that it allows the user to interact with map layers and get information about the coordinates of locations shown in the PDF. For instance, a GeoPDF could contain multiple layers of data that the user could turn on and off (such as annotation for road names or a separate layer of the roads themselves) as in GIS (see Chapter 13 for more usages of GeoPDFs with other maps).

> **GeoPDF** a format that allows for maps to be exported to a PDF format, yet contain geographic information or multiple layers.

Chapter Wrapup

Once GIS data is created, compiled, or analyzed, the results are usually best communicated using a map, and the ability to create a map of data is a standard feature in GIS software packages such as ArcGIS or ArcExplorer Java Edition for Educators. However, for producing a useful, readable, well-balanced map, there are several choices that go into map design, the form of the layout, and the presentation of data, colors, and symbols.

This chapter presented an overview of several cartographic and map-design concepts, and this chapter's lab will have you take the Esri data you've been analyzing in Chapters 5 and 6 and create a professional-quality map from it. Next chapter, we're going to examine some specific types of maps—road network maps—and how GIS and geospatial technologies are used to design street maps that can be used for locating addresses and computing shortest paths (and directions) between destinations.

Important note: The references for this chapter are part of the online companion for this book and can be found at http://www.whfreeman.com/shellito1e.

Key Terms

cartography (p. 185)
map (p. 185)
geographic scale (p. 186)
map scale (p. 186)
RF (p. 187)
large-scale map (p. 187)
small-scale map (p. 187)
scale bar (p. 189)
north arrow (p. 189)
legend (p. 189)
type (p. 190)
label (p. 190)
fonts (p. 190)
layout (p. 191)
map template (p. 191)
reference map (p. 192)

thematic map (p. 192)
graduated symbols (p. 193)
choropleth map (p. 193)
data classification (p. 193)
Natural Breaks (p. 193)
Quantile (p. 194)
Equal Interval (p. 194)
Standard Deviation (p. 195)
normalized (p. 196)
RGB (p. 197)
CMYK (p. 197)
color ramp (p. 197)
JPEG (p. 198)
TIFF (p. 198)
DPI (p. 198)
GeoPDF (p. 199)

7.1 Geospatial Lab Application

GIS Layouts: AEJEE Version

This lab will introduce you to the concepts of taking GIS data and creating a print-quality map from it. This map should contain the following:

- The contiguous United States (48 states without Alaska and Hawaii) population per square mile, set up in an appropriate color scheme
- The data in a different projection than the default GCS one
- An appropriate legend (make sure your legend items have normal names and that the legend is not called "legend")
- An appropriate title (make sure that your map title doesn't include the word "map" in it)
- A north arrow
- A scale bar
- Text information: your name, the date, and the source of the data
- Appropriate borders, colors, and design layout (your map should be well designed instead of map elements thrown on at random)

Important note: At the end of this lab, there is a checklist of items to aid you in making sure the map you make is complete and of the best quality possible.

Similar to the geospatial lab applications in Chapters 5 and 6, two versions of this lab are provided. The first version (*7.1 Geospatial Lab Application: GIS Layouts: AEJEE Version*) uses the free ArcExplorer Java Edition for Educators (AEJEE). The second version (*7.2 Geospatial Lab Application: GIS Layouts: ArcGIS Version*) provides the same activities for use with ArcGIS 10.

Objectives

The goals for you to take away from this lab are:

- Familiarize yourself with the layout functions of AEJEE.
- Arrange and print professional-quality maps from geographic data using the various layout elements.

Obtaining Software

The current version of AEJEE (2.3.2) is available for free download at http://edcommunity.esri.com/software/aejee.

Important note: Software and online resources sometimes change fast. This lab was designed with the most recently available version of the software at the time of writing. However, if the software or Websites have significantly changed between then and now, an updated version of this lab (using the newest versions) is available online at **http://www.whfreeman. com/shellito1e.**

Lab Data

There is no data to copy in this lab. All data comes as part of the AEJEE sample data that gets installed with the software.

Localizing This Lab

The dataset used in this lab is Esri sample data for the entire United States, and you'll be creating a map layout of the United States. However, the layout tools can be used to make a map of whatever dataset you desire—rather than creating a layout map of the entire United States, focus on your home state and create a layout of that instead (see Section 7.3 for how to focus the layout on one state instead of the whole United States). If you're only going to work with one state, use the counties.shp file instead and make a map of county population by square mile instead.

7.1 Initial Pre-Layout Tasks

1. Start AEJEE (the default install folder is called AEJEE). After AEJEE opens, do the following (if needed, refer back to *5.1 Geospatial Lab Application: GIS Introduction: AEJEE Version* for specifics on how to do these things):

2. Add the **states** shapefile from the usa sample data folder.

3. Pan and zoom the View so that the lower 48 states are shown filling up the View.

4. Change the projection in the View from the default GCS to something more visually pleasing.

   Important note: 5.1 Geospatial Lab Application: GIS Introduction: AEJEE Version used the Lambert Conformal Conic projection, but there are many more to choose from—keep in mind that your final map will be of the United States if selecting a regional or world projection.

5. Leave the states symbology alone for now—you'll change it in the next step.

7.2 Graduated Symbology

1. AEJEE gives you the ability to change an object's symbology from a single symbol to multiple symbols or colors and allows for a variety of different data classification methods.

2. Right-click on the **states** shapefile in the TOC and select **Properties**. Click on the **Symbol** tab. To display the states as graduated symbols, use the following settings:

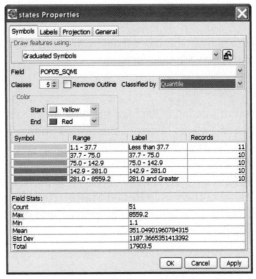

(Source: Esri® ArcGIS ArcExplorer graphical user interface Copyright © Esri.)

3. Under "Draw features using:" select **Graduated Symbols**.

4. The Field to use (in this exercise) is the **POP05_SQMI** (the states' population per square mile from the year 2005).

5. Use **5** for the number of Classes.

6. Use **Quantile** for the Classified by option.

7. For the color scheme, select an appropriate choice for the Start and End choices. (The default should be yellow for Start and red for End, thus giving you a range from yellow to orange to red for the five classes.)

8. When you have things arranged how you want them, click **Apply** to make the changes. Take a look at the map and make any color changes you think are needed.

9. Click **OK** to close the dialog box.

10. Now, the symbology of the states has changed in the View and the values that make up each one of the breaks can be seen in the TOC.

7.3 Layouts in AEJEE

To begin laying out the print-quality version of the map, you'll need to begin working in the Layout View. This mode of AEJEE works like a blank canvas, allowing you to construct a map using various elements. Note that whatever appears in the Map View will also appear in the Layout View. For instance, if you zoom the Map View to only show Ohio, and then switch to Layout View, the layout will only show Ohio.

1. To begin, select **Layout View** from the **View** pull-down menu.

2. In the Layout View, the black border around the white page represents the border of the printed page of an 8½ × 11 piece of paper, so be careful when working near the edges of the page and keep all elements of the map within that border.

3. The toolbar will switch to a new set of tools—the navigation tools are as follows:

4. Starting at the left and moving right are the following:

   a. The cursor that is used to select map elements

   b. The printer icon that is used when printing (see later in the exercise)

   c. The plus and minus magnifying glasses that are used to zoom in and out of the layout

   d. The hand icon that is used to pan around the map

5. And also from the right and moving left are the following:

   a. The 1:1 icon that is used to zoom the layout to 100%

   b. The white pages with the arrows that are used to (from right to left) zoom to see the entire page, zoom out a fixed amount, and zoom in a fixed amount

6. The other tools are used for the addition of map elements to the layout, and we'll examine them individually.

7.4 Map Elements

1. Again, think of the layout as a blank sheet of paper that you'll use to construct your map. There are numerous map elements that can be added, including scale bars, north arrows, and a legend. Each element has properties (such as creating borders, filling colors, or fixing size and position) that can be accessed by selecting the **Map Elements Properties icon** on the toolbar:

2. The default element that appears on the layout is the Data Frame that contains all of the GIS data. All of the visible layers in the TOC under the word "Layers" will appear in this frame. You can't manipulate individual layers (for instance, you can't click and drag a state somewhere else), but you can treat the entire Frame as if it's a map element. Click and drag the **Frame** around the map or resize it as you see fit (using the eight blue tabs at its corners and midpoints), or access its properties (like any regular element) to create borders around the map or fix it in a particular position.

3. Map elements can be resized and moved by selecting them with the mouse and resizing like you would the data frame.

4. Map elements can be deleted by selecting them with the cursor and pressing the delete key on the keyboard, or by selecting the **Remove icon** on the toolbar:

7.5 Layer Properties and Layouts

Each layer you use in AEJEE (such as the states) has a set of layer properties. Anything changed in the layer properties will be reflected in changes to the layout. For instance, if you change the symbology of a layer in its properties, it will be changed in the layout elements such as the map legend. However, to change the name of how the layer will appear in something like a map legend, the shapefile must be renamed in the TOC.

1. To do this, right-click on the name you want to change (the **states** layer) and select **Rename**.

2. In the TOC itself, you can type a new name for the layer (like "Population Per Square Mile"). Click the **enter** key when you've typed in a new name.

7.6 Inserting a Scale Bar

1. To add a scale bar to the map, select the **scale bar icon** from the toolbar:

2. A number of options for scale bars will appear in a new Scale Bar Selector dialog box:

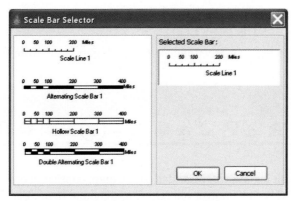

(Source: Esri® ArcGIS ArcExplorer graphical user interface Copyright © Esri.)

3. Select an appropriate scale bar and click **OK**. Use the cursor to position or resize the scale bar on the layout.

7.7 Inserting a North Arrow

1. To add a north arrow to the map, select the **north arrow icon** from the toolbar:

2. A number of options for north arrows will appear in a new North Arrow Selector dialog box:

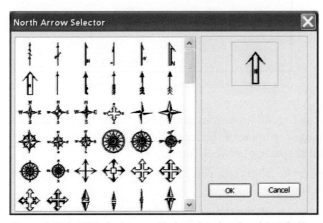

(Source: Esri® ArcGIS ArcExplorer graphical user interface Copyright © Esri.)

3. Select an appropriate north arrow and click **OK**. Use the cursor to position or resize the north arrow on the layout.

7.8 Inserting a Legend

1. To add a legend to the map, select the **legend icon** from the toolbar:

2. A default legend will be added to the map, consisting of all the layers in the TOC with whatever names are assigned to them. Use the cursor to move and resize the legend.

3. To make changes to the default legend, right-click on the **legend** itself, then select **Properties** from the new menu options.

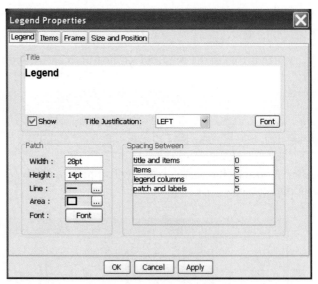

(Source: Esri® ArcGIS ArcExplorer graphical user interface Copyright © Esri.)

4. In the Legend Properties dialog box, under the Legend tab, you can change the name of the legend, its font, justification, and appearance, as well as spacing between items.

5. Under the **Items** tab, you can select what layers from the TOC appear in the legend.

6. Under the **Frame** tab, you can change the legend's color, border, and fill.

7. Under the **Size and Position** tab, you can manually change the size and position of the legend inside the layout.

8. Click **Apply** to have your legend changes take effect and **OK** to close the dialog box.

7.9 Adding Text to the Layout

1. To add text to the map (such as a title or any other text you may want to add), select the **Add Text icon** from the toolbar:

2. Note that when text is added, a small box called "right click this text" is placed near the center of the map.

3. Do what AEJEE asks and right-click the text box and select **Properties** from the new menu items. The Text Properties dialog box will appear.

(Source: Esri® ArcGIS ArcExplorer graphical user interface Copyright © Esri.)

4. Type the text you want to appear in the box. To alter the font, color, and other properties of the text, click the "**Change Properties . . .**" option.

More options will appear—note that you can alter each text box separately, to create larger fonts for titles, smaller fonts for type, and so on.

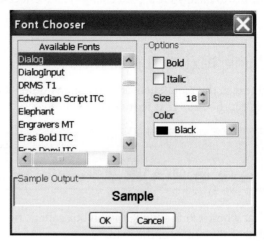

(Source: Esri® ArcGIS ArcExplorer graphical user interface Copyright © Esri.)

7.10 Other Map Elements

1. Though they're not used in this lab, two other map elements can be added:

    a. An overview map—this allows you to add a second map to the layout to show a larger context of the area. Use this icon:

    b. An image—this allows you to add graphics to your map, including downloaded images, digital camera images, or other graphics. Use this icon:

7.11 Printing the Layout

1. When you have constructed the map the way you want, choose **Print** from the **File** pull-down menu (or select the **Print icon** from the toolbar):

2. In the Print dialog box under the **Page Setup** tab, you can make additional changes, such as moving the printed map from Landscape to Portrait or adding margins for printing.

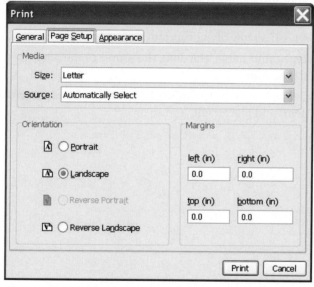

(Source: Esri® ArcGIS ArcExplorer graphical user interface Copyright © Esri.)

3. When the layout is ready the way you want it, click **Print** and AEJEE should print to your computer's default printer.
4. Exit AEJEE by selecting **Exit** from the **File** pull-down menu.

Closing Time

The geospatial lab application in Chapter 8 will keep using AEJEE for analysis—you'll be using a Web service to take a series of addresses and plot them on a map, then using those results in AEJEE for some simple analysis.

Final Layout Checklist

_____ The contiguous 48 states (without Alaska and Hawaii) United States population per square mile, classified, and displayed with an appropriate color scheme

_____ States shown in a different (and appropriate) projection than the default

_____ Appropriate map legend, with all items listed with normal names (the legend should not be called "legend" and should be appropriate size, font, number of columns, and so on)

_____ Scale bar

_____ Appropriate title (don't use the words "map" or "title" in your title) and appropriate size and font

_____ A north arrow of appropriate size and position

_____ Type (your name, the date, the source of the data) in an appropriate size and font

_____ Overall map design (appropriate borders, color schemes, balance of items and placement, and so on)

Geospatial Lab Application

GIS Layouts: ArcGIS Version

This lab will introduce you to the concepts of taking GIS data and creating a print-quality map from it. This map should contain the following:

- The contiguous United States (48 states without Alaska and Hawaii) population per square mile, set up in an appropriate color scheme
- The data in a different projection than the default GCS one
- An appropriate legend (make sure your legend items have normal names and that the legend is not called "legend")
- An appropriate title (make sure that your map doesn't include the word "map" in it)
- A north arrow
- A scale bar
- Text information: your name, the date, and the source of the data
- Appropriate borders, colors, and design layout (your map should be well designed instead of map elements thrown on at random)

Important note: At the end of this lab, there is a checklist of items to aid you in making sure the map you make is complete and of the best quality possible.

The previous *7.1 Geospatial Lab Application, GIS Layouts: AEJEE Version* uses the free ArcExplorer Java Edition for Educators (AEJEE); however, this lab provides the same activities for use with ArcGIS 10.

Objectives

The goals for you to take away from this lab are:

- Familiarize yourself with the layout functions of ArcGIS.
- Arrange and print professional-quality maps from geographic data using the various layout elements.

Obtaining Software

The current version of ArcGIS (10) is not freely available for use. However, instructors affiliated with schools that have a campus-wide software license may request a 1-year student version of the software online at http://www. esri.com/industries/apps/education/offers/promo/index.cfm.

Important note: Software and online resources sometimes change fast. This lab was designed with the most recently available version of the software at the time of writing. However, if the software or Websites have significantly changed between then and now, an updated version of this lab (using the newest versions) is available online at http://www.whfreeman.com/shellito1e.

Lab Data

There is no data to copy in this lab. All data comes as part of the AEJEE sample data that gets installed with the software.

Important note: In order to keep the data and results similar, both the AEJEE and ArcGIS 10 portions of the lab use the same setup sample data that comes with AEJEE. Thus, if you're using ArcGIS 10 for the lab, please download and install AEJEE in order to use its sample data.

Localizing This Lab

The dataset used in this lab is Esri sample data for the entire United States. You'll be creating a map layout of the United States. However, the layout tools can be used to make a map of whatever dataset you desire—rather than creating a layout map of the entire United States, focus on your home state and create a layout of that instead (see Section 7.4 for how to focus the layout on one state instead of the whole United States). If you're only going to work with one state, use the counties.shp file instead and make a map of county population by square mile instead.

7.1 Getting Data for Mapping

1. Start ArcMap. After ArcMap opens, do the following (if needed, refer back to the *5.2 Geospatial Lab Application: GIS Introduction: ArcGIS Version* for specifics on how to do these things):

2. Add the **states** shapefile from the USA sample data folder.

3. Pan and zoom the View so that the lower 48 states are shown filling up the View.

4. Change the projection in the View from the default GCS to something more visually pleasing.

    Important note: 5.2 Geospatial Lab Application: GIS Introduction: ArcGIS Version used the Lambert Conformal Conic projection, but there are many more to choose from—keep in mind that your final map will be of the United States when selecting a regional or world projection.

5. Leave the states symbology alone for now—you'll change it in the next step.

7.2 Setting Graduated Symbology

ArcMap gives you the ability to change an object's symbology from a single symbol to multiple symbols or colors and allows for a variety of different data classification methods.

1. Right-click on the **states** shapefile in the TOC and select **Properties**. Click on the **Symbology** tab. To display the states as graduated symbols, use the following settings:

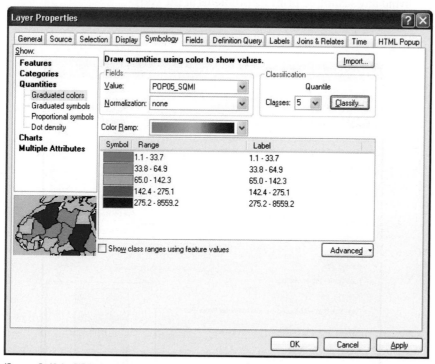

(Source: Esri® ArcGIS ArcMap ArcInfo graphical user interface Copyright © Esri.)

2. Under the **Show:** options, select **Quantities**, then select **Graduated colors**.

3. The Field to use (in this exercise) is the **POP05_SQMI** (the states' population per square mile from the year 2005).

4. For the color ramp, select an appropriate choice from the available options.

5. Use **5** for the number of Classes.

6. To change the classification method, press the **Classify** button. From the new options available in the pull-down menu, choose **Quantile** for the Method.

7. When you have things arranged how you want them, click **Apply** to make the changes. Take a look at the map and make any color changes you think are needed.

8. Click **OK** to close the dialog box.

9. Now, the symbology of the states has changed in the View, and the values that make up each one of the breaks can be seen in the TOC.

7.3 Page Setup and Layouts

1. Before starting the layout process, first set up the properties of the map page, as they'll fit with the printer you'll be using to print the map (to avoid any scaling or printing problems at the end).

2. From the **File** pull-down menu, select **Page and Print Setup**.

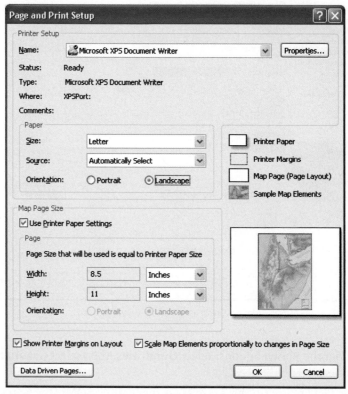

(Source: Esri® ArcGIS ArcMap ArcInfo graphical user interface Copyright © Esri.)

3. Under **Name**, select the printer you'll be using.

4. For **Orientation**, choose how you want to compose the layout—using Portrait (a vertical orientation for the map) or Landscape (a horizontal orientation).

5. Place check marks in the following boxes:

   a. Use Printer Paper Settings

   b. Show Printer Margins on Layout

   c. Scale Map Elements proportionally to changes in Page Size

6. Click **OK** when all settings are correct.

7.4 Layouts in ArcMap

1. To begin laying out the print-quality version of the map, you'll need to begin working in the Layout View. This mode of ArcMap works like a blank canvas, allowing you to construct a map using various elements.

2. Note that whatever appears in the Data View will also appear in the Layout View. For instance, if you zoom the Map View to only show Ohio, and then switch to Layout View, the layout will only show Ohio.

3. To begin, select **Layout View** from the **View** pull-down menu.

    a. You can switch back to the Data View by selecting Data View from the View pull-down menu.

    b. You can also switch between Layout and Data Views by using the icons at the bottom left-hand corner of the View.

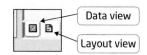

4. In the Layout View, the black border around the white page represents the border of the printed page of an 8½ × 11 piece of paper, so be careful when working near the edges of the page and keep all elements of the map within that border.

5. The toolbar will switch to a new set of tools—the navigation tools are as follows:

6. Starting at the left and moving right are the following:

    a. The plus and minus magnifying glasses that are used to zoom in and out of the layout

    b. The hand icon that is used to pan around the map

    c. The 1:1 icon that is used to zoom the layout to 100%

    d. The white pages with the four black arrows that are used for zooming in and out from the center of the page

    e. The white pages with the blue arrows that are used for returning to previous scales

7.5 Map Elements

1. Again, think of the layout as a blank sheet of paper that you'll use to construct your map. There are numerous map elements that can be added, including scale bars, north arrows, and a legend. Each element has properties (such as creating borders, filling colors, or fixing size and position) that can be accessed individually (using the black 'select elements' arrow on the **Tools** toolbar).

2. The default element that appears on the layout is the Data Frame, which contains all of the GIS data. All of the visible layers in the TOC under the word "Layers" will appear in this frame. You can't manipulate parts of individual layers (for instance, you can't click and drag a state somewhere else), but you can treat the entire Frame as if it's a map element. Click and drag the **Frame** around the map or resize it as you see fit (using the eight blue tabs at its corners and midpoints), or access its properties (like any regular element) to create borders around the map or fix it in a particular position.

3. Map elements can be resized and moved by selecting them with the mouse and resizing like you would the data frame.

4. Map elements can be deleted by selecting them with the cursor and pressing the delete key on the keyboard.

7.6 Layer Properties and Layouts

Each layer you use in ArcMap (such as the states) has a set of layer properties. Anything changed in the layer properties will be reflected in changes to the layout. For instance, if you change the symbology of a layer in its properties, it will be changed in the layout elements such as the map legend. However, to change the name of the layer and how it will appear in something like a map legend, the shapefile must be renamed in its Layer Properties.

1. To do this, return to the **Layer Properties** and choose **General**. You can type a new name for the layer (like "Population Per Square Mile"). Click **OK** when you've typed in a new name.

7.7 Inserting a Scale Bar

1. To add a scale bar to the map, choose the **Insert** pull-down menu and select **Scale Bar**.

2. A number of options for scale bars will appear in a new Scale Bar Selector dialog box:

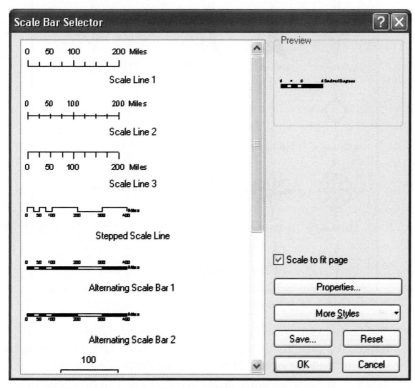

(Source: Esri® ArcGIS ArcMap ArcInfo graphical user interface Copyright © Esri.)

3. Additional scale bars are available by clicking the **More Styles** button. You can alter the appearance of the scale bar (adjusting the number of divisions on the bar, for the display units shown on the scale bar, etc.) by selecting the **Properties** button.

4. Select an appropriate scale bar and click **OK**. Use the cursor to position or resize the scale bar on the layout.

7.8 Inserting a North Arrow

1. To add a north arrow to the map, choose the **Insert** pull-down menu and select **North Arrow**.

2. A number of options for north arrows will appear in a new North Arrow Selector dialog box:

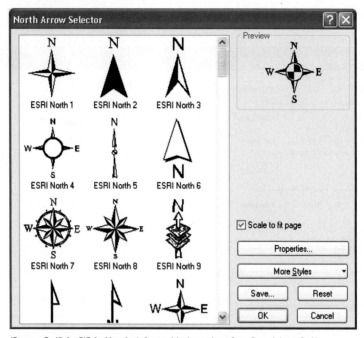

(Source: Esri® ArcGIS ArcMap ArcInfo graphical user interface Copyright © Esri.)

3. Additional north arrows are available by clicking the **More Styles** button. You can alter the appearance of the north arrow (adjusting the color, rotation angle, etc.) by selecting the **Properties** button.

4. Select an appropriate north arrow and click **OK**. Use the cursor to position or resize the north arrow on the layout.

7.9 Inserting a Legend

1. To add a legend on the map, choose the **Insert** pull-down menu and select **Legend**.

2. The Legend Wizard will open—this is a set of menus to help you properly set up the legend for the map.

    Important note: At any step, you can see what the legend will look like on the map by clicking the **Preview** button in the wizard. By pressing **Preview**, the legend will be added to the map, and you can see what it looks like at the current stage of legend design. If you're satisfied, you can click **Finish** in the Legend Wizard to add the legend. If you want to keep going with the steps in the Legend Wizard, press **Preview** again and the legend will disappear, so you can advance to the next Wizard step.

3. The first screen will allow you choose the layers that will appear in the legend, along with the number of columns you want the legend to be.

Choose the **states** layer for this map and select one column. Click **Next** to advance to the second menu.

4. The second menu will allow you to give the legend a name—give it something more appropriate besides "legend." You can also adjust the color, size, font, and justification of the legend itself. Once you've set these values, click **Next** to advance to the third menu.

5. The third menu will allow you to adjust the color, thickness, and appearance of the legend's border, background, and drop-shadow effects. Examine various options for the legend's appearance and, when ready, click **Next** to advance to the fourth menu.

6. The fourth menu will allow you to alter the symbol patch of the legend items (in this case, the states layer). When ready, click **Next** to advance to the fifth menu.

7. The fifth and final menu will allow you to adjust the appearance and spacing of items in the legend itself. When you're satisfied with how the legend will look, click **Finish**. The legend will then be added to the layout.

8. Once the legend is created, you can make changes to it by right-clicking on the legend itself in the layout and selecting **Properties**. For instance, selecting the **Items** tab under **Properties** allows you to access a "Style" option that you can use to further modify the appearance of the legend.

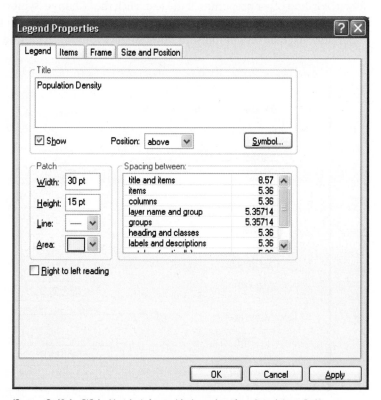

(Source: Esri® ArcGIS ArcMap ArcInfo graphical user interface Copyright © Esri.)

(Source: Esri® ArcGIS ArcExplorer graphical user interface
Copyright © Esri.)

7.10 Adding Text and a Title to the Layout

1. To add text to the map (such as your name, the date, or any other text you may want to add), select **Text** from the **Insert** pull-down menu.

2. Note that when text is added, a small box called "Text" is placed near the center of the map. You'll have to move each text box to somewhere else in the layout.

3. Right-click the text box and select **Properties** from the new menu items. The Properties dialog box will appear.

4. Under the Text tab, type the text you want to appear in the box. To alter the font, color, and other properties of the text, click the "**Change Symbol . . .**" button. More options will appear—note that you can alter each text box separately, to create larger fonts for titles, smaller fonts for type, and so on.

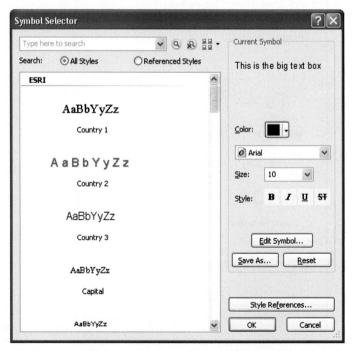

(Source: Esri® ArcGIS ArcMap ArcInfo graphical user interface Copyright © Esri.)

7.11 Other Map Elements

1. Though they're not used in this lab, other map elements can be added from the **Insert** pull-down menu:

a. Dynamic Text allows you to insert information about the current date, current time, the date the map was saved, and so on.

b. Neatline allows you to draw a line around map elements.

c. Scale Text allows you to insert verbal statements of scale rather than only using a scale bar for representation.

d. Picture allows you to add graphics to your map, including downloaded images, digital camera images, or other graphics.

e. Object allows you to insert things from other software applications, including video clips, Microsoft PowerPoint slides, or sound clips.

7.12 Printing the Layout

1. When you have constructed the map the way you want, choose **Print Preview** from the **File** pull-down menu. You'll see how your map will appear on the printed page. Pay careful attention to borders, lines, and placement of objects when comparing the screen appearance to what they will look like on paper. Make whatever adjustments are necessary to the final draft of the layout before printing.

2. When the map's ready, select **Print** from the **File** pull-down menu—
    ArcMap will use this dialog box to send the layout to the printer.

3. Exit ArcMap by selecting **Exit** from the **File** pull-down menu. There's no
    need to save any data in this lab.

Closing Time

The geospatial lab application in Chapter 8 will keep using ArcGIS for
analysis—you'll be using a Web service to take a series of addresses and plot
them on a map, then use those results in ArcGIS for some simple analysis.

Final Layout Checklist

_____   The contiguous 48 states (without Alaska and Hawaii) United States
         population per square mile, classified, and displayed with an appro-
         priate color scheme

_____   States shown in a different (and appropriate) projection than the
         default

_____   Appropriate map legend, with all items listed with normal names
         (the legend should not be called "legend" and should be appropriate
         size, font, number of columns, and so on)

_____   Scale bar

_____   Appropriate title (don't use the words "map" or "title" in your title)
         and appropriate size and font

_____   A north arrow of appropriate size and position

_____   Type (your name, the date, the source of the data) in an appropriate
         size and font

_____   Overall map design (appropriate borders, color schemes, balance of
         items and placement, and so on)

8

Getting There Quicker with Geospatial Technology

Vehicle Navigation Systems, Road Maps in a Digital World, Creating a Street Network, Geocoding, Shortest Paths, and Street Networks Online

Vehicle navigation systems (like those made by companies such as Garmin, Magellan, or Tom-Tom) are really revolutionary technologies. One small device mounted on the dashboard will find your precise location, plot it on a map, determine where the nearest gas stations are, then compute the quickest route to get you there, all with turn-by-turn directions that announce names of streets and the distance to the turn (**Figure 8.1**). The position determination is straightforward—the device has a GPS receiver in it that finds your location on Earth's surface using the methods discussed in Chapter 4. In fact, most of these devices are simply referred to as a "GPS," as in "punch our destination into the GPS" or "what does the GPS say the shortest route to get there is?" However, it's doing a disservice to these things to simply call them

vehicle navigation system a device used to plot the user's position on a map, using GPS technology to obtain the location.

FIGURE 8.1 A Garmin GPS vehicle navigation system. (Source: Edward J. Bock III/Dreamstime.com)

a "GPS" since it's obvious they do so much more than what a regular GPS receiver does.

These devices rely on a GIS-style system at their core—hence their ability to handle spatial data in the form of road-network maps, use those maps for routing, determine the shortest path between two (or more) points, and match an address to a spatial location. This same type of data is used to route emergency vehicles to the site of a 911 emergency phone call or manage a fleet of delivery vehicles. The same sort of system is at the heart of online mapping applications such as MapQuest or Google Maps and in the location and mapping apps of a smartphone. This type of technology is changing fast and improving as time goes on.

A March 2009 article in *USA Today* described an incident where a vehicle navigation device instructed the driver to turn off a road and follow a snowmobile trail toward a destination, which ended with the car stuck in the snow and the state police being called for emergency help. Similarly, a March 2009 article of the *Daily Mail* relates a story of a vehicle navigation system that directed a driver along a walking footpath, which ended in a sheer cliff drop of 100 feet (luckily, the driver stopped in time). When these kinds of errors occur, it's usually not because the GPS receiver is finding the incorrect position from the satellite information. Rather, it is more likely there are problems with the base network data itself. These types of systems are only as accurate as the base network data they have available to them. This chapter delves into how these types of geospatial technology applications function, how they're used, what makes them tick, and why they may sometimes lead you astray.

Thinking Critically with Geospatial Technology 8.1

What Happens When the Maps Are Incorrect?

How many times has this happened to you when using some sort of mapping service—either the directions take you to the wrong place, they indicate you should turn where you can't, or they can't find where you want to go (or the road you want to go on)? The GPS location is probably correct, but the maps being used for reference are likely either outdated or have errors in them. How much does the usefulness of this aspect of geospatial technology rely on the base data? If the base maps are incomplete or not updated, then how useful is a mapping system to the user?

In addition, many vehicle navigation systems will recompute a new route on the fly if you miss a turn or start to take another route. However, if you have been taken on an improper route, how useful will the system be in getting you back to where you want to go? A key purpose of mapping systems is using them to navigate through unfamiliar territory—but if they don't properly fulfill their function, what do travelers rely on? Are there any sorts of liability issues that could result from inaccurate mapping systems?

How Do You Model a Network for Geospatial Technology?

Back in Chapter 5, we discussed how real-world items are modeled or represented using GIS. Any type of **network** is going to involve connections between locations, whether streams, power lines, or roads. Thus (to use Esri terminology), in GIS a network in its most basic form is represented by a series of **junctions** (point locations or nodes) that are connected to each other by a series of **edges** (lines or links). For instance, in a road network, junctions might be the starting and ending points of a road or the road intersections, while the edge would be the line representing the road itself. When designing a road network, keep in mind that there may be many types of edges and junctions to represent. For example, a city's road network would have edges that represent streets, highways, railroads, light-rail systems, subway lines, or walking paths, while junctions may represent not only the starting and ending of streets, but also highway entrances and exits, freeway overpasses and underpasses, subway stops, or rail terminals.

When dealing with all these different types of edges and junctions, the **connectivity** of the network in GIS is essential when modeling it. With proper connectivity, all junctions and edges should properly connect to one another, while things that should not connect, do not connect. For example, if a freeway crosses over a road via an overpass, the network connectivity should not show that as a valid intersection allowing the street to turn onto the freeway at that junction. If your vehicle navigation system leads you to this point, then instructs you to "turn right onto the highway," it's impossible for you to do so, but the device thinks you should be able to because of how the network data is set up. In the same way, a railroad line may intersect with a street, but the network should not have a connection showing that the street could continue along the rail line. If this kind of connection was built into the data, you could conceivably be routed to turn onto the railroad line and continue on it toward your destination. It sounds silly to think of driving your car on the railroad tracks, but due to incorrect network data, this line would simply represent the next road to take to get to your destination.

Thinking along these lines, other features of a road network must also be included in the model. For instance, some streets may be one way, or some junctions may not allow left-hand turns, or U-turns may not be permitted. These types of features need to be properly modeled for the system to be an accurate, realistic model of the road network. Although you may be able to see the "one way" street sign when you're driving, if that feature has not been properly set up in the network, the system would have no way of knowing not to try to route cars in both directions along the street. As discussed previously, an overpass or underpass should not show up as being connected to the road network (to be a viable option for a turn—if a device

network a series of junctions and edges connected together for modeling concepts such as streets.

junction a term used for the nodes (or places where edges come together) in a network.

edge a term used for the linkages of a network.

connectivity the linkages between edges and junctions of a network.

or GIS instructs you to make a right-hand turn onto the freeway that you're currently driving under, then something's gone wrong with some aspect of the technology).

When a network is being modeled, each edge is considered to be a separate entity, not necessarily each individual street. A long city road may be modeled in the GIS as several **line segments**, with each segment representing a section of the road. A major urban street may be made up of more than 100 line segments—each segment being a different section of the road (with each segment being delineated by roads that intersect it). For example, Figure 8.2 shows a geospatial road network of Virginia Beach, Virginia. The road cutting through the center of the city (highlighted in blue) is Virginia Beach Boulevard, a major multi-lane city street, with numerous intersecting roads and street lights. Although we think of Virginia Beach Boulevard as one big, long street, the system models it as 129 line segments, with each line segment representing a portion of the road.

Breaking a road up into individual line segments allows the modeling of different attributes for each segment. Attributes such as the name of the road, the address ranges for each side of the street, the suffix of the road name (whether it's a Drive, Avenue, Boulevard, etc.), the type of road (such as residential street, interstate, highway, etc.), and the speed limit are all examples of the types of values that can be assigned to individual segments. Thus, an attribute table of this layer in a GIS would consist of 129 records, each with multiple attributes.

Several different types of geospatial road network files are available. A **street centerline** file is a file modeling each city road as a line, containing the different types of roads. The U.S. Census Bureau also regularly issues this type

---

line segment a single edge of a network that corresponds to one portion of a street (for instance, the edge between two junctions).

street centerline a file containing line segments representing roads.

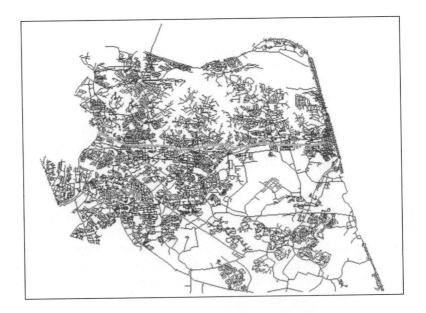

FIGURE 8.2 Virginia Beach Boulevard shown highlighted in a GIS network. Although it is only one road, it consists of 129 line segments. (Source: Esri® ArcGIS ArcMap ArcInfo graphical user interface. Copyright © Esri.)

Hands-on Application 8.1

TIGER Files Online

TIGER files are made freely available from the Census Bureau via the Web. Open your Web browser and go to **http://www.census.gov/geo/www/tiger**–this is the part of the U.S. Census Bureau's Website for downloading TIGER/Line files. There's also full documentation of TIGER files available in PDF format on the Website. Files for U.S. counties can be downloaded in shapefile format (see Chapter 5 for more info about shapefiles) to be used in GIS products such as ArcGIS or ArcExplorer Java Edition for Educators. You can also download several other types of TIGER files besides road-network data, including census block information, hydrography, landmarks, and American Indian reference data–check to see what types of datasets are available for download.

Esri also makes TIGER 2000 data freely available–this data has already been separated into its various components (such as roads, railroads, etc.) for download so that each layer can be downloaded as its own shapefile. Open your Web browser and go to **http://arcdata.esri.com/data/tiger2000/tiger_download.cfm**. All data is downloaded in a zipped shapefile format (same as the U.S. Census Bureau's Website) for use in ArcGIS or AEJEE. Select your county to see what TIGER data is available for you to use. This chapter's lab uses a TIGER dataset downloaded from the Esri Website.

of road network data in a format usable by geospatial technology software. The data is provided in the **TIGER/Line** files. TIGER stands for Topologically Integrated Geographic Encoding Referencing and the files delineate different boundaries throughout the United States (such as congressional districts), in addition to containing road-network data. (See *Hands-on Application 8.1: TIGER Files Online* for more information.)

Each record in a TIGER/Line file represents a segment of a road network, and thus each segment (record) can have multiple attributes (fields) assigned to it. Figure 8.3 on page 228 again shows a TIGER/Line file of Virginia Beach and a portion of the attribute table of those selected segments that make up Virginia Beach Boulevard. Note how many attributes there are (information that gets encoded into each road segment, and the entire Virginia Beach TIGER/Line file is made up of over 19,000 segments). The TIGER/Line file contains the following standard attributes (among others):

> **TIGER/Line** a file produced by the U.S. Census Bureau that contains (among other items) the line segments that correspond with roads across the United States.

- FEDIRP: This is the prefix direction of the road (*N.* Smith St. or *W.* Broad Ave.).
- FENAME: This is the name of the road (N. *Smith* St. or W. *Broad* Ave.).
- FETYPE: This is the type of road (N. Smith *St.* or W. Broad *Ave.*).
- FEDIRS: This is the suffix direction of the road (Cherry Lane *S.* or Canal Street *E.*).
- CFCC: This is the Census Feature Class Code, a standardized encoding used to separate different kinds of roads (such as residential street, highways, or interstates).

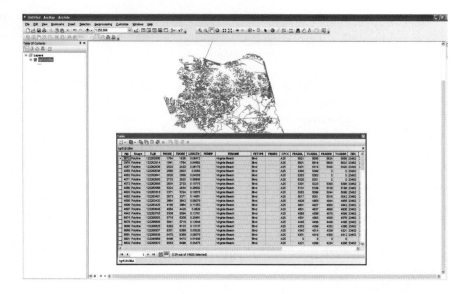

⦿ FRADDL: This is the start of the address range on the left side of the street
   (for instance, if the street addresses go from 101 to 199, this value would
   be 101).

⦿ TOADDL: This is the end of the address range on the left side of the street
   (for instance, if the street addresses go from 101 to 199, this value would
   be 199).

⦿ FRADDR: This is the start of the address range on the right side of the
   street (for instance, if the street addresses go from 100 to 198, this value
   would be 100).

⦿ TOADDR: This is the end of the address range on the left side of the street
   (for instance, if the street addresses go from 100 to 198, this value would
   be 198).

⦿ ZIPL: This is the zip code used for addresses on the left side of the
   street.

⦿ ZIPR: This is the zip code used for addresses on the right side of the
   street.

These attributes define the characteristics of each road segment, and
similar attributes would be found in road-network data, such as other street
centerline files. If these attributes are incorrect, then the base network map
will be incorrect. If the vehicle navigation system gives you incorrect street
names or calls a road "east" when it's really "west," it's likely that there are
incorrect values in the base network data's attributes.

The TIGER file attributes (or similar base road-network data created by
others) concerning specific address ranges, zip-code information, and detailed data for the names of roads can be used as a base map source for other

applications, such as pinpointing specific addresses on a road. It's this source data that allows for a match of a typed street address to a map of the actual location.

How Is Address Matching Performed?

Whenever you use a program like MapQuest or Google Maps to find a map of a location, you're typing in something (like "1600 Pennsylvania Avenue, Washington, D.C.") and somehow the Website translates this string of characters into a map of a spatial location (like the White House). The process of taking a bunch of numbers and letters and finding the corresponding location that matches up with them is called **address matching** or **geocoding** (Figure 8.4). Although the process seems instantaneous, there are several steps involved in geocoding that are happening "behind the scenes" when you use an address-matching system (like those in GIS).

First, you need to have some sort of **reference database** in place—this is a road network that the addresses will be matched to. A TIGER/Line file, another type of street centerline file, or some other road-network data (like those created by commercial companies) is needed here. What's essential is that the line segments contain attributes such as those found in a TIGER/Line file—for example, street direction, name of the street, address ranges on the left and right sides of the street, street suffix, and zip codes on the left and right sides of the street. This information will be used as the source to match addresses to as well as a source for the final plotted map.

address matching another term for geocoding.

geocoding the process of using the text of an address to plot a point at that location on a map.

reference database the base network data used as a source for geocoding.

FIGURE 8.4 A map generated by querying MapQuest for "1600 Pennsylvania Avenue, Washington, D.C." (Source: Map(s) and data © 2011 by MapQuest, Inc., Navteq, and Intermap as applicable. MapQuest and the MapQuest logo are trademarks of MapQuest, Inc. Used with permission.)

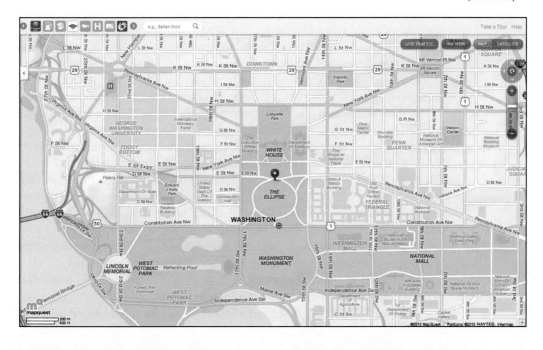

TABLE 8.1   Addresses that have been parsed into their component parts and standardized.

Location	Address	Prefix	Number	Street name	Street type	Suffix
White House	1600 Pennsylvania Avenue NW		1600	Pennsylvania	AVE	NW
National Gallery of Art	401 Constitution Avenue NW		401	Constitution	AVE	NW
U.S. Capitol	1 1st Street NE		1	1st	ST	NE
National Archives and Records	800 North Capitol Street NW	N	800	Capitol	ST	NW
Washington National Cathedral	3101 Wisconsin Avenue NW		3101	Wisconsin	AVE	NW
United States Holocaust Memorial Museum	100 Raoul Wallen-berg Place SW		100	Raoul Wallenberg	PL	SW

parsing breaking an address up into its component parts.

address standardization setting up the components of an address in a regular format.

Next, the address information is **parsed** into its component pieces, and **address standardization** is performed to set up data in a consistent format. The geocoding process needs to standardize addresses to properly match a location using its appropriate attributes in the reference database. Table 8.1 shows a number of locations in the Washington, D.C., area with their addresses, as well as these addresses parsed and standardized. For instance, in the National Gallery of Art's address, the street name is "Constitution." When the address matches, the system refers to line segments with a name attribute of "Constitution" and those segments with a street-type attribute of "AVE" (rather than ST, BLVD, LN, or others) and a suffix direction attribute of NW (instead of some other direction). The street number, 401, is used to determine which road segments match an address range (on the left or right side of the street, depending on whether the number is odd or even).

After the address has been parsed and standardized, the matching takes place. The geocoding system will find the line segments in the reference database that are the best match to the component pieces of the address and (in ArcGIS) rank them. For instance, in trying to address match the National Gallery of Art, a line segment with attributes of Name = "Constitution," Type = "AVE," Suffix Direction = "NW," and an address range on the left side of "401-451" would likely be the best (or top-ranked) match. A point corresponding with this line segment is placed at the approximate location along the line segment to match the street number. For instance, our address of 401 would have a point placed near the start of the segment, while an address of 425 would get placed close to the middle. The method used to plot a point at its approximate distance along the segment is called **linear interpolation**.

linear interpolation a method used in geocoding to place an address location among a range of addresses.

Keep in mind that the plotting is an approximation of where a specific point should be. For instance, if a road segment for "Smith Street" has an address range of 302 through 318, an address of 308 would be placed near the

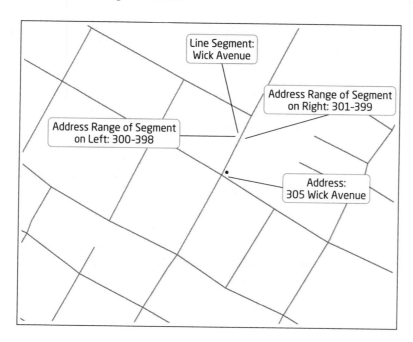

FIGURE 8.5 Plotting an
address point on a line
segment in GIS. (Source:
Esri® ArcGIS ArcMap ArcView
graphical user interface.
Copyright © Esri.)

middle. However, if the actual real-world location of house number 308 was closer to the end of the street, then the placement of the plotted point would not necessarily match up with the actual location of the house. In cases of streets that contain only a handful of houses that correspond with the address range in the reference file, plotted locations may be estimated incorrectly. See Figure 8.5 for an example of plotting a geocoded point on a road network in GIS.

In GIS, or in a vehicle-navigation system or smartphone equipped with geospatial technology, whenever you specify an address, the system will match the address and fix it as a destination point to be found. The process of geocoding multiple addresses at once is referred to as **batch geocoding**. For instance, in batch geocoding, you could have a list of the addresses of all coffee shops in Seattle, and the GIS would match all addresses on the list, rather than you having to input one at a time (for an example, see *Hands-on Application 8.2: Geocoding Using Online Resources* on page 232). If no match can be found, or if the ranking is so poor as to be under a certain threshold for a match, sometimes a point will not be matched for that address (or it may be matched incorrectly or placed at something like the center of a zip code). You will sometimes be prompted to recheck the address or to try to interactively match the address by manually stepping through the process.

Finally, when an address is plotted, the system may have the capability to calculate the *x* and *y* coordinates of that point and return those values to the user. For example, the street address of the Empire State Building in New York City is "350 5th Ave., New York, NY 10018." From address matching using the free online gpsvisualizer.com utility (also used in this chapter's lab), the GIS coordinates that fit that address are computed to

batch geocoding
matching a group of
addresses together
at once.

Hands-on Application 8.2

Geocoding Using Online Resources

Many online mapping resources (see *Hands-on Application 8.4: Online Mapping and Routing Applications and Shortest Paths* for details on what they are and how they're used) will geocode a single address (or a pair of addresses used for calculating directions between the two). You can use GIS software to geocode multiple addresses, or you can also use a resource like the BatchGeo Website. Open your Web browser and go to **http://www.batchgeo.com**—this is a free online service that allows you to geocode an address (or multiples in batches), and then view the plotted points on a map. Set up a batch of three or more addresses (such as home addresses for a group of family members, a group of friends, or several workplaces) and run the batch geocode utility. The Website will give you examples of how the addresses need to be formatted. From there, you can view your results on the Google map that the Website generates.

This chapter's lab will utilize other functions of the BatchGeo Website (and other resources)—you'll first geocode a set of points, and then use GIS to map them and conduct some analysis.

be: latitude 40.74807 and longitude −73.984959 (Figure 8.6). Thus, geocoded points have spatial reference attached to them for further use as a geospatial dataset.

Geocoding is a simple yet powerful process, but it's not infallible. There are several potential sources of error that can give an incorrect result and plot an address in the incorrect location. Since the addresses you're entering will be parsed to match up with the segments in the reference database, the result can sometimes be an error in matching. For instance, the address of the White House in Table 8.1 is listed as "1600 Pennsylvania Avenue NW." The line segments that match up with this street should have the name listed with "Avenue" (or "Ave.") as a suffix. If you input different things like "1600 Pennsylvania Street" or "1600 Penn Avenue," the location could potentially

FIGURE 8.6 The geocoded result for the Empire State Building in New York City, with its corresponding latitude and longitude calculation. (Source: GPSVisualizer.com/ Map data © 2011 Google, Sanborn.)

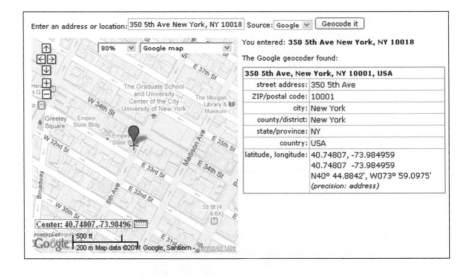

receive a lower ranking or end up plotted elsewhere or otherwise not properly matched. With more complete information (such as "1600 Pennsylvania Avenue NW, Washington, D.C. 20003") the system should be better able to identify a more accurate match.

Also, the geocoding system can only properly identify locations if the line segments are in the reference database. If you're searching for an address established after the reference database network file was put together, the system will not be able to properly match the address. If the point is plotted on the correct street but at an incorrect location on the street, it's likely a problem with address range data in the reference database and how it reflects the real world (for instance, a house at #50 Smith Street is not necessarily halfway down a street segment that begins with 2 and ends with 100). The geocoding will usually be only as accurate as the base data it is being matched to. If the reference database does not contain all line segments (for instance, subdivisions, streets, or new freeway bypasses that haven't been mapped and added to the reference database), or if its attributes contain inaccurate address ranges or incorrect or missing attributes, the geocoding process will likely be unable to accurately match the addresses or may plot them in the wrong location.

New methods for geocoding addresses to get a more accurate match are being developed. Rather than using a line segment and interpolating the address location, point databases are being created in which a point represents the center of a parcel of land (for instance, a house or a commercial property). When geocoding with this point data, address information can get matched to the point representing the parcel, and the address location can be found for the road immediately next to the parcel.

Once locations are geocoded, the system (or GIS) can begin to examine the routes between locations to determine the shortest path from one location to another. With a vehicle navigation system, you enter the address of the destination you want to travel to, and the system will match that address. The device's current position is determined using GPS and plotted on the network map (and this will be the origin). With two points, the system will then compute the shortest route between the origin and destination across the network. The same holds true for an online system to find directions—it has a matched origin and destination and will compute what it considers the best route for you to follow between the two points. With so many different ways to get from the origin to the destination, the system now needs a way to determine the "shortest path" between these locations.

How Are Shortest Paths Found?

When you leave your home to go to work, you likely have several different ways you can go. Some of them are very direct and some of them are very roundabout, but you have plenty of options available. If you want to take the "shortest" path from home to work, you'd likely focus on some of the more

shortest path the route that corresponds to the lowest cumulative transit cost between stops in a network.

direct routes and eliminate some of the longer or more circuitous routes. However, a **shortest path** can mean different things. For instance, driving through city streets may be the shortest physical driving distance (in terms of mileage), but if those streets have lots of stop lights and traffic congestion, then this "shortest path" will likely take longer in terms of the time spent driving, rather than traveling a longer distance on a highway that does not have these impediments. In this case, if you wanted to minimize the time spent driving, the highway route would likely get you to work faster, but you'd actually be driving a longer distance. Major city streets may take a long time to traverse at rush hour, but you might sail through quickly if you're driving on them late at night.

All of these are things to consider when figuring out the shortest path (or "best route") to take when traveling from an origin to a destination. People use their own decision-making criteria when determining the shortest path they're going to take—things like "always use main roads" or "try to use highways whenever possible" or "don't make a left turn." For this reason, vehicle navigation systems often offer multiple options, such as "shortest distance" or "shortest driving time" or "avoid highways" to compute the best route between points.

transit cost a value that represents how many units (of time or distance) are used in moving across a network edge.

Within a vehicle-navigation system (or GIS), each line segment has a **transit cost** (or impedance) assigned to it. The transit cost reflects how many units (of things like distance or travel time) it takes to traverse that edge. For example, the transit cost may reflect the actual distance in miles from one junction to another along the edge. The transit cost could also be the equivalent driving time it takes to traverse that particular segment. Whatever transit cost is used, that value will be utilized in the shortest path computation. Other impedance attributes can be modeled as well—segments could have a different transit cost under certain conditions (such as heavy traffic or construction). These types of impedance factors can help in making a network more realistic for use.

algorithm a set of steps used in a process (for example, the steps used in computing a shortest path).

Dijkstra's Algorithm an algorithm used in calculating the shortest path between an origin node and other destination nodes in a network.

The shortest path is then calculated using an **algorithm**, or a set of steps used in determining the overall lowest transit cost to move along the network from a starting point to a destination. There are various types of shortest path algorithms, including **Dijkstra's Algorithm** (see *Hands-on Application 8.3: Solve Your Network Problems with Dijkstra* for more information on how to use this algorithm), which will compute the path of lowest cost to travel from a starting point to each destination in the network. For instance, say you have three different destinations you plan to travel to (work, the pizza shop, and the grocery store). Dijkstra's Algorithm will evaluate the overall cost from your home to each destination and find the shortest path from your home to work, from your home to the pizza shop, and from your home to the grocery store.

Whatever type of algorithm is used, the system will compute the shortest path, given the constraints of the network (such as transit cost or directionality of things like one-way streets). The system can then generate directions for you by translating the selected path into the various turns and names of

Hands-on Application 8.3

Solve Your Network Problems with Dijkstra

The inner workings of Dijkstra's Algorithm are beyond the scope of this book, but there's an excellent free online resource that allows you to construct a sample network, then run the Dijkstra Algorithm to find the shortest path. The algorithm will walk through the shortest path step-by-step and describe the actions it's taking (and how the shortest paths are created). Open your Web browser and go to **http://www.dgp.toronto.edu/people/** **JamesStewart/270/9798s/Laffra/DijkstraApplet. html**. On this Website, you can set up a series of nodes (junctions) and links (edges), assign weights (transit costs) and directions to them, and run the algorithm to find the shortest path between the origin and all destinations on the network. Use the interactive interface to construct a sample network (or use the pre-made example) and use Dijkstra to set up the shortest paths for you.

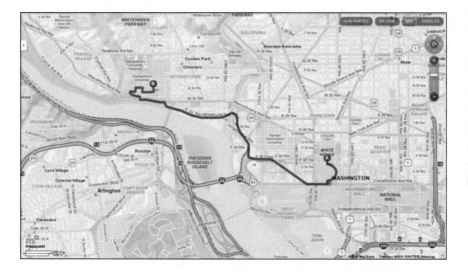

FIGURE 8.7 The "shortest path" (the blue line) computed from MapQuest from the White House to Georgetown University in Washington, D.C. (Source: Map(s) and data © 2011 by MapQuest, Inc., Navteq, and Intermap as applicable. MapQuest and the MapQuest logo are trademarks of MapQuest, Inc. Used with permission.)

streets that will be traversed on that path. See Figure 8.7 for an example of using an online geospatial technology utility to compute the shortest path or best route between two locations. Also check out *Hands-on Application 8.4: Online Mapping and Routing Applications and Shortest Paths* on page 236 for some other Web tools for online directions and routing.

Of course, sometimes you have more than one destination to visit—say you have three places you want to stop at (shoe store, music store, and book store) and you want to drive the overall shortest route to hit all three places. When using geospatial technology to compute a shortest route between several of these **stops,** there are two types of scenarios to choose from:

1. Finding the shortest path when visiting stops in a pre-defined order: This means you have to stop at the shoe store first, the music store second, and the book store last. In this case, you'd want to find the shortest path

stops destinations to visit on a network.

Hands-on Application 8.4

Online Mapping and Routing Applications and Shortest Paths

There are many online applications available for creating a map of an address and then generating a shortest path and directions from that address to another one. Examine the functionality of some of the following online mapping sites:

1. Google Maps: **http://maps.google.com**
2. MapQuest: **http://www.mapquest.com**
3. Yahoo! Maps: **http://maps.yahoo.com**
4. Bing Maps: **http://www.bing.com/maps**

Try inputting the same pair of locations (origin and destination) into each Web service and compare the shortest paths and routes they generate (some of them may be similar, some may be different, and some will likely give you more than one option for the "shortest path"). In addition, the services will also allow you to alter the route interactively by moving and repositioning junctions of the path, so you can tailor the route more to your liking. Set up a shortest route, then position your cursor at junctions along the path—you can move the nodes and reset the routes. How do they change in terms of time and distance by remaking them?

Also, for each service, note the source of the maps being generated (usually in small text at the bottom of the map). What is the copyrighted source of the maps?

from your house to the shoe store, then the shortest path from the shoe store to the music store, and finally the shortest path from the music store to the book store.

2. Finding the shortest path when you can arrange the order of visiting the stops: For instance, if the book store and the shoe store are near each other, it makes sense to rearrange your travels, so you visit them one after the other, and then drive to the music store.

Geospatial technology applications can help determine some solutions to both of these scenarios. To visit stops in a pre-determined order, the system will find the shortest route (of all possible routes) from the origin to the first stop, then choose the shortest route to travel from the first stop to the second stop, and so on. An example of this in vehicle-navigation systems is being able to set up a "Via Point" (another stop) between your origin and final destination—the device will then calculate the route from the origin (or current location) to the Via Point, then from the Via Point to the final destination.

With the ability to rearrange the order of stops, a system will evaluate options by changing the order of visiting stops to produce the overall shortest route. See Figure 8.8 for an example of how the "shortest path" changes between visiting stops in order as opposed to being able to rearrange the order of visiting the stops. Additional constraints can be placed on the problem—for instance, you may have to return home (the starting point) after visiting all of the stops. When rearranging stops, the starting point then becomes the last stop and may affect how the reordering is done. Conversely, you may want

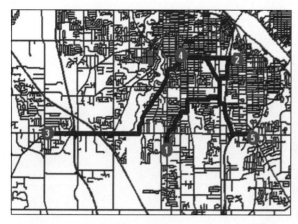

FIGURE 8.8 Two different shortest paths involving multiple stops (selected public libraries in Youngstown, Ohio): First, stops are visiting in numerical order, and in the second image, stops are rearranged to find the shortest route (without a fixed starting or ending point). (Source: Esri® ArcGIS ArcMap ArcView graphical user interface. Copyright © Esri.)

to have your starting point and ending point different from one another, and these types of parameters would have to be placed on the problem. Keep in mind, with rearranging the order of stops, the program will likely give a decent solution to the problem; however, the only way to truly identify which configuration is the best would be for the program to work through every possible arrangement—something that would be impossible with a larger number of stops.

How Are Networks Used in Geospatial Technology?

All of the things discussed in this chapter are used for a variety of applications with geospatial technology. Through GIS, these types of network data can be created and utilized in various ways. The vehicle navigation systems integrate many of these concepts together—using GPS to pinpoint the device's location on a map, then using network base data, geocoding, and shortest paths to navigate through locales worldwide. Other options on these systems involve utilizing real-time data broadcast into the device to determine areas of congestion, high traffic, construction, or road accidents. This data can then be used in the shortest-path process to route you around these types of barriers or things that would slow down your travel. These same maps and technology are being integrated into smartphones to put GPS locations, mapping, geocoding, shortest-paths, and real-time routing information capabilities into the palm of your hand (Figure 8.9). As noted, however, often these devices are only as accurate as the base maps they're utilizing, so companies provide a means for consumers to obtain regular map updates or give users the option to make their own updates.

These same types of network base maps are used online for services like MapQuest, Bing Maps, Yahoo! Maps, or Google Maps (see *Hands-on Application 8.4: Online Mapping and Routing Applications and Shortest Paths*), which allow you to perform geocoding and obtain directions for a shortest

FIGURE 8.9 A smartphone running a GPS mapping application. (Source: Pharos Science & Applications, Inc.)

FIGURE 8.10 The U.S. Capitol Building as seen from Google Maps Street View. (Source: © 2011 Google)

Street View a component of Google Maps and Google Earth that allows the viewer to see 360-degree imagery around an area on a road.

path between points. **Street View** allows a Google Maps (or Google Earth) user to examine 360 degrees of photography of a location along a street, as if your car was stopped at that location. With this function, you can examine canned photography of an address or destination when planning a stop (Figure 8.10). Cars equipped with special cameras that can capture a 360-degree view travel along roads, taking imagery along the way (see Figure 8.11 and *Hands-on Application 8.5: Examining Google Street View*). Google's even extending this Street View onto places inaccessible by cars— by attaching the same type of camera equipment to bicycles and traversing hiking and biking trails.

With this level of data availability, and geospatial networks being integrated into so many different applications, mapping (and providing accurate and up-to-date base maps) has become big business. Companies such as Tele Atlas and NAVTEQ produce these maps, and their products are often what you're accessing on an online service or from a vehicle-navigation system. Next time you access one of these kinds of services, look for the copyright

FIGURE 8.11 An example of a Google Street View car. (Source: Bob Bobster/Wikimedia)

Hands-on Application 8.5

Examining Google Street View

Google Street View is a very cool application that gives you a look at what the street network would look like if you were driving or riding past. Open your Web browser and go to Google Maps at **http://maps.google.com**, and then enter a particular address. For example, type in the address of the Rock and Roll Hall of Fame in Cleveland, Ohio (1100 Rock and Roll Blvd., Cleveland, OH) and zoom in to the map until you can see the distinctive building of the Rock Hall and its surrounding streets. Above the zoom slider on the left side of the map is an orange icon that looks like a person (called the "pegman"). Grab that with the mouse and place it on top of the streets in front of the Rock Hall (E. 9th Street). Streets that have been covered by Google Street View will be highlighted in dark blue. Place the icon on one of the available streets and the view will shift to photography of that area (see Figure 8.10 for an example). Use the mouse to move the view about for 360-degree imagery of the area. You'll also see some white lines and arrows superimposed on the road—clicking on them will move you along the street and you'll see the imagery change. Move around until you can get a good street-level view of the Rock and Roll Hall of Fame.

Examine some local areas near you—see if Google Street View is available for roads near where you live, work, or attend school. If the imagery is available, use Google Street View to take a "virtual tour" of some of the main streets of your town, or to examine street-level photographs of nearby places.

data somewhere in the map to see which company is producing the map data you're using. NAVTEQ, for instance, sends teams out in high-tech cars to collect data on the location and attributes of new roads, housing developments, and other items for updates of maps (as well as photography equipment to produce items like the Street View scenes).

Thinking Critically with Geospatial Technology 8.2

What Kind of Issues Come with Google Street View?

Open your Web browser and go to Google Maps at **http://maps.google.com**, and then enter the address of the White House (1600 Pennsylvania Avenue, Washington, D.C., 20006). When you try to move the Street View symbol to view the roads, you'll see that the streets immediately surrounding the White House are unavailable to view in Street View, presumably for security purposes. However, most of the surrounding streets can be viewed using Street View, allowing you to look at the exteriors of shops, residences, and other government buildings. Can the images being collected by Street View pose a security risk, not just for government areas, but private security of homes or businesses?

An article in the May 31, 2008, issue of the *Star Tribune* (available online at **http://www.startribune.com/lifestyle/19416279.html**) describes how the community of North Oaks, Minnesota, demanded that Google remove the Street View images obtained of the privately owned roads in the area, citing laws against trespassing. Is Street View encroaching too far onto people's private lives or not (since you could just view or record the same things by just driving down a street)?

Chapter Wrapup

Networks, geocoding, and routing are all powerful tools for use in geospatial technology. With GIS, these concepts are used in a variety of applications and businesses today. Today 911 operators can geocode the address of a call and emergency services can determine the shortest route to a destination. Delivery services can use geocoding and routing applications to quickly determine locations and reduce travel time to shortest paths. Also, the use and development of these types of techniques in vehicle navigation systems represents a rapidly changing technology that keeps getting better.

Geospatial Lab Applications 8.1 and 8.2 will use online geocoding applications, as well as Google Earth and GIS software for investigating shortest paths and TIGER/Line mapping uses. In the next chapter, we're going to start looking at a whole different aspect of geospatial technology—remote sensing. All of these overhead images that you can see on applications like Google Maps or MapQuest have to come from somewhere to get incorporated into the program, and we'll start looking at the methods behind remote sensing in the next chapter.

Important note: The references for this chapter are part of the online companion for this book and can be found at http://www.whfreeman.com/shellito1e.

Key Terms

vehicle navigation system (p. 223)

network (p. 225)

junction (p. 225)

edge (p. 225)

connectivity (p. 225)

line segment (p. 226)

street centerline (p. 226)

TIGER/Line (p. 227)

address matching (p. 229)

geocoding (p. 229)

reference database (p. 229)

parsing (p. 230)

address standardization (p. 230)

linear interpolation (p. 230)

batch geocoding (p. 231)

shortest path (p. 234)

transit cost (p. 234)

algorithm (p. 234)

Dijkstra's Algorithm (p. 234)

stops (p. 235)

Street View (p. 238)

Geocoding and Shortest Path Analysis: AEJEE Version

This chapter's lab will introduce you to the concepts of calculating a shortest path between stops along a network as well as generating directions for the path using Google Earth and Google Maps. You'll also be performing geocoding using a Web service, examining the geocoding results, then performing some basic spatial analysis of the results and a TIGER file using Google Earth and GIS.

Similar to some of the geospatial lab applications in previous chapters, two versions of this lab are provided. The first version (*Geospatial Lab Application 8.1: Geocoding and Shortest Path Analysis: AEJEE Version*) uses the free ArcExplorer Java Edition for Educators (AEJEE). The second version (*Geospatial Lab Application 8.2: Geocoding and Shortest Path Analysis: ArcGIS Version*) provides the same activities for use with ArcGIS 10.

Objectives

The goals for you to take away from this lab are:

◉ Familiarizing yourself with the shortest path and directions functions of Google Earth.

◉ Using Google Maps to alter the shortest path to account for route changes.

◉ Utilizing an online geocoding service to geocode a series of addresses, then examining the results.

◉ Using the AEJEE software to plot the results of a geocoding operation.

◉ Examining some basic spatial analysis of locations and their relation to a road network (TIGER file) in AEJEE.

Obtaining Software

◉ The current version of Google Earth (6.0) is available for free download at http://earth.google.com.

◉ The current version of AEJEE (2.3.2) is available for free download at http://edcommunity.esri.com/software/aejee.

Important note: Software and online resources sometimes change fast. This lab was designed with the most recently available version of the software at the time of writing. However, if the software or Websites have significantly changed between then and now, an updated version of this lab (using the newest versions) is available online at http://www.whfreeman.com/shellito1e.

Lab Data

Copy the folder 'Chapter8'—it contains:

- A Microsoft Excel spreadsheet featuring a series of addresses to be geocoded
- A second Microsoft Excel spreadsheet containing latitude and longitude coordinates for each address
- A .kmz file containing library locations in a format readable by Google Earth
- A TIGER 2000 line file in shapefile format

Localizing This Lab

The datasets in this lab focus on the locations of public libraries and the road network in Virginia Beach, Virginia. However, this lab can be modified to examine your local area by doing the following:

- Use your local county library's Website (or the phone book) as a source of names and addresses of local libraries and use Microsoft Excel to create a file like the one in the 'Chapter8' folder (where each column has a different attribute of data). If there are not enough local libraries around, use the addresses of something else, like coffee shops, pizza shops, or drugstores.
- A TIGER file (in shapefile format) of the roads of your county can be downloaded free from Esri at: http://arcdata.esri.com/data/tiger2000/tiger_download.cfm. Unzip the file and the shapefile will be ready to go (this was the source of the Virginia Beach TIGER file used in this lab). Coordinates are in NAD83 decimal degrees.

8.1 Google Earth's Shortest-Path Functions

1. Start **Google Earth (GE)** and "Fly To" Virginia Beach, Virginia.
2. Let's start with this scenario: You're on vacation in Virginia Beach, visiting the Old Coast Guard Museum (which fronts on the oceanfront boardwalk). You have tickets to see a show at the Norfolk Scope, located in nearby downtown Norfolk. You want to take the shortest route to get from your beachfront location to the arena. Similar to what you did in *Geospatial Lab Application 1.1*, select the **Directions** tab. Use the following addresses for directions:
   a. From: 2400 Atlantic Avenue, Virginia Beach, VA 23451 (this is the Museum)
   b. To: 201 E. Brambleton Avenue, Norfolk, VA 23510 (this is the Norfolk Scope)

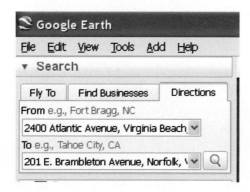

3. Click the **Begin Search** button for Google Earth to compute the shortest route between the two points.

4. The shortest path will be highlighted in purple.

5. Make sure the **Roads** layer is turned on in the Layers box.

Question 8.1 Without transcribing directions, what is the general route that Google Earth has created for the shortest route between the oceanfront museum and the arena?

6. The bottom of the search box will give information about the length of the route.

Question 8.2 What is the distance (and estimated driving time) of the route using this shortest path?

8.2 Google Maps' Shortest-Path Functions

1. Keeping in mind the shortest path that Google Earth calculated, open your Web browser and navigate to the Google Maps Website at **http:// maps.google.com**.

2. Type **"Virginia Beach, VA"** and click **Search Maps**.

3. When the map of Virginia Beach appears, select **Get Directions** and use the Old Coast Guard Museum address for option A and the Norfolk Scope address for option B.

4. Select **"By Car"** for your method of travel (the car icon) and click on **Get Directions**.

5. The shortest route between the two points will appear, highlighted in purple. It should be the same route that Google Earth calculated for you. It may give an alternate route to take as well.

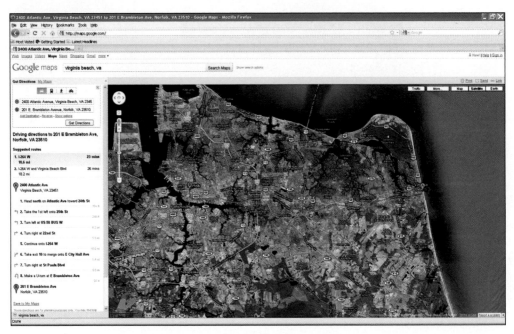

(Source: © 2010 Google, Imagery Copyright 2011 TerraMetrics, Inc. www.terrametrics.com)

6. Zoom in closer to the original point A. Scroll the mouse over the purple shortest path line and you'll see a white circle appear along the path. This lets you change the calculated route to account for any variety of factors (such as travel preference, known congestion or construction areas, rerouting to avoid areas or known delays, and so on).

7. Start by placing the mouse over the turn onto I-264 and drag the circle up to state route 58 (also known as Laskin Road or Virginia Beach Boulevard—the other major east-west road just above and parallel to I-264). The route will change by first having you drive north on Atlantic Avenue, turning west on 58, then merging back onto I-264 again a little later in the route. Even though it's a small change, you'll see that the driving distance and estimated time have changed. Answer Question 8.3. When you're done, return the circle back to its original starting point and the route will go back to how it originally was.

Question 8.3 What is the new distance (and estimated driving time) of the route using this new path? What would account for this?

8. Scroll across the map until you see where the route crossed I-64 (about two-thirds of the way between the two points). I-64 is the major highway into the Hampton Roads area. Let's throw another change into the scenario—say, for instance, that there is heavy construction on I-264 west of the I-64 intersection. At the intersection, move the circle off the

route and north onto I-64 far enough that the remainder of the route to Norfolk Scope diverts your shortest path onto I-64 headed north and not so that it returns you to I-264 west (you may have to move other circles as well if Google Maps diverts your path back onto I-264).

Question 8.4 What is the new distance (and estimated driving time) of the route using this new path?

Question 8.5 Without transcribing directions, what is the general route that Google Maps has created for the new route between the museum and the arena (taking this detour into account)?

9. Reset the route back to its original state. Now, change the route so you would have to travel south on I-64 instead of north (and still avoiding I-264).

Question 8.6 What is the new distance (and estimated driving time) of the route using this new path?

Question 8.7 Without transcribing directions, what is the general route that Google Maps has created for the new route between the oceanfront resort and the arena (taking this detour into account)?

10. All of these options can be used to model alternative routes or potential barriers or user-defined choices during shortest-path calculations. Each of these routes is a "best route," just based on the parameters (such as simulating barriers) given to the system.

8.3 Geocoding and Google Earth

Before a shortest path could be calculated, the system first had to match the addresses of the starting and ending points to their proper location. The next section of this lab will introduce a method of doing this address matching (geocoding) to create a new point layer for use.

1. Open a new tab or window with your Web browser and go to the following URL: http://batchgeo.com/.

   Important note: BatchGeo is a free online tool that will let you match multiple (or single) addresses to their road-network locations and create a point layer marking each of those address locations. In this portion of the lab, you'll still be working with the Virginia Beach area, focusing on the locations of the city's public libraries.

2. Open the **VBLibraries** file in the 'Chapter8' folder (it's in Microsoft Excel format). This lists the name and address of ten different public library venues in Virginia Beach.

3. Use the mouse to select all rows and columns in the Excel file that contain data in them (including the column headers like "Name," "Address," and so on). Copy this data.

(Source: Batchgeo.com)

4. On the BatchGeo Website, click the mouse in the **Step 1** box to select all of the data listed in it and delete it. Then paste the data from the Excel file in place of it. The Step 1 box will now show all of the data from the Excel file but separated by large spaces.

5. Click on the **Validate and Set Options** button. This will ensure that the data is in the proper format that batchgeo.com needs it to be. If an error message results, delete the information in the Step 1 box, then re-copy and paste the Excel data.

6. In Step 2, BatchGeo needs the names of the fields identified for it to begin the geocoding process (in other words, it wants to know which column contains the Address data, which column contains the Zip Code data, and so on).

7. Make sure the following fields are selected:
    a. Address – should be ADDRESS
    b. City – should be CITY

    c.   State/Province – should be STATE

    d.   Zip/Postal Code – should be ZIP

8. All other fields can stay at their defaults.

9. In Step 3, click on the **Make Google Map** button.

10. You'll see a Google Map generated of Virginia Beach that contains symbols indicating the libraries' locations. This indicates the geocoding process is complete.

11. Pan and Zoom around the map to get a feel for where the libraries were geocoded to. If you wanted, you could save this map for your own use. However, at the time if this writing, you cannot access coordinate data for the points or download a Google Earth compatible file. In order to get both of these things, you'll use a different Web resource.

12. Open your Web browser and navigate to: http://www.gpsvisualizer. com/geocoding.html.

13. On the Website, click the link for **Geocode Multiple Addresses**. A new Web interface will appear.

14. Return to Excel and re-copy all of the data, including the headers for each column.

15. Back on the GPS Visualizer Website, paste all of the Excel data in the Input box.

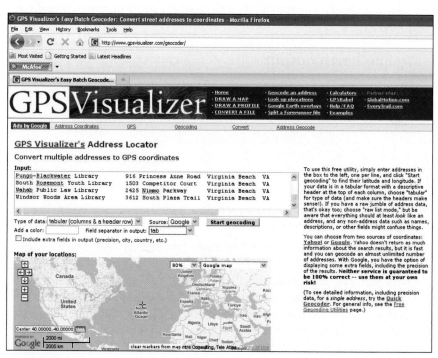

(Source: GPSVisualizer.com/Map data © 2011 Google)

16. For the various settings, use the following:
    a.   For Type of data, choose **tabular (columns & header row)**.
    b.   For Source, choose **Google**.
    c.   For Field separator in output, choose **tab**.

17. When the settings are correct, press the **Start geocoding** button.

18. You'll see a new map appear with markers showing the locations of the libraries.

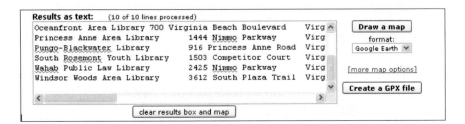

19. In the format pull-down menu under the "Draw a map" button, select **Google Earth**. This will enable the GPS Visualizer Website to create a file of the geocoded addre s that is compatible with Google Earth.

20. After choosing that option, click on the **Draw a map** button (located under the map).

21. A new Web page will open and a blue and white "KMZ" icon will be shown with a link next to it (the link will be a string of numbers ending in the letters .kmz).

Google Earth output

Your GPS data has been processed. Here's your KML or KMZ file:

 1298404026-25572-150.134.80.140.kmz

22. Click on the link and if prompted choose to **Open With Google Earth**. Click **OK** in the dialog box.

23. Also, do NOT close the browser window with the geocoded data. We'll return to it later in the lab.

24. In the Places box, there should be a new heading for Temporary Places, and under that should be a new checkbox for a .kmz file called GPS data. Turning this option on and off will toggle the display of the points representing your geocoded libraries. Basically, the GPS Visualizer Website converted the geocoded locations into a file that could be read by Google Earth. Note that a copy of this GPS data .kmz file is included with the 'Chapter8' data, just in case you're unable to access the Web resources used to generate it.

25. Once the geocoded results are added to Google Earth, in the Google Earth view you'll see each of the libraries represented by a point.

26. To make your layer easier to see, right-click on the **.kmz file** and select **Properties**. Under the **Style, Color** tab you can alter the color and size of both the icons and their labels.

27. Rolling the mouse over the top of a point will display its name. Clicking on a point will open a box with its name and address. Right-clicking on a point will allow you to access options for Directions to be calculated to that point or from that point.

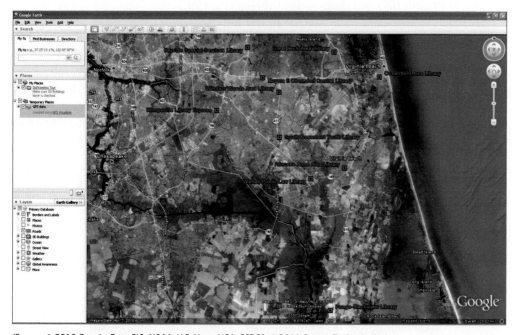

(Source: © 2010 Google, Data SIO, NOAA, U.S. Navy, NGA, GEBCO. © 2011 Europa Technologies)

28. Select the **Great Neck Area Library**, right-click on it, and select the **directions from here** option.

29. In the Search box, you'll see the Directions From option switch to say "Great Neck Area Library" with its latitude and longitude.

30. Next, locate the Princess Anne Area Library, right-click on it, and select the **directions to here** option.

31. In the Search box, the To directions will switch to the new Princess Anne destination. A new shortest path from the Great Neck Area Library to the **Princess Anne Area Library** will be calculated and displayed in purple.

Question 8.8 What is the distance (and estimated driving time) of the route between the Great Neck Area Library and the Princess Anne Area Library?

32. Locate both the South Rosemont Youth Library and the Oceanfront Area Library.

Question 8.9 What is the distance (and estimated driving time) of the route between the South Rosemont Youth Library and the Oceanfront Area Library?

8.4 Preparing Geocoded Results for GIS Analysis

The next step of the lab will investigate how to use the GIS road files (like a TIGER file) and these geocoded results in GIS.

Important note: ArcGIS 10 has full geocoding capabilities, allowing you to start with a table of addresses and create a point layer from them (without having to use a Website to perform the geocoding), but the free ArcExplorer Java Edition for Educators does not. However, you can use the results from the **gpsvisualizer.com** Website with AEJEE by converting the address points into latitude/longitude coordinates and mapping those as points in AEJEE. The **gpsvisualizer.com** Website will report back the latitude and longitude coordinates for all geocoded points. If so, it would be easy enough to copy the full contents of the "Results as text" box into an Excel spreadsheet. You would then have the addresses and their coordinates together in a file. Also, the latitude and longitude coordinates for each point are stored in the .kmz file you loaded into Google Earth.

1. Return to Google Earth, right-click on the **GPS data** .kmz file, and select **Copy**.

2. Open a text editor utility like Notepad (in the Accessories folder in Windows), and **Paste** the .kmz file there.

You'll see the code that makes up the file, including the latitude and longitude coordinates for each geocoded point. You could make two new columns in the original Excel table—one for latitude and one for longitude—and copy and paste the coordinate information for each one into its proper place in the table. It's a longer way of doing things, but in the end you'll have coordinates for each geocoded address available.

For the purposes of this lab, a separate Excel file called VBCoords is available within the 'Chapter8' folder that has all of this done for you.

Important note: The latitude and longitude computed by **gpsvisualizer. com** are decimal degree values.

8.5 Using TIGER Files and Geocoded Results in AEJEE

1. Open the **VBCoords.xls** file in Excel. Use Excel's **Save As** option to save the file not as a regular Excel file, but as a Comma Delimited (.csv) file (in Excel 2007, this is available under the "Other Formats" options in Save As). Call this new file **GISlibraries.csv** and save it on your computer in the 'Chapter8' folder.

2. GISlibraries.csv now contains all the library address information, plus a latitude and longitude location value for each library. This information will be used to plot their locations in AEJEE.

3. Start **AEJEE**. From the 'Chapter8' folder, add the **tgr51810lka.shp** shapefile to the map (see *Geospatial Lab Application 5.1* for the basics of starting AEJEE and adding data to it). This shapefile is a TIGER 2000 file of the Virginia Beach road network.

4. Open the TIGER file's attribute table. You'll see it contains 19,026 records, each representing a link of the city's road network. You'll also see the standard TIGER file information of address ranges, census feature class codes, and so on assigned to each link. You can close the attribute table for now.

5. Select the **View** pull-down menu and choose **Add Event Theme**.

6. Choose the **GISlibraries.csv** file as the Table (AEJEE will use this as the source of the latitude/longitude values).

7. Select **LONG** for the X Field.

8. Select **LAT** for the Y Field.

9. Select the **'Chapter8'** folder as the Output Dir (where AEJEE will place its new shapefile of points).

10. Choose an appropriate Style, Color, and Size for the appearance of the resulting points.

11. Click **OK** when all settings are ready.

12. A new point layer will be created (called GISlibraries), with each point representing the location of a library in Virginia Beach. Open this new layer's attribute table and you'll see that all of the data from the file with the addresses has been importing into attribute table format.

8.6 Analysis of Geocoded Results in AEJEE

In this section, you will start performing some basic spatial analysis to examine the relationship between the library locations and the road network (see *Geospatial Lab Application 6.1* for details on how to build queries and buffers using AEJEE).

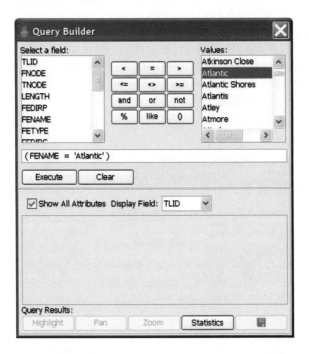

1. First, use AEJEE to build a query selecting all road segments that have their name (FENAME) equal to 'Atlantic' (Atlantic Avenue is a main north-south road that runs parallel to the boardwalk along the Virginia Beach oceanfront). Even though there are more than 100 records, have AEJEE display all the unique names. Keep in mind that to build a query, you will first have to click on the file (**tgr51810lka**) in AEJEE's Table of Contents.

2. If AEJEE asks, indicate that you want to see all records.

3. You'll see the road segments that make up Atlantic Avenue appear in the yellow color of selected AEJEE objects.

4. Next, build a 2-mile buffer around the selected features of the road. Use the buffer to select libraries that are within the buffer.

Question 8.10 How many libraries are within 2 miles of Atlantic Avenue? Which libraries are these?

5. Clear the selected features from the roads and libraries and then select all road segments with their name equal to 'Virginia Beach.' This will

select the segments that comprise Virginia Beach Boulevard, a major east-west corridor that runs through the center of Virginia Beach from the oceanfront on the east straight through into Norfolk on the west.

6. Construct a 2-mile buffer around the selected features of road and use the buffer to select libraries that are within the buffer.

Question 8.11 How many libraries are within two miles of Virginia Beach Boulevard? Which libraries are these?

7. Clear the selected features and build a final query to select all road segments with their name equal to 'Independence.' This will select all the segments that comprise Independence Boulevard, a key north-south road in the middle of Virginia Beach.

8. Construct a 2-mile buffer around the selected features of road and use the buffer to select libraries that are within the buffer.

Question 8.12 How many libraries are within two miles of Independence Boulevard? Which libraries are these?

Closing Time

This lab demonstrated several types of features associated with geospatial network data, including calculating a shortest path, geocoding, and using geocoded results in conjunction with a road network file in GIS. Chapter 9 will switch gears and introduce some new concepts dealing with remote sensing of Earth (and the roads built on top of it).

- Exit Google Earth by selecting **Exit** from the **File** pull-down menu.
- Also, exit AEJEE by selecting **Exit** from the **File** pull-down menu.
- There's no need to save any data in this lab.

Geocoding and Shortest Path Analysis: ArcGIS Version

This chapter's lab will introduce you to the concepts of calculating a shortest path between stops along a network as well as generating directions for the path using Google Earth and Google Maps. You'll also be performing geocoding using a Web service, examining the geocoding results, then performing some basic spatial analysis of the results and a TIGER file using Google Earth and ArcGIS.

The previous version of this lab (*8.1 Geospatial Lab Application: Geocoding and Shortest Path Analysis: AEJEE Version*) uses the free ArcExplorer Java Edition for Educators (AEJEE). However, this lab provides the same activities for use with ArcGIS 10.

Objectives

The goals for you to take away from this lab are:

◉ Familiarizing yourself with the shortest path and directions functions of Google Earth.

◉ Using Google Maps to alter the shortest path to account for route changes.

◉ Utilizing an online geocoding service to geocode a series of addresses, then examining the results.

◉ Using ArcGIS to plot the results of a geocoding operation.

◉ Examining some basic spatial analysis of locations and their relation to a road network (TIGER file) in ArcGIS.

Obtaining Software

◉ The current version of Google Earth (6.0) is available for free download at http://earth.google.com.

◉ The current version of ArcGIS (10) is not freely available for use. However, instructors affiliated with schools that have a campus-wide software license may request a 1-year student version of the software online at http://www.esri.com/industries/apps/education/offers/promo/index.cfm.

Important note: Software and online resources sometimes change fast. This lab was designed with the most recently available version of the software at the time of writing. However, if the software or Websites have significantly changed between then and now, an updated version of this lab (using the newest versions) is available online at: http://www.whfreeman.com/shellito1e.

Lab Data

Copy the folder 'Chapter8'—it contains:

- A Microsoft Excel spreadsheet featuring a series of addresses to be geocoded
- A second Microsoft Excel spreadsheet containing latitude and longitude coordinates for each address
- A .kmz file containing library locations in a format readable by Google Earth.
- A TIGER 2000 line file in shapefile format

Localizing This Lab

The datasets in this lab focus on the locations of public libraries and the road network in Virginia Beach, Virginia. However, this lab can be modified to examine your local area by doing the following:

- Use your local county library's Website (or the phone book) as a source of names and addresses of local libraries and use Microsoft Excel to create a file like the one in the 'Chapter8' folder (where each column has a different attribute of data). If there are not enough local libraries around, use the addresses of something else, like coffee shops, pizza shops, or drugstores.
- A TIGER file (in shapefile format) of the roads of your county can be downloaded for free from Esri at http://arcdata.esri.com/data/tiger2000/tiger_download.cfm. Unzip the file and the shapefile will be ready to go (this was the source of the Virginia Beach TIGER file used in this lab). Coordinates are in NAD83 decimal degrees.

8.1 Google Earth's Shortest-Path Functions

1. Start **Google Earth (GE)** and **"Fly To"** Virginia Beach, Virginia.
2. Let's start with this scenario: You're on vacation in Virginia Beach, visiting the Old Coast Guard Museum (which fronts on the oceanfront boardwalk). You have tickets to see a show at the Norfolk Scope, located in nearby downtown Norfolk. You want to take the shortest route to get from your beachfront location to the arena. Similar to what you did in *Geospatial Lab Application 1.1*, select the **Directions** tab. Use the following addresses for directions:

    a. From: 2400 Atlantic Avenue, Virginia Beach, VA 23451 (this is the Museum)

    b. To: 201 E. Brambleton Avenue, Norfolk, VA 23510 (this is the Norfolk Scope)

3. Click the **Begin Search** button for Google Earth to compute the shortest route between the two points.

4. The shortest path will be highlighted in purple.

5. Make sure the **Roads** layer is turned on in the Layers box.

Question 8.1 Without transcribing directions, what is the general route that Google Earth has created for the shortest route between the oceanfront museum and the arena?

6. The bottom of the search box will give information about the length of the route.

Question 8.2 What is the distance (and estimated driving time) of the route using this shortest path?

8.2 Google Maps' Shortest-Path Functions

1. Keeping in mind the shortest path that Google Earth calculated, open your Web browser and navigate to the Google Maps Website at **http:// maps.google.com**.

2. Type "**Virginia Beach, VA**" and click **Search Maps**.

3. When the map of Virginia Beach appears, select **Get Directions** and use the Old Coast Guard Museum address for option A and the Norfolk Scope address for option B.

4. Select "**By Car**" for your method of travel (the car icon) and click on **Get Directions**.

5. The shortest route between the two points will appear, highlighted in purple. It should be the same route that GE calculated for you. It may give an alternate route to take as well.

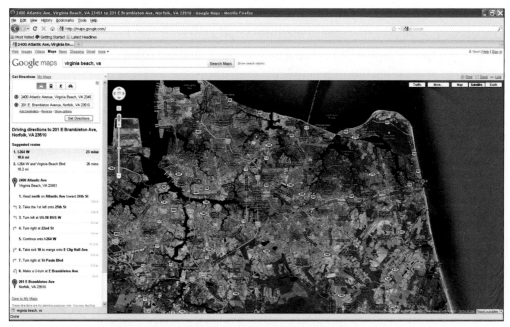

(Source: © 2010 Google, Imagery Copyright 2011 TerraMetrics, Inc. www.terrametrics.com)

6.   Zoom in closer to the original point A. Scroll the mouse over the purple shortest path line and you'll see a white circle appear along the path. This lets you change the calculated route to account for any variety of factors (such as travel preference, known congestion or construction areas, rerouting to avoid areas or known delays, and so on).

7.   Start by placing the mouse over the turn onto I-264 and drag the circle up to state route 58 (also known as Laskin Road or Virginia Beach Boulevard—the other major east-west road just above and parallel to I-264). The route will change by first having you drive north on Atlantic Avenue, turning west on 58, then merging back onto I-264 again a little later in the route. Even though it's a small change, you'll see that the driving distance and estimated time have changed. Answer Question 8.3. When you're done, return the circle back to its original starting point and the route will go back to how it originally was.

Question 8.3   What is the new distance (and estimated driving time) of the route using this new path? What would account for this?

8.   Scroll across the map until you see where the route crossed I-64 (about two-thirds of the way between the two points). I-64 is the major highway into the Hampton Roads area. Let's throw another change into the scenario—say, for instance, that there is heavy construction on I-264 west of the I-64 intersection. At the intersection, move the circle off the route and north onto I-64 far enough that the remainder of the route to Norfolk Scope diverts your shortest path headed north onto I-64 and not

so that it returns you to I-264 west (you may have to move other circles as well if Google Maps diverts your path back onto I-264).

Question 8.4 What is the new distance (and estimated driving time) of the route using this new path?

Question 8.5 Without transcribing directions, what is the general route that Google Maps has created for the new route between the museum and the arena (taking this detour into account)?

9.   Reset the route back to its original state. Now, change the route so you would have to travel south on I-64 instead of north (and still avoiding I-264).

Question 8.6 What is the new distance (and estimated driving time) of the route using this new path?

Question 8.7 Without transcribing directions, what is the general route that Google Maps has created for the new route between the oceanfront resort and the arena (taking this detour into account)?

10. All of these options can be used to model alternative routes or potential barriers or user-defined choices during shortest-path calculations. Each of these routes is a "best route," just based on the parameters (such as simulating barriers) given to the system.

8.3 Geocoding and Google Earth

Before any type of shortest path could be calculated, the system first had to match the addresses of the starting and ending points to their proper location. The next section of this lab will introduce a method of doing this address matching (geocoding) to create a new point layer for use.

1.   Open a new tab or window with your Web browser and go to the URL **http://batchgeo.com/**.

Important note: BatchGeo is a free online tool that will let you match multiple (or single) addresses to their road-network locations and create a point layer marking each of those address locations. In this portion of the lab, you'll still be working with the Virginia Beach area, focusing on the locations of the city's public libraries.

2.   Open the **VBLibraries** file in the 'Chapter8' folder (it's in Microsoft Excel format). This lists the name and address of ten different public library venues in Virginia Beach.

3.   Use the mouse to select all rows and columns in the Excel file that contain data in them (including the column headers like "Name," "Address," and so on). Copy this data.

4.   On the BatchGeo Website, click the mouse in the Step 1 box to select all of the data listed in it and delete it. Then paste the data from the Excel

file in place of it. The Step 1 box will now show all of the data from the
Excel file but separated by large spaces.

(Source: BatchGeo.com)

5. Click on the **Validate and Set Options** button. This will ensure that the
   data is properly in the format that batchgeo.com needs it to be in. If an
   error message results, delete the information in the Step 1 box, then re-
   copy and paste the Excel data.

6. In Step 2, BatchGeo needs the names of the fields identified for it to
   begin the geocoding process (in other words, it wants to know which
   column contains the Address data, which column contains the Zip Code
   data, and so on).

7. Make sure the following fields are selected:
   a. Address – should be ADDRESS
   b. City – should be CITY
   c. State/Province – should be STATE
   d. Zip/Postal Code – should be ZIP

8. All other fields can stay at their defaults.

9. In Step 3, click on the **Make Google Map** button.

10. You'll see a Google Map generated of Virginia Beach that contains symbols indicating the libraries' locations. This indicates the geocoding process is complete.

11. Pan and Zoom around the map to get a feel for where the libraries were geocoded to. If you wanted, you could save this map for your own use. However, at the time if this writing, you cannot access any coordinate data for the points or download a Google Earth compatible file. In order to get both of these things, you'll use a different Web resource.

12. Open your Web browser and navigate to: http://www.gpsvisualizer. com/geocoding.html.

13. On the Website, click the link for **Geocode Multiple Addresses**. A new Web interface will appear.

14. Return to Excel and re-copy all of the data, including the headers for each column.

15. Back on the GPS Visualizer Website, paste all of the Excel data in the Input box.

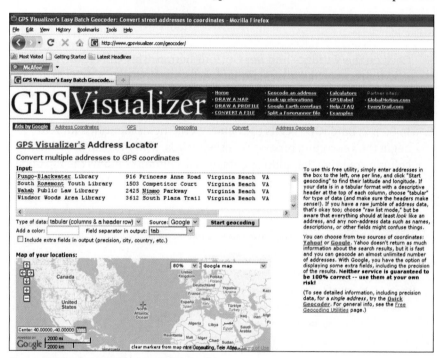

(Source: GPSVisualizer.com/Map data © 2011 Google)

16. For the various settings, use the following:

    a. For Type of data, choose **tabular (columns & header row)**.

    b. For Source, choose **Google**.

    c. For Field, separator in output choose **tab**.

17. When the settings are correct, press the **Start geocoding** button.
18. You'll see a new map appear with markers showing the locations of the libraries.

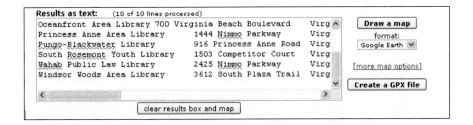

19. In the format pull-down menu under the "Draw a map" button, select **Google Earth**. This will enable the GPS Visualizer Website to create a file of the geocoded addresses that is compatible with Google Earth.
20. After choosing that option, click on the **Draw a map** button (located under the map).
21. A new Web page will open and a blue and white "KMZ" icon will be shown with a link next to it (the link will be a string of numbers ending in the letters .kmz).

Google Earth output

Your GPS data has been processed. Here's your KML or KMZ file:

 1298404026-25572-150.134.80.140.kmz

22. Click on the link and if prompted choose to **Open With Google Earth**. Click **OK** in the dialog box.
23. Also, do NOT close the browser window with the geocoded data. We'll return to it later in the lab.
24. In the Places box, there should be a new heading for Temporary Places, and under that should be a new checkbox for a .kmz file called GPS data. Turning this option on and off will toggle the display of the points representing your geocoded libraries. Basically, the GPS Visualizer Website converted the geocoded locations into a file that could be read by Google Earth. Note that a copy of this GPS data .kmz file is included with the 'Chapter8' data, just in case you're unable to access the Web resources used to generate it.
25. Once the geocoded results are added to Google Earth, in the Google Earth view you'll see each of the libraries represented by a point.

26. To make your layer easier to see, right-click on the **.kmz file** and select **Properties**. Under the **Style, Color** tab you can alter the color and size of both the icons and their labels.

27. Rolling the mouse over the top of a point will display its name. Clicking on a point will open a box with its name and address. Right-clicking on a point will allow you to access options for Directions to be calculated to that point or from that point.

(Source: ©2010 Google, Data SIO, NOAA, U.S. Navy, NGA, GEBCO. © 2011 Europa Technologies)

28. Select the **Great Neck Area Library**, right-click on it, and select the **directions from here** option.

29. In the Search box, you'll see the Directions From option switch to say "Great Neck Area Library" with its latitude and longitude.

30. Next, locate the **Princess Anne Area Library**, right-click on it, and select the **directions to here** option.

31. In the Search box, the To directions will switch to the new Princess Anne destination. A new shortest path from the Great Neck Area Library to the Princess Anne Area Library will be calculated and displayed in purple.

Question 8.8 What is the distance (and estimated driving time) of the route between the Great Neck Area Library and the Princess Anne Area Library?

32. Locate both the South Rosemont Youth Library and the Oceanfront Area Library.

Question 8.9 What is the distance (and estimated driving time) of the route between the South Rosemont Youth Library and the Oceanfront Area Library?

8.4 Preparing Geocoded Results for GIS Analysis

The next step of the lab will investigate how to use the GIS road files (like a TIGER file) and these geocoded results in GIS.

Important note: ArcGIS has full geocoding capabilities, allowing you to start with a table of addresses and create a point layer from them (without having to use a Website to perform the geocoding). However, you can use the results from the **gpsvisualizer.com** Website with ArcGIS by converting the address points into latitude/longitude coordinates and mapping those as points in ArcGIS. The **gpsvisualizer.com** Website will report back the latitude and longitude coordinates for all geocoded points. If so, it would be easy enough to copy the full contents of the "Results as text" box into an Excel spreadsheet. You would then have the addresses and their coordinates together in a file. Also, the latitude and longitude coordinates for each point are stored in the .kmz file you loaded into Google Earth.

1. Return to Google Earth, right-click on the **GPS data** .kmz file, and select **Copy**.

2. Open a text editor utility like Notepad (in the Accessories folder in Windows), and **Paste** the .kmz file there.

You'll see the code that makes up the file, including the latitude and longitude coordinates for each geocoded point. You could make two new columns in the original Excel table—one for latitude and one for longitude—and copy and paste the coordinate information for each one into its proper place in the table. It's a longer way of doing things, but in the end you'll have coordinates for each geocoded address available.

For the purposes of this lab, a separate Excel file called VBCoords is available within the 'Chapter8' folder that has all of this done for you.

Important note: The latitude and longitude computed by **gpsvisualizer. com** are decimal degree values.

8.5 Using TIGER Files and Geocoded Results in ArcGIS

1. Start **ArcMap**. From the 'Chapter8' folder, add the **tgr51810lka.shp** shapefile to the map (see *Geospatial Lab Application 5.2* for the basics of starting ArcMap and adding data to it). This shapefile is a TIGER 2000 file of the Virginia Beach road network.

2. Open the TIGER file's attribute table. You'll see it contains 19,026 records, each representing a link of the city's road network. You'll also see the standard TIGER file information of address ranges, census feature class codes, and so on assigned to each link. You can close the attribute table for now.

3. Open **ArcToolbox** by selecting its icon from the Standard toolbar (it's the small red box icon):

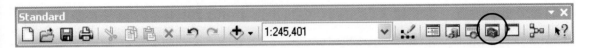

4. ArcToolbox is a dockable window that contains a multitude of useful tools for ArcGIS. From the **Data Management** tools, select **Layers and Table Views**, then select **Make XY Event Layer**.

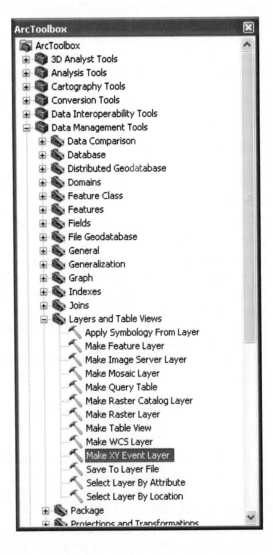

5. An Event Layer will be a temporary layer created from a set of coordinates. In this case, you'll be plotting points based on the latitude and longitude coordinates of the geocoded addresses.

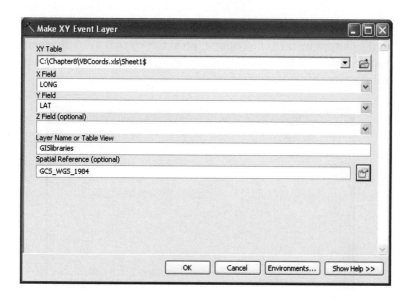

6. Choose the **VBCoords.xls** file (the **Sheet1$** part of it) as the XY Table (that ArcMap will use as the source of the latitude/longitude values).

7. Select **LONG** for the X Field.

8. Select **LAT** for the Y Field.

9. Leave the Z Field blank.

10. Type **GISlibraries** for the Layer Name.

11. From the Spatial Reference choices, select **Geographic Coordinate Systems**, then select **World**, then select **WGS84.prj**.

12. Click **OK** when all settings are ready.

13. A new point layer will be created called GISlibraries, with each point representing the location of a library in Virginia Beach. Open this new layer's attribute table and you'll see that all of the data from the file with the addresses has been imported into ArcGIS attribute table format.

14. In order to use this layer for querying, it will first have to be converted over to a format like a shapefile for use. To do this, first right-click on the **GISlibraries** file and select **Data**, then **Export Data**. A new Export Data dialog will appear.

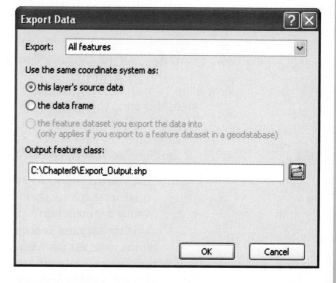

15. Use the same coordinate system as **this layer's source data.**

16. In the Output feature class option, press the **open** button. By default, ArcGIS will want to name the new shapefile Export_Output.shp, but you will change that in the next step.

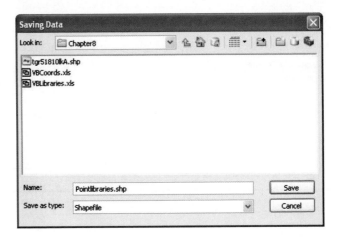

17. In the Saving Data dialog, navigate to the 'Chapter8' folder and call the new file you'll be creating **Pointlibraries** instead of Export_Output.

18. In the Save as type: option, use the pull-down menu to select **Shapefile.**

19. Click **Save** when everything's ready. This will return you to the Export Data dialog, except your Output feature class should now be Pointlibraries.shp. If everything's okay, click **OK** to continue. ArcGIS will export your event layer into a shapefile called Pointlibraries.

20. Click **Yes** when prompted if you want to add the exported data as a new layer. Pointlibraries will appear in the Table of Contents.

21. Turn off the old **GISlibraries** event layer. You'll be working with the Pointlibraries shapefile in the next part.

8.6 Some Basic Analysis of Geocoded Results in ArcMap

In this part, you will start performing some basic spatial analysis to examine the relationship between the library locations and the road network (see *Geospatial Lab Application 6.2* for details on how to build queries based on attributes and queries based on location using ArcMap).

1. First, use the **Select By Attributes** tool to build a query selecting all road segments that have their name (FENAME) equal to 'Atlantic' (Atlantic Avenue is a main north-south road that runs parallel to the boardwalk along the Virginia Beach oceanfront). Press the **Get Unique Values** button to access the names of all the roads. You'll see the road segments that make up Atlantic Avenue appear in the default cyan color of selected ArcMap features.

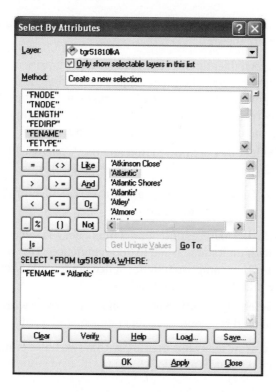

2. Next, use **Select By Location** to find all library points (the Pointlibraries) within 2 miles of the selected road segments of Atlantic Avenue.

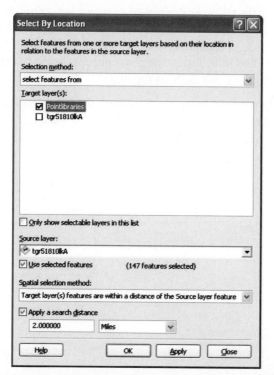

Question 8.10 How many libraries are within 2 miles of Atlantic Avenue? Which libraries are these?

3. Clear the selected features from the roads and libraries and then select all road segments with their name equal to 'Virginia Beach.' This will select the segments that comprise Virginia Beach Boulevard, a major east-west corridor that runs through the center of Virginia Beach from the oceanfront on the east straight through into Norfolk on the west.

4. Next, use **Select By Location** to find all library points within 2 miles of the selected road segments of Virginia Beach Boulevard.

Question 8.11 How many libraries are within two miles of Virginia Beach Boulevard? Which libraries are these?

5. Clear all selected features and build a final query to select all road segments with their name equal to 'Independence.' This will select all the segments that comprise Independence Boulevard, a key north-south road in the middle of Virginia Beach.

6. Next, use **Select By Location** to find all library points within 2 miles of the selected road segments of Independence Boulevard.

Question 8.12 How many libraries are within two miles of Independence Boulevard? Which libraries are these?

Closing Time

This lab demonstrated several types of features associated with geospatial network data, including calculating a shortest path, geocoding, and using geocoded results in conjunction with a road-network file in GIS. Chapter 9 will switch gears and introduce some new concepts dealing with remote sensing of Earth (and the roads built on top of it).

◉ Exit Google Earth by selecting **Exit** from the **File** pull-down menu. **Exit** ArcGIS as well.

◉ There's no need to save any data in this lab.

9

Remotely Sensed Images from Above

Where Aerial Photography Came from, Color Infrared Photos, Orthophotos, Oblique Photos, Visual Image Interpretation, and Photogrammetric Measurements

Whenever you take a picture of something with a digital camera, somehow the image is captured and stored as a file on a memory card. Taking a picture at a birthday party or on vacation is actually a form of remote sensing—acquiring data without being in direct contact with the subject. For instance, you don't shove the camera right on top of the birthday cake to take a picture of it, you stand back a distance to do it. In essence, the camera is "sensing" (taking the picture) while you're some distance ("remotely") away from the target. In this same way, your eyes function as "remote sensing" devices—when you're looking at this book, you don't place your eyeball directly on the page. Instead, your eyes are "sensing" the data and they're "remotely" located about a foot away from the page. Both a camera and your eyes are performing "remote sensing" as they're acquiring information or data (in this case, visual information) from a distance away without actually making contact with the item.

However, when it comes to **remote sensing** in geospatial technology, things are a little more specific. In remote sensing, the data being acquired is information about the light energy being reflected off of a target. In this case, your eyes can still function like remote sensing devices, as what you're actually seeing is light being reflected off objects around you processed by your eyes. A digital camera does the same thing by capturing the reflection of light from whatever you're taking a picture of. In 1826, a French inventor named Joseph Niepce took the first photograph, capturing an image of the courtyard of his home. Photography has come a long way since those days (when it took 8 hours to expose the film) to the point where digital cameras

remote sensing the process of collecting information related to the reflected or emitted electromagnetic energy from a target by a device a considerable distance away from that target from an aircraft or spacecraft.

are now commonplace and it's difficult to find a cell phone without a built-in camera. In many ways, you can think of a camera as the original device for remote sensing—it captures data from the reflection of visible light nearly instantaneously from a distance away from the target.

Taking pictures of the ground with such an instrument started not long after cameras became available, leading to the development of **aerial photography**. Capturing images of the ground from the sky has been going on for over 150 years—first using balloons and kites, then by utilizing aircraft and rockets, then using satellites to get images from space. When dealing with remote sensing in geospatial technologies, we're concerned with acquiring information from a platform on some type of aircraft or spacecraft. Currently, these types of imagery are obtained in both the public and private sector.

aerial photography
taking photographs of objects on the ground from an airborne platform.

Whether the use has been for military reconnaissance, surveillance, studies of the landscape, planning purposes, or even the good old "hey, I can see my house," people have been acquiring images of Earth from above for a long time. Today's aerial photographs are used for all manner of applications—for instance, aerial imagery serves as a base source for creation and updating of geospatial data and maps (such as USGS topographic maps or road networks). Some of the extremely crisp high-resolution imagery you see on Google Earth is taken from aerial photography sources. In this chapter, we'll focus on how aerial images are captured and analyzed, as well as getting used to looking at Earth from above.

How Did Aircraft Photography Develop?

A French photographer named Gaspar Felix Tournachon (also known as "Nadar") took the first aerial photograph in 1858. Tournachon captured an image of the landscape outside Paris while tethered above the ground in a balloon. This first aerial photograph no longer exists, but this was clearly the beginning of something big. In 1860, aerial photographs were taken of Boston from a photographer in a balloon above the city (Figure 9.1). However, development times, problems with landing at the proper location, and interference from the balloon's gas with the photo plates all contributed to balloon photography not being the best medium for capturing aerial photographs.

By the end of the nineteenth century, photographic equipment was attached to kites (or a series of kites) and used to remotely photograph the landscape while the photographer was safely on the ground. A famous example of kite photography comes from 1906, when a photographer named George Lawrence used a set of kites 1000 feet above San Francisco Bay to capture an aerial image he termed "San Francisco in Ruins" which showed the aftermath of the 1906 earthquake (Figure 9.2). These early uses of taking images from the air set the stage for the use of aircraft to take aerial photographs.

FIGURE 9.1 The earliest surviving aerial photograph, taken of Boston in 1860. (Source: Boston Public Library)

Flying with an aircraft started with the first flight of the Wright Brothers at Kitty Hawk, North Carolina, in 1903. In 1908, a passenger aboard an aircraft was able to capture aerial imagery for the first time over France. By the time of World War I, aerial photography from an airplane was commonly used for mapmaking and planning of military tactics. Aerial photography of the battlefields allowed for the patterns of trenches to be determined, troop movements and encampments to be plotted, and the location of artillery or

FIGURE 9.2 "San Francisco in Ruins" kite aerial photo taken in 1906. (Source: Library of Congress, Prints and Photographs Division [LC-USZC4-3870 DLC])

FIGURE 9.3 World War I aerial photography showing trench formations. (Source: National Air and Space Museum, National Air and Space Museum Archives)

supplies to be found. Countless aerial photos were shot and developed over the course of the war, and aerial photography greatly contributed to military intelligence during World War I. See Figure 9.3 for an example of trench development captured by an aerial camera.

Aerial photography continued to be extremely important for tactical planning and reconnaissance during World War II. Both the Allies and the Axis forces depended heavily on aerial photography for obtaining vital information about locations prior to military action at those places, and it became a key reconnaissance technique during the war. For instance, a V-2 rocket base in Peenemunde, Germany, was the subject of British aerial photography in June of 1943, which allowed for bombing strikes to destroy the base less than two months later. Similarly, aerial reconnaissance was used in obtaining photographs of dams prior to bombing missions sent to destroy them. A huge amount of aerial photography was performed during the war, and a good source of historical imagery is The Aerial Reconnaissance Archives (TARA), currently held at the Royal Commission on the Ancient and Historical Monuments of Scotland (RCAHMS), which hosts a large archive of European World War II aerial photos (see *Hands-on Application 9.1: World War II Aerial Photography Online* for a glimpse of these historic photos).

Hands-on Application 9.1

World War II Aerial Photography Online

The National Collection of Aerial Photography is a great online resource for viewing aerial photography of Europe during World War II. Open your Web browser and go to **http://aerial.rcahms.gov.uk** and select the option for TARA, which contains imagery from several countries. Select a location (such as France or Germany), then select a region. Check out some of the countless historic photos that TARA has archived online.

Post–World War II aerial photography remained critical for military applications, especially for spy planes. The 1950s saw the development of the U-2 spy plane, capable of flying at 70,000 feet and photographing the ground below. The U-2 was used for taking aerial photos over the Soviet Union, but also enabled the acquisition of imagery during the Cuban Missile Crisis, which showed the placement of missile launchers at key points in Cuba in October 1962 (**Figure 9.4**). Spy plane technology continued to advance with the development of the Lockheed SR-71 Blackbird, which could fly at 80,000 feet or higher, while traveling at more than three times the speed of sound (**Figure 9.5** on page 274), thus enabling it to go higher and faster than the U-2. The Blackbird was an essential aerial photography plane through the Cold War and beyond, until it was retired in the late 1990s.

Today, digital aerial photography (among many other types of remotely sensed data) is obtained by unmanned aerial vehicles (**UAVs**). These planes

UAV unmanned aerial vehicle—a reconnaissance aircraft that is piloted from the ground via remote control.

FIGURE 9.4 Aerial imagery of Cuba from 1962, showing missile facilities during the Cuban Missile Crisis. (Source: U.S. Air Force Photo)

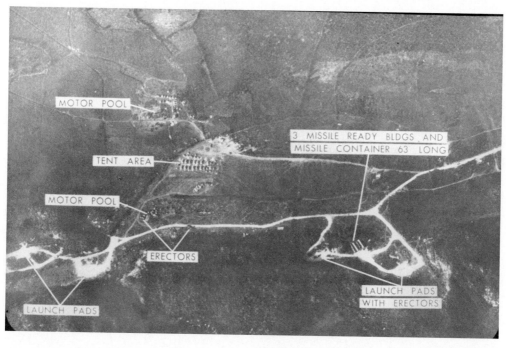

FIGURE 9.5 The SR-71
Blackbird in action. (Source:
USAF/Judson Brohmer)

are piloted from the ground via remote control and are currently being used by the military for aerial reconnaissance in Iraq and Afghanistan. There are numerous types of UAVs, ranging from smaller planes like the Predator to larger ones such as the RQ-4 Global Hawk (Figure 9.6). Outside of military uses, aerial photography is extensively used by state and local governments or private companies for data collection (see *Hands-on Application 9.2: The National Aerial Photography Program* for an example of an online government aerial photo archive).

FIGURE 9.6 An RQ-4
Global Hawk Unmanned
Aerial Vehicle. (Source: U.S.
Air Force photo/Stacey Knott)

Thinking Critically with Geospatial Technology 9.1

How Can UAVs Be Used for Security Purposes?

UAVs are not limited to military operations. Aircraft that can be piloted via remote control and equipped with multiple types of aerial surveillance equipment have a variety of other uses. For example, an article in the January 2010 issue of *Popular Science* (available online at **http://www.popsci.com/technology/article/2010-01/british-police-monitor-civilians-uavs-2012**) indicates that the British police intend to use UAVs for security purposes, especially during the 2012 Olympics in London. How else can UAVs be utilized for functions such as border security or law enforcement? How could police or federal agents utilize UAV capabilities for gathering intelligence about dangerous situations or acquiring reconnaissance before entering an area? Also, if UAVs are being used to remotely monitor civilian conditions, what potential is there for abuse of these types of data-collection platforms? Is UAV surveillance in a domestic urban environment an invasion of privacy or not?

Hands-on Application 9.2

The National Aerial Photography Program

The National Aerial Photography Program (NAPP) was a federal program sponsored by several agencies (including the U.S. Department of Agriculture and the U.S. Department of the Interior) to provide regular aerial photography of the United States. NAPP imagery consists of 1-meter-resolution photography and is now available online via the USGS's EarthExplorer utility (also used back in *Hands-on Application 5.3: USGS Digital Line Graphs*). Open your Web browser and go to **http://edcsns17.cr.usgs.gov/NewEarthExplorer/** to begin (make sure your computer can accept cookies for the Website to load properly).

Using Earth Explorer is a four-step process. First, enter your search criteria, such as a place name (for example, try New York City, NY). Next, select the Data Sets tab and choose the datasets you want to search (in this case, expand the Aerial Photography option and select NAPP). In the third option, Additional Criteria, you could add other information to your search request to help narrow it down (if needed). Lastly, click on Results to see what NAPP imagery is available for the place you've searched for. A thumbnail preview of the images will be available that you can click on to expand so that you can view the image. The USGS requires users to be logged in with an account before you can access download options for the photos. Use EarthExplorer to view what NAPP imagery is available for your local area or home address, and then examine the results.

What Are the Different Types of Aerial Photos?

When you're flying in a plane and look out the window, you're looking down on the landscape below. If you had a camera and took a picture straight down from under the plane, you'd be capturing a **vertical photo**. Take a look back at Figure 9.4—the Cuban landscape as seen by the U-2 is photographed like you

vertical photo an aerial photo in which the camera is looking down at a landscape.

nadir the location under the camera in aerial photography.

panchromatic black-and-white aerial imagery.

CIR photo color infrared photo—a photo where infrared reflection is shown in shades of red, red reflection is shown in shades of green, and green reflection is shown in shades of blue.

were looking down from the plane. Many aerial photos are vertical (although there may be some minor tilting from the camera). The spot directly under the camera is referred to as **nadir**.

Aerial photos can also be **panchromatic** or color. Panchromatic imagery is only capturing the visible portion of light in its entirety. As a result, panchromatic aerial photos will be grayscale (that is, black and white—like the Cuban aerial photo). Color imagery is capturing the three main bands of visible light—red, green, and blue—and the colors are composited together in the digital imagery (or film) as a true color composite (we'll discuss much more about light and composites in Chapter 10).

A distinctive type of color aerial photo is a color infrared **(CIR) photo**. A CIR image is captured with a special type of film sensitive to infrared light. Infrared energy is invisible to the naked eye, but special film processes can capture it. In order for us to see infrared energy reflection, a special filter is applied to the film and the end result is a CIR photo, wherein near-infrared (NIR) reflection is displayed in the color red, red reflection is displayed with the color green, and green reflection is displayed in the color blue. Blue reflection is not shown in the image, as it is blocked by the filter and displayed as black (**Figure 9.7**).

Being able to examine CIR imagery is very useful in environmental studies, for instance, to use the infrared reflection to examine the relative health of vegetation. For example, examine the CIR aerial photo of Burley, Idaho, in **Figure 9.8.** The bright red sections of the image are indicative of areas with a high amount of NIR reflection (since NIR reflection is being shown with the color red on the image), likely grass, trees, or healthy agricultural fields (all elements that would be reflecting large amounts of NIR light). The river appears black since water heavily absorbs all types of light, except the color blue

FIGURE 9.7 How the actual reflection of blue, green, red, and NIR light appears in its corresponding colors in a CIR image.

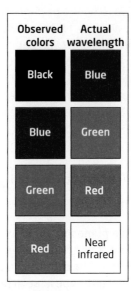

Observed colors	Actual wavelength
Black	Blue
Blue	Green
Green	Red
Red	Near infrared

FIGURE 9.8 A CIR aerial image of Burley, Idaho. (Source: NASA Airborne Science Program image acquired by NASA ER-2 aircraft 26 August 1983)

(which is then blocked in the CIR photo and shows up as black). Developed areas appear as a white or cyan color, the same with barren lands that have been harvested or not yet planted.

No matter what type of aerial photo is being used, the big problem is that the photo cannot be used as a map. Even though it shows all the features you might need (like roads or footprints of buildings) you can't rely on the photo to be a map for one reason—the photo doesn't have the same scale at every spot on the image. As a plane flies over the surface, some terrain features will be closer to the camera and some will be further away—and thus, some areas will have a larger or smaller scale than others. When looking down, the center of the photo is referred to as its **principal point**. In aerial photos, you'll see tall objects (such as terrain relief, towers, or buildings) having a tendency to "lean" away from this center point toward the edges of the photo—this effect is called **relief displacement**. See Figure 9.9 on page 278 for an example of this—if you look at the photo's center, it's like looking straight down at that area. However, if you look outward from the principal point, you'll see that the tall buildings seem to be "leaning" away from the center of the photo (rather than looking straight down on all the buildings in the image).

Regular aerial photos do not have uniform scale, but a different type of photos called **orthophotos** do. To create an orthophoto, a process called **orthorectification** is performed on a regular aerial photo to remove the

principal point the center point of an aerial photo.

relief displacement the effect seen in aerial imagery where tall items appear to "bend" outward from the photo's center toward the edges.

orthophoto an aerial photo with uniform scale.

orthorectification a process used on aerial photos to remove the effects of relief displacement and give the image uniform scale.

FIGURE 9.9 The effect of relief displacement in an aerial photo. (Source: aerial photograph by VTN Consolidated, Inc. courtesy of J. Van Eden)

effects of terrain and relief displacement and enforce the same scale across the photo. A regular aerial photo can't be used as a map since it doesn't have uniform scale everywhere, but an orthophoto can. Even with uniform scale, some of the side appearances of objects can still be seen since, when the photo was taken, some of the sides of the building may have been hidden due to relief displacement and not visible in the photo. This effect is removed in a **true orthophoto**, which gives the appearance of looking directly straight down on all objects in the image. This is achieved by using several images of the area for filling in the missing information, while keeping constant scale in the final photo.

Orthophotos are used to create special products called Digital Orthophoto Quads, or DOQs. A **DOQ** is orthophotography that covers 3.75 minutes of latitude and 3.75 minutes of longitude of the ground, the same area as one-fourth of a 7.5 minute USGS topographic map (see Chapter 13 for more about topographic maps). DOQs (sometimes referred to as DOQQs, or Digital Orthophoto Quarter Quads) have been georeferenced (see Chapter 3 for more information on georeferencing) so that they can be matched up with other geospatial datasets. Also note that the USGS produces certain other DOQ products that cover the area of the entire quad. See *Hands-on Application 9.3: Using MSR Maps for Viewing Aerial Images* for an online application that is used for viewing aerial images.

true orthophoto an orthophoto where all objects look as if they're being seen from directly above.

DOQ a Digital Orthophoto Quad. Orthophotos that cover an area of 3.75 minutes of latitude by 3.75 minutes of longitude, or one-fourth of a 7.5 minute USGS quad.

Hands-on Application 9.3

Using MSR Maps for Viewing Aerial Images

MSR Maps (formerly TerraServer-USA) is an online utility sponsored by the USGS and Microsoft for viewing aerial imagery. Open your Web browser and go to **http://msrmaps.com** to get started. MSR Maps allows you to search for a specific place, an address, or a set of latitude and longitude coordinates and view available aerial imagery for that location. Alternatively, some landmarks around the world have been pre-selected as "Famous Places"—select that option and view aerial/DOQ imagery of places like Alcatraz, Three Mile Island, or Walt Disney World. When you're done perusing aerial images of famous locales, search for your home address or other local places and see what kinds of aerial imagery MSR Maps has to offer.

In an **oblique photo,** the camera is tilted so that it's not positioned directly at nadir, but rather at an angle. Take a look at Figure 9.2 again for an example of this—the image of San Francisco taken from the kites. Unlike a vertical photo, you can see the sides of some of the buildings, and you're really seeing the city from a perspective view instead of an overhead one. Oblique images allow for very different views of a target rather than from straight overhead and can capture other types of information (such as the sides of buildings or mountains). Law enforcement agencies could make use of oblique photos of a building to obtain detailed information of all sides of a building (allowing a view of entrances, exits, windows, and so on) before sending police into a potentially dangerous situation. Companies such as Pictometry extensively collect high-resolution oblique aerial imagery for sale. You can view oblique images of several locations online using Microsoft's free Bing Maps utility if its "Bird's eye" imagery is available (see Figure 9.10 and *Hands-on Application 9.4: Oblique Imagery on Bing Maps* for examples of oblique images created on Bing Maps).

oblique photo an aerial photo taken at an angle.

FIGURE 9.10 Oblique imagery of the Lincoln Memorial, Washington, D.C., as shown in Bing Maps' "Birds's eye view" feature. (Source: Pictometry Bird's Eye © 2010 Pictometry International Corp © AND © 2010 NAVTEQ © 2011 Microsoft Corporation)

Hands-on Application 9.4

Oblique Imagery on Bing Maps

Bing Maps is a great source of viewing online oblique imagery supplied via pictometry. Like Google Earth and Google Maps, Bing Maps allows you to view aerial or satellite images of areas but also gives you the option to view high-resolution oblique images of areas (where oblique imagery is available). To start, open your Web browser and go to **http://www.bing.com/maps** and type the name of a location (for instance, Atlantic City, New Jersey). In the Atlantic City map, zoom in until you can see the ocean and boardwalk. From the Road pull-down menu, select the option for "Bird's eye" and be sure there's a checkmark in the "show angled view" box. The view will switch to oblique photography of At-lantic City. You can zoom in and out and pan the im-agery (when you move past the edge of a photo, a new tile of oblique imagery should load).

Reposition the images so you can see the boardwalk and oceanfront at an oblique angle and then move down the boardwalk to see the casinos, piers, and developments. In addition, Bing Maps provides oblique imagery from mul-tiple angles—click on one of the arrows pointing downward to the left and right of the compass rose (around the letter "N") to rotate the direc-tion and load oblique imagery of the same loca-tion but shot from a different angle. You can use this function to examine a spot from multiple viewpoints. Once you get used to navigating and examining "Bird's eye" oblique images of the New Jersey shore, search for your local area with Bing Maps to see if Bird's eye imagery is available to view your home, school, or workplace.

How Can You Interpret Objects in an Aerial Image?

Understanding what you're looking at when viewing Earth from the sky in-stead of the ground takes a completely new skill set. Objects that you're used to looking at head-on have a whole different appearance from above. For ex-ample, Figure 9.11 shows two different views of the Stratosphere Hotel and Casino in Las Vegas, Nevada—one a regular oblique photograph showing details of the tower and its buildings and the second photo showing an over-head aerial view. Obviously, the complex looks completely different from the ground than from directly above. If you look closely, you can see where areas in the two photos match up, such as the tower on the right-hand side of the image and the remainder of the hotel and casino on the left. How-ever, if you were only given the aerial image and asked which building (of all possible structures in the world) this was, it would be a whole different ballgame.

From the aerial image alone, you'd have to search for clues to determine what was really being shown. For instance, based on the size of the complex (roughly a city block) you might guess that it was some sort of hotel, resort, entertainment attraction, or museum. By looking at its location amidst other large buildings and multiple-lane roads, you could guess that the building

(a)

(b)

was in a big city. Looking closely, you can make out a different colored object (the light blue) on the main complex and determine that it's a swimming pool, further narrowing down the building to some sort of hotel or resort. Although the height of the tower on the right-hand side of the image isn't really discernible (since the image was taken looking down onto the tower itself), the huge shadow being cast by the tower (stretching right to left) is visible. Based on the length of the shadow (and the shape of the shadow being cast), you could determine that object was a tower (or a similar very large, very tall object). Putting all of these clues together (city-block-sized hotel/resort in a big city with a massive tower in front of it), you would be able to use some outside information (like examining some travel books or a short Web search) to quickly determine that the image is of the Stratosphere Hotel in Las Vegas.

When you try to interpret features in an aerial photo or other remotely sensed satellite image (for instance, objects in developed areas or physical features on the landscape), you act like a detective searching for clues in the image to figure out what you're really looking at. Like in the Stratosphere Hotel example, clues like the size and shape of objects or the shadows being cast aid in determining what you're really looking at. **Visual image interpretation** is the activity of identifying features in an aerial (or other remotely sensed) image based on several distinct elements as follows:

⦿ **Pattern**: This is the physical arrangement of objects in an image. The pattern of how objects are lined up (or disarrayed) will often aid in interpreting an image. A large series of cars uniformly set up in a parking lot around a relatively small building would likely be a clue indicating a car dealership rather than another type of shopping area. Evenly spaced airplanes or jets may be a clue to identifying a military base, while a haphazard arrangement of aircraft could indicate a group of statics on display at an aircraft or military museum.

FIGURE 9.11 The Stratosphere Hotel and Casino in Las Vegas, Nevada: (a) an overhead aerial view of the building and (b) an oblique image showing the tower and the hotel layout. (Sources: (a) Bing Map © 2011 Microsoft Corporation Image courtesy of the Nevada State Mapping Advisory Committee. Esri® ArcGIS ArcExplorer graphical user interface Copyright © Esri. (b) Tim Jarrett/Wikipedia)

visual image interpretation the process of discerning information to identify objects in an aerial (or other remotely sensed) image.

pattern the arrangement of objects in an image. An element of image interpretation.

site and association the information referring the location of objects and their related attributes in an image. Used as elements of image interpretation.

- ⊙ **Site and Association:** Site represents the location characteristics of an item, while association represents relating an object in an image to other nearby features in the image. For example, a football field itself has enough distinctive features to identify it, but the related phenomena you could see in the image (the number of bleachers, the number and distribution of the seats, and perhaps the amount of nearby parking) would help in determining if you're looking at a field used by a high school team, a 1-AA college team, or a professional NFL team.

size the physical dimensions (such as length and width) of objects. An element of image interpretation.

- ⊙ **Size:** This is information about the length and width of objects in the image. The relative size of objects in an image can offer good clues in visual image interpretation. For instance, the average length of a car is about 15 feet. If a car is present in an image, you could compare the length of a car to the length of other objects to quickly gain information if a structure is the size of a house or a shopping center. Similarly, some sizes remain constant throughout images. If objects such as a baseball diamond or a football field are present, elements in them (the 90 feet between bases or the 100 yards between goal posts) can be compared to other objects in a photo to gain relative size information.

shadow the shadings in an image caused by a light source. An element of image interpretation.

- ⊙ **Shadow:** This is the shading cast by light shining onto an object. Shadows also help provide information about the height or depth of the objects that are casting the shadows. For instance, in the Stratosphere Hotel example, the height of the tower itself was partially hidden due to the nature of being photographed from above, but the fact that the object was a very tall structure was evident from the shadow cast by the tower. Shadows can also help in determining objects that would be near-indistinguishable from looking down on them, such as the shadows cast by things like railings or telephone poles.

shape the form of objects. An element of image interpretation.

- ⊙ **Shape:** This is the form of objects in an image. The distinctive shapes of objects in an aerial photo can greatly aid in their interpretation. The diamond shape of a baseball field will help to quickly identify it, or the circular shape of crops may indicate the presence of center-pivot irrigation in fields. A racing track has a distinctive oval shape, and even an abandoned horse race track may still show evidence of the oval shape on the landscape.

texture repeating tones in an image. An element of image interpretation.

- ⊙ **Texture:** This refers to the differences of a certain tone throughout parts of the image. The texture of objects can be identified as coarse or smooth. For instance, different types of greenery can be quickly distinguished based on texture—a forest of trees and a field of grass may have the same tone, but the trees appear very rough in an image, while grass will look very smooth. The texture of a calm lake will look very different from the rocky beach surrounding it.

tone the grayscale levels (from black to white) or ranges of a color for objects present in an image. An element of image interpretation.

- ⊙ **Tone:** This is the grayscale (black to white) or intensity of a particular color of objects in an image. The tone of an object can help discern important information about items in a photo. Like in the Stratosphere example, the light-blue color of the swimming pool made identifying it easy. Similarly, a wooden walkway extending into a sandy beach will have different tones that help in distinguishing them from one another in a photo.

FIGURE 9.12 An overhead view of an area containing several different objects. (Source: www.satimagingcorp. com, Quickbird Satellite Sensor, February 2002 © 2007–DigitalGlobe All rights reserved.)

These same elements can be used to identify items in satellite images as well. Figure 9.12 is an overhead view of an area containing several objects that all add up to represent one thing. We'll apply these elements to this image to determine what is being shown in the image.

- Pattern: The three main objects are arranged in a definite diagonal line.

- Texture: The area around the three objects is of very fine texture, compared with some of the rougher textures nearby, especially in the east.

- Tone: The area around the three objects is much lighter than the area of rougher texture (and that area's probably a town or city of some sort, given the pattern of the objects in this range). Given the tone and texture of the ground around the three objects, it's likely not smooth water or grass or concrete, so sand is a likely choice.

- Site and association: The three objects are located in a large sandy area, adjacent to a city, with what resembles roads leading to the objects.

- Size: Comparing the size of the objects to the size of the buildings indicates that the objects are very large—even the smallest of the three is larger than many buildings, while the largest is greater in size than blocks of the city.

- Shadow: The shadows cast by the objects indicate that not only are they tall, but they have a distinctive pointed shape at their top.

- ◉ Shape: The objects are square at the base, but have pointed, triangular tops, making them pyramid shaped.

Bringing all of these elements of interpretation together, we can find three pyramids of different sizes (although still very large), in a definite fixed arrangement, in sand near a busy city. All of these clues combine to identify the Great Pyramid of Giza in Cairo, Egypt. Aerial images often contain clues to try and determine the identity of the objects they contain, or to at least narrow down the options to a handful that could be determined through some brief research through other collateral material (like investigating groupings of large desert pyramids in that particular arrangement).

Visual image interpretation skills enable the viewer to discern more information about the nature of objects within a remotely sensed image. The elements of interpretation can be used for closer investigation to identify (for instance) the specific type of fighter jet that can be seen on a runway based on the size and shape of its features, wingspan, engines, and armaments. A forestry expert could use visual image interpretation to determine what type of tree stand is being examined in a photo based on features associated with the trees.

How Can You Make Measurements from an Aerial Photo?

photogrammetry the process of making measurements using aerial photos.

Once identification of objects in an image has been positively made, these objects can be used for making various types of measurements. **Photogrammetry** is the process of obtaining measurements from aerial photos. Photogrammetric techniques can be used for determining things like the height and depth of objects in an aerial photo. For instance, the lengths of visible features in a photo or heights of buildings seen in the photo can be calculated. There are a lot of possible photogrammetric measurements, and the following are two simple examples of how these types of measurements can be made.

photo scale the representation used to determine how many units of measurement in the real world are equivalent to one unit of measurement on an aerial photo.

Just like the map scale discussed in Chapter 7, every photo has a **photo scale**, listed as a representative fraction. For example, in a 1:8000 scale aerial photo, one unit of measurement in the photo is equivalent to 8000 of those units in the real world. Using the photo scale, you can determine the real-world size of features. For instance, say you measure a length of rail on an aerial photo to be a half-inch and the photo scale is 1:8000. Thus, 1 inch in the photo is 8000 inches in the real world, and so a measurement of 0.5 inches in the photo would be 4000 inches (0.5 times 8000), or 333.33 feet, in length.

The scale of a photo relies on the focal length of the camera's lens and the altitude height of the plane when the image is taken. The problem comes when you're trying to make measurements from an orthophoto but you don't know the photo scale. Without knowing how many real-world units equate to one aerial photo unit, measurements can't be accurately made. An unknown photo scale (of a vertical photo taken over level terrain) can be determined by using a secondary source that has a known scale, so long as an item is visible in both the photo and the source with the known scale (and that you would be able to measure the item in both). A good secondary source would be a topographic map (see Chapter 13) with a known scale since it would contain many features that could also be clearly seen in an aerial photo (such as a road where the beginning and ending are visible). Be cautioned when making these types of measurements, as a regular aerial photo will not have the same scale everywhere in the photo, whereas an orthophoto does have uniform scale across the image.

By being able to make the same measurement on the map (where the scale is known) and on the aerial photo (where the scale is unknown), you can determine the photo scale. This works as follows—your photo could be 1:6000 or 1:12000 or 1: "some number." Just like the representative fraction (RF) discussed in Chapter 7, the photo scale can be written as a fraction—1:24000 can be written as 1/24000. So, assume your unknown photo scale is the RF—this is equal to the distance measured on the photo, known as the photo distance (PD), divided by the real-world distance measured on the ground, known as the ground distance (GD), or:

$$RF = \frac{PD}{GD}$$

Say, for example, you have a 1:12000 scale map showing an oceanfront boardwalk and you also have an orthophoto of unknown scale showing the same region. For the photo to be useful, you have to determine its scale. You can find the same section of boardwalk in both the map and the photo. By measuring the boardwalk section on the map, you find it is 0.59 inches. However, that's not the ground distance, or how long that section of the boardwalk is in the real world—because of the map scale, every one inch measured on the map translates into 12,000 inches in the real world. So, 0.59 inches on the map is actually 7080 inches in the real world. This measure is the GD variable in the equation above (the ground distance):

$$RF = \frac{PD}{7080 \text{ in}}$$

You can measure the same section of boardwalk on the photo and find that it's 1.77 inches. This measurement is the PD variable in the equation (the photo distance):

$$RF = \frac{1.77 \text{ in}}{7080 \text{ in}}$$

Doing some quick division, you find that RF is equal to 1/4000.

$$RF = \frac{1}{4000}$$

So the photo scale is 1:4000. One unit measured on the photo is equal to 4000 units in the real world.

A second type of measurement that you can make from the elements found in an aerial photo is the ability to accurately calculate the height of an object in the photo simply by examining its shadow in the photo. At first blush, it seems you'd need to know all sorts of other information—where the photo was taken, what time of day it was taken, the date on which it was taken— all variables related to the relative location of the Sun and how the shadows would be cast. Chances are you wouldn't be able to easily get your hands on a lot of this type of information, so you're probably thinking that there must be a better way to do this.

You'd be right—photogrammetric measurements give you a much simpler way of determining the heights of objects in a photo from their shadows without needing all that other data. It relies on three things: (1) knowing the scale of the photo (which we just figured out); (2) being able to clearly see the full shadow (from the top of the object) on level ground of all objects whose heights you want to measure; and (3) already knowing the height of one object with a shadow you can measure. With these things, measuring heights is a snap.

Let's take that hypothetical 1:4000 photo from the last example and assume that it's got a number of large hotels casting shadows on the boardwalk. You know the height of one building (115 feet) and can measure its shadow in the photo (from the base to the top) to be 0.10 inches. You can use this information to calculate the angle of the Sun, which is casting the shadows in the photo as follows:

$$\tan a = \frac{h}{L}$$

In this equation, a is the angle of the Sun, h is the real-world height of the object, and L is the real-world length of the shadow. From basic trigonometry, the tangent of a right angle ($\tan a$) is equal to its opposite value (h) divided by its adjacent value (L). See Figure 9.13 for a diagram of how this works. You already know the height, h (115 feet). The length, L, can be found by taking the height of the shadow measured in the photo (0.10 inches) and multiplying by the photo scale (1:4000). It turns out that 0.10 inches on the photo would be 400 inches in the real world, or 33.33 feet. This means if you were to measure the building's photo by actually going to the boardwalk, its shadow would be 33.33 feet long. Plugging these numbers into the formula, we find that:

$$\tan a = \frac{115 \text{ ft}}{33.33 \text{ ft}}$$

and that the tangent of angle a (or "tan a") is equal to 3.45.

Now, since an aerial photo represents a single snapshot in time, we can assume that the angle of the Sun is going to remain the same across the photo for all of the buildings casting shadows. So, the measure for "tan a" will be the same value when applied to all heights we're trying to determine. We can use

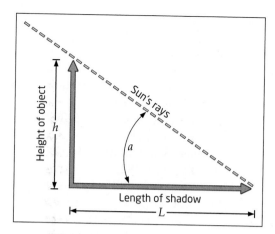

FIGURE 9.13 The relationship between the height of an object (h), the length of its shadow (L), and the angle of the Sun's rays.

this information to calculate the height of any object in the photo casting a shadow we can measure. Say you measure the shadow cast by a second building in the photo on the boardwalk and find it to be 0.14 inches in length. Using the 1:4000 photo scale, this shadow in the real-world would be 560 inches long, or 46.67 feet in length. The only thing that's unknown now is the actual height of the building. Using the equation:

$$3.45 = \frac{h}{46.67 \text{ ft}}$$

and solving for h, we can find that the height of the new building is about 161 feet. This type of measurement can be applied to other objects casting shadows, allowing one to quickly determine the heights of multiple objects in a single photo. Again, be cautious on how you employ these techniques—for finding heights because of shadows, the object must be straight up and down while casting a full shadow on level ground.

Chapter Wrapup

Aerial photography has a wide variety of uses and is an integral part of geospatial technology. Aerial photos are used for interpretation, photogrammetry, and additional data sources for other aspects of geospatial technology, such as GIS. Despite its myriad of uses and applications, aerial photography is just one part of remote sensing. The next chapter will delve into how the whole remote sensing process actually works.

This chapter's lab will put you to work with several remotely sensed images and applying elements of visual image interpretation to items in the images.

Important note: The references for this chapter are part of the online companion for this book and can be found at **http://www.whfreeman.com/shellito1e.**

Key Terms

remote sensing (p. 269)

aerial photography (p. 270)

UAV (p. 273)

vertical photo (p. 275)

nadir (p. 276)

panchromatic (p. 276)

CIR photo (p. 276)

principal point (p. 277)

relief displacement (p. 277)

orthophoto (p. 277)

orthorectification (p. 277)

true orthophoto (p. 278)

DOQ (p. 278)

oblique photo (p. 279)

visual image interpretation (p. 281)

pattern (p. 281)

site and association (p. 282)

size (p. 282)

shadow (p. 282)

shape (p. 282)

texture (p. 282)

tone (p. 282)

photogrammetry (p. 284)

photo scale (p. 284)

Geospatial Lab Application

Visual Imagery Interpretation

This chapter's lab will get you started about thinking about how objects appear from the sky rather than the ground. You'll be examining a series of images in terms of the elements of visual image interpretation from the chapter—think of it like being a detective and searching the image for clues to figure out just what it is you're looking at. You should be able to figure out what, exactly, you're looking at, or to at least narrow down the objects to a short list before using other data to help in determining what you're looking at.

Although the items you'll be examining are all large and prominent objects, buildings, or other features, the application of visual image interpretation elements for these simple (and hopefully, fun) examples will help get you started with looking at the world from above.

Objectives

The goals for you to take away from this exercise are:

◉   Thinking of how objects look from an aerial perspective.

◉   Applying the elements of visual image interpretation (as described in the chapter) in order to discern the identity of objects in the images.

Obtaining Software

There is no special software used in this lab, aside from whatever program your computer uses to view graphics and images. You may find Google Earth useful during the lab, however.

Lab Data

Copy the folder 'Chapter9'—it contains a series of JPEG images (.jpg files) that contain items you will be attempting to interpret. These are numbered, and tend to increase in difficulty of interpretation as the numbers go up.

Localizing This Lab

The images used in this lab were taken from a variety of locations around the United States. You could use Google Earth, MSR Maps, or Bing Maps to locate good overhead views of nearby areas (showing prominent developed or physical features) to build a dataset of local imagery to use for visual image interpretation.

9.1 Applying Elements of Visual Image Interpretation

Take a look at each one of the images (there are questions related to each further down) and try to determine just what it is you're looking at. Although everyone's application of the elements may vary, there are some guidelines that you may find useful in interpreting the images.

Several images may contain multiple items, but all of them work together to help define one key solution.

1. We'll start with a sample image (a digital copy of this sample image is also in the 'Chapter 9' folder labeled "sample"):

(Source: Bing Map © 2011 Microsoft Corporation. Esri® ArcGIS ArcExplorer graphical user interface Copyright © Esri.)

2. First, look at the image as a whole for items to identify:

    a. The central object is obviously some sort of structure, and a large one at that, judging from the shadow being cast. The relative size of the structure (when compared to the size of several of the cars seen throughout the image) helps to likely identify it as a building and not a monument of some kind.

    b. It's set on a body of water (as seen by the tone of the water being different from the tone of the nearby concrete or greenery). Thus, the site can be fixed.

    c. One of the most notable parts is the large triangular shape of the main portion of the building as well as its texture being very distinctive from the rest of the building (the three white sections).

    d. The shape of the entranceway is also very distinctive, being large and round with a couple of concentric circles.

3. Second, take the interpretation of the initial items and start looking for specifics or relationships between the items in the image:

    a. There's something about the pattern of those concentric circles at the entrance plaza that's striking. Two outer rings and a third inner ring with some sort of a design placed in it. Judging from the relative size of a person (seen by picking out shadows near the plaza or on the walkway along the water), the plaza is fairly large. The pattern sort of resembles a giant vinyl record album (and the concept solidifies further with the shape of the large curved area that follows the plaza on its right that looks like the arm of an old vinyl record player).

b. The texture of the triangular portions of the building looks like they could be transparent, as if that whole portion of the structure was made of glass.

c. The association of the area shows no parking lot right next to the building, indicating that you'd have to park somewhere else and walk there. Likewise, there are no major roads nearby, again indicating this is some sort of destination you can't just drive up to and park in front of. Situated on an (apparently) large body of water gives a feel that this is in a large city somewhere.

4. Third, put the clues together and come up with an idea of what the object or scene could be:

a. The building's distinctive enough to be a monument, but too large to probably be one. It's likely some sort of museum, casino, shopping center, or other attraction. However, the building's too small to be a casino or resort and lacks the necessary parking to be a casino, office, or shopping center.

b. The very distinct "record" motif of the front plaza seems to indicate that music (and more specifically older music, given the whole "vinyl record" styling) plays a large part in whatever the building is.

c. Thus, from the clues in the image, this could be some sort of music museum, like the Rock and Roll Hall of Fame, located in Cleveland, Ohio, right on the shore of Lake Erie.

5. Fourth and final, use a source to verify your deduction, or to eliminate some potential choices.

a. By using Google Earth to search for the Rock and Roll Hall of Fame, the image would be confirmed. Doing an image search on Google can also turn up some non-aerial pictures of the site to help verify your conclusion as well.

Not all images will be this extensive, and not all may make use of all elements—for instance, there may be one or two items in the image that will help you make a quick identification of it. However, there are enough clues in each image to figure out what you're looking at—or to at least narrow the choices down far enough that some research using other sources may be able to narrow it down.

For instance, by putting all of the clues together, you could start looking up large, prominent music museums (with triangular glass structures) and the Rock and Roll Hall of Fame would be determined in very short order.

9.2 Visual Image Interpretation

1. To examine an image, open the '**Chapter9**' folder and double-click on the appropriate image (.jpg) file. Whatever program you have on your computer for viewing images should open with the image displayed. You may find it helpful to zoom in on parts of an image as well.

2. For each image, answer the questions presented below, and then explain which of the elements of visual image interpretation led you to this conclusion (and how they did). Writing "shadow and shape helped identify traits of the building" would be unacceptable. However, writing something that would answer the question "What was so special about the shadows in the scene or the shape of items that helped?" would be much better.

3. When identifying the items, be very specific (that is, not just "a baseball stadium," but "Progressive Field" in Cleveland, Ohio).

   Important note: All images used in this exercise are from areas within the United States.

4. You may want to consult some outside sources for extra information, such as Websites, search engines, books, or maps. Lastly, when you finally think you have the image properly identified, you may want to use something like Google Earth or Bing Maps to obtain a view of the object in question to help verify your answer.

Question 9.1 Examine '**image1**.' There are several items in this image, but there's one that's the most prominent. What, specifically, is this image showing? What elements of visual image interpretation lead you to draw this conclusion?

Question 9.2 Examine '**image2**.' What (specifically) is being displayed in this image? What elements of visual image interpretation lead you to draw this conclusion?

Question 9.3 Examine '**image3**.' What (specifically) is being displayed in this image? What elements of visual image interpretation lead you to draw this conclusion?

Question 9.4 Examine '**image4**.' There are several items in this image, but there's one that's the most prominent. What, specifically, is this image showing? What elements of visual image interpretation lead you to draw this conclusion?

Question 9.5 Examine '**image5**.' There are many similar objects in this image, but you'll notice some differences related to some of them. What (specifically) is being displayed in this image, and what location is this an image of? What elements of visual image interpretation lead you to draw this conclusion?

Question 9.6 Examine **'image6.'** There are several items in this image but they all add up to one specific thing. What, specifically, is this image showing, and what location is this an image of? What elements of visual image interpretation lead you to draw this conclusion?

Question 9.7 Examine **'image7.'** What (specifically) is being displayed in this image? What elements of visual image interpretation lead you to draw this conclusion?

Question 9.8 Examine **'image8.'** There are several items in this image, but they all add up to one specific thing. What, specifically, is this image showing (and what is its geographic location)? What elements of visual image interpretation lead you to draw this conclusion?

Question 9.9 Examine **'image9.'** What feature is prominent in the image? Be specific as to what the area is and what geographic location is being shown here. What elements of visual image interpretation lead you to draw this conclusion? For instance, what are all those white objects throughout the image (and how do they help identify the area)?

Question 9.10 Examine **'image10.'** What feature is prominent in the image? Be specific as to what the object is and what geographic location being shown here. What elements of visual image interpretation lead you to draw this conclusion? For instance, several airplanes are prominently displayed—how do they help identify the feature?

Closing Time

Now that you've gotten your feet wet with some visual image interpretation and have hopefully gotten used to viewing Earth from above, Chapter 10 will delve into satellite imagery and its exercise will have you doing (among other things) some more interpretation tasks based on satellite capabilities.

Once you've identified all of the images and explained your application of the elements of image interpretation, you can close the **'Chapter9'** folder and any other programs or resources you have open.

10

How Remote Sensing Works

Electromagnetic Energy, the Remote Sensing Process, Spectral Reflectance, NDVI, Digital Imagery, and Color Composites

When you use Google Earth to get an overhead view of your house, Google had to get those images from somewhere, but there are a number of steps that occur before that image of your house can be acquired and finally turned into a picture you can view on the computer screen. In Chapter 9, we discussed how what's really being captured via **remote sensing** is reflected light—this is a form of electromagnetic energy and a sensor is measuring the amount of energy that reflects off a target. In addition, in remote sensing the device being used to obtain this information about energy reflectance will be on some type of airborne or spacecraft platform that will be a considerable distance away from a target on the ground below. Some remote sensing devices measure the energy emitted from objects on the ground (for instance, as a means of measuring the heat or thermal properties of an item); however, this chapter focuses on measuring the reflection of energy by a sensor.

If remote sensing is measuring the reflectance of energy, there needs to be a source for that energy. Luckily, we have one nearby—our solar system's Sun provides the source of the electromagnetic energy that is being sensed. The Sun is almost 93 million miles away from Earth, and sunlight takes about 8.3 minutes to reach Earth from the Sun. This energy from the Sun radiates through space at the speed of light (almost 300 million meters per second) to reach Earth. This energy passes through and interacts with Earth's atmosphere and then reaches objects on Earth's surface. The energy then interacts with those objects, and some of the energy is reflected back into space. That reflected energy again passes through the atmosphere a second time on its way back off Earth and finally makes it to a sensor either on an airborne or spaceborne platform, where the energy is recorded (**Figure 10.1**, page 296).

These measurements made at the sensor are being captured as remotely sensed data. Thus, the remotely sensed information you see on things like Google Earth is actually the reflection of energy off targets on the ground

> **remote sensing** the process of collecting information related to the reflected or emitted electromagnetic energy from a target by a device a considerable distance away from the target onboard an airborne or spacecraft platform.

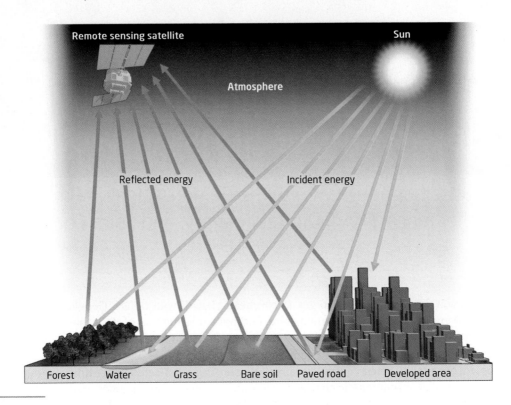

Remote sensing satellite

Sun

Atmosphere

Reflected energy

Incident energy

| Forest | Water | Grass | Bare soil | Paved road | Developed area |

FIGURE 10.1 The remote sensing process: Energy radiates from the Sun, interacts with the atmosphere, interacts with items on the ground, then some energy reflects upward to be measured by the sensor.

that has been measured, processed, and turned into imagery to view. We'll go through each of these stages in more detail to see how you can start with energy from the Sun and end up with a crisp satellite image of a target on the ground (also see *Hands-on Application 10.1: Viewing Remotely Sensed Imagery Online* for various examples of the finished product of remote sensing).

Hands-on Application 10.1

Viewing Remotely Sensed Imagery Online

An excellent tool for viewing different types of remotely sensed data in your Web browser (without using a separate program like Google Earth) is Flash Earth. Like the name implies, make sure you have Flash installed on your computer—then open your Web browser and go to **http://www.flashearth.com** to get started. When Flash Earth starts, you can pan around Earth and zoom in on an area (or search by location). To get started, look at the United States and search for New York City. Zoom in toward Manhattan, and examine what types of imagery become available from the Map Source options—you can select from Bing Maps or Yahoo! Maps imagery—the closer you zoom in, new imagery will load for different scales. Examine some of the areas around Manhattan to see what types of remotely sensed images are available at each option. When you're done looking around New York, type your home location (or nearby local areas) into the Search box to see what remotely sensed imagery you can view of where you are now.

Note that this process (and this chapter) describes passive remote sensing—where the sensor doesn't do anything but simply measures and records reflected or emitted energy. A separate branch encompasses active remote sensing, wherein the sensor generates its own energy, casts it at a target, then measures the return of that form of energy. Radar is a good example of active remote sensing—a device on a plane throws radar waves (microwaves) at a target, those waves interact with the target, and the device measures the backscatter of the returning radar waves.

What Is Remote Sensing Actually Sensing?

The information that is being sensed is the reflection of electromagnetic (EM) energy off a target. There are two ways of thinking of light energy, as either a particle or as a wave. For our purposes, we'll stick to the concept of light energy as a wave. When the Sun radiates this energy, it radiates in the form of waves. Think of this like sitting on the beach and watching the waves come in from the ocean to crash on the shore. The distance between the crests of two waves is called the **wavelength** (λ), with some wavelengths being very long and some being very short. If it's a calm day at the ocean, you would see waves less frequently than you would if it was a rough day, thus there will be a longer distance between waves (and a longer wavelength). If it's a rough day at the ocean, you'll see waves a lot more frequently and the distance between waves will be shorter (that is, a shorter wavelength).

Thus, there's a definite connection between wavelengths and the frequency of those wavelengths. With a longer wavelength, you'll see waves far less frequently, and with a shorter wavelength, you'll see the waves a lot more frequently (Figure 10.2). If you sat there on the beach and counted the waves, you'd count a higher number of waves when the wavelength between them was small, and consequently you would count fewer waves when the wavelength between them was larger.

wavelength the distance between the crests of two waves.

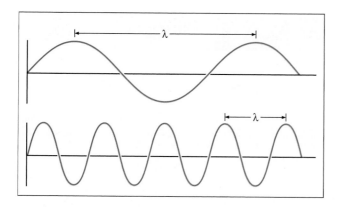

FIGURE 10.2 The relationship between wavelength and frequency: Long wavelengths equate to a low frequency while short wavelengths equate to a high frequency.

Waves of energy work on a similar principle. Energy is moving at a constant speed, that is, the speed of light (c), which is about 300 million meters per second. If you multiply the frequency times the wavelength, it is always equal to the speed of light. When the value for the wavelength of light energy is high, the frequency is low, and vice versa. So if the value for the frequency is high, the value for the wavelength has to be lower to make the product of the two values always come out to be the same number.

Electromagnetic energy at different wavelengths also has different properties. Light energy with very long wavelengths will have different characteristics than light that has very short wavelengths. The **electromagnetic spectrum** is used to examine the properties of light energy in relation to the wavelengths of energy. The values of the wavelengths are usually measured in extremely small units of measurement called **micrometers** (μm), which is one-millionth of a meter (roughly the thickness of bacteria), or an even smaller measurement called **nanometers** (nm), which is one-billionth of a meter. Forms of electromagnetic energy with very short wavelengths are cosmic rays, gamma rays, and x-rays, while forms of electromagnetic energy with very long wavelengths are microwaves and radio waves (**Figure 10.3**).

Something to remember is that the Sun is radiating all of these types of energy, but they're invisible to the human eye. For instance, when you put some cold pizza in the microwave oven to heat it, a light turns on in the oven, but you don't actually see the microwaves bombarding the pizza. In the same way, you can't see the x-rays that a doctor uses in an exam, or you can't see the radio waves as they're picked up by your car's antenna. All of this energy is out there, but you can't actually see it. This is because the composition of your eyes (that allow you to see reflected energy) is only sensitive to the reflection of light that has a wavelength between 0.4 and 0.7 micrometers. This portion of the electromagnetic spectrum is called the **visible light spectrum** or "visible spectrum" (see Figure 10.3 for where the visible light portion falls relative to other forms of electromagnetic energy). If your eyes could see the reflection of slightly shorter wavelengths of energy, you'd be

electromagnetic spectrum the light energy wavelengths and the properties associated with them.

micrometer a unit of measurement equal to one-millionth of a meter. Abbreviated μm.

nanometer a unit of measurement equal to one-billionth of a meter. Abbreviated nm.

visible light spectrum the portion of the electromagnetic spectrum with wavelengths between 0.4 and 0.7 micrometers.

FIGURE 10.3 The electromagnetic spectrum, showing the wavelengths of energy and the properties they correspond with. The visible light portion of the spectrum occurs at wavelengths between 0.4 and 0.7 micrometers.

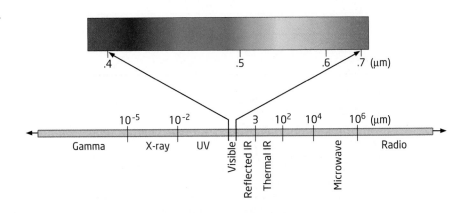

able to see ultraviolet (**UV**) light (like bees can) and if you could see slightly longer wavelengths of energy, you'd be able to see infrared light. However, the human eye is only sensitive to electromagnetic energy in this very small range of wavelengths.

The visible spectrum can be broken down further to the colors that it represents. Shorter visible light wavelengths (between 0.4 and 0.5 micrometers) represent the blue portion of the spectrum. Medium visible light wavelengths (between 0.5 and 0.6 micrometers) represent the green portion of the spectrum. Longer visible light wavelengths (between 0.6 and 0.7 micrometers) represent the red portion of the spectrum. For instance, when you see a shirt with a bright red color, your eyes are actually sensing the reflection of light with a wavelength measurement somewhere between 0.6 and 0.7 micrometers.

Each of these narrow portions of wavelengths of the electromagnetic spectrum that represents a different form of light is referred to as a **band** of energy in remote sensing. For example, the shorter wavelengths of 0.4 to 0.5 micrometers are referred to as the **blue band** of the electromagnetic spectrum since the properties of those wavelengths of light correspond to the characteristics of blue energy. Similarly, the section of wavelengths between 0.5 and 0.6 micrometers make up the **green band** of the electromagnetic spectrum, and the range of wavelengths of 0.6 to 0.7 micrometers make up the **red band.**

Keep in mind that while the human eye is restricted to seeing only these narrow bands of the electromagnetic spectrum, remote sensing devices have the capability to sense energy from other portions of the spectrum. For instance, the infrared (**IR**) light portion of the electromagnetic spectrum extends to wavelengths between 0.7 and 100 micrometers in length. A sensor could be tuned to only measure the reflection of a narrow infrared band of energy between 0.7 and 0.9 micrometers, and even though this energy is not visible to the human eye, a remote sensing device can easily measure this reflection of energy. As we'll see, there's an enormous amount of information that can be gained from examining the reflection of these other forms of energy off objects. When doing remote sensing of infrared light, there are a few wavelength ranges that are commonly utilized:

◉ 0.7 to 1.3 micrometers—Bands of this portion are referred to as "near infrared" (**NIR**), which is commonly used by satellite remote sensing (see Chapter 11) and also in infrared photography (see Chapter 9).

◉ 1.3 to 3.0 micrometers—Bands of this portion are referred to as "middle infrared" (**MIR**), which is also utilized by different satellite sensors. For instance, measurements of bands of MIR energy are used in studies related to water content of plants.

◉ 3.0 to 14.0 micrometers—Bands of this portion are referred to as "thermal infrared" (**TIR**), which is used for measuring heat sources or radiant heat energy.

UV (ultraviolet) the portion of the electromagnetic spectrum with wavelengths between 0.01 and 0.4 micrometers.

band a narrow range of wavelengths being measured by a remote sensing device.

blue band the range of wavelengths between 0.4 and 0.5 micrometers.

green band the range of wavelengths between 0.5 and 0.6 micrometers.

red band the range of wavelengths between 0.6 and 0.7 micrometers.

IR (infrared) the portion of the electromagnetic spectrum with wavelengths between 0.7 and 100 micrometers.

NIR (near infrared) the portion of the electromagnetic spectrum with wavelengths between 0.7 and 1.3 micrometers.

MIR (middle infrared) the portion of the electromagnetic spectrum with wavelengths between 1.3 and 3.0 micrometers.

TIR (thermal infrared) the portion of the electromagnetic spectrum with wavelengths between 3.0 and 14.0 micrometers.

What Is the Role of the Atmosphere in Remote Sensing?

Earth's atmosphere acts like a shield around the planet and the energy from the Sun has to pass through this medium before it reaches Earth's surface. As a result, a considerable portion of the Sun's electromagnetic energy never actually makes it to the ground. Consequently, if it never makes it to the ground, it's never going to be reflected back up to the remote sensing device (and is not part of the whole remote sensing process). Earth's atmosphere contains a variety of gases (including carbon dioxide and ozone) that serve to absorb numerous types of electromagnetic wavelengths. For instance, ozone absorbs wavelengths of energy that correspond to ultraviolet light, the "UV rays" that can harm you. Thus, a considerable amount of ultraviolet radiation gets trapped in the atmosphere and doesn't pass through to Earth. Very short wavelengths and some very long wavelengths are also absorbed by the atmosphere.

Those wavelengths of energy that pass through the atmosphere (rather than being absorbed) are referred to as **atmospheric windows**. Remote sensing is done of the wavelengths that make up these "windows," as they are the ones that transmit most of their energy through the atmosphere to reach Earth. An example of atmospheric windows would be the visible light wavelengths of the electromagnetic spectrum (wavelengths between 0.4 and 0.7 micrometers). Most of the energy at these wavelengths is not absorbed by the atmosphere and instead transmits to Earth to reflect off objects to be viewed by our eyes (as well as cameras and satellite sensors). Other windows used in remote sensing include portions of the infrared and thermal infrared sections of the spectrum (see **Figure 10.4** for a diagram of the atmospheric windows). Sensors are set up to measure the energy at the wavelengths of the windows. It wouldn't do any good whatsoever to have a sensor measuring the wavelengths that were completely absorbed by the atmosphere, as there would be no later energy reflection to measure (such as the entire ultraviolet portion of the spectrum).

atmospheric windows those wavelengths of electromagnetic energy in which most of the energy passes through Earth's atmosphere.

FIGURE 10.4 The absorption of various wavelengths by different gases in Earth's atmosphere, culminating in the atmospheric windows used in remote sensing.

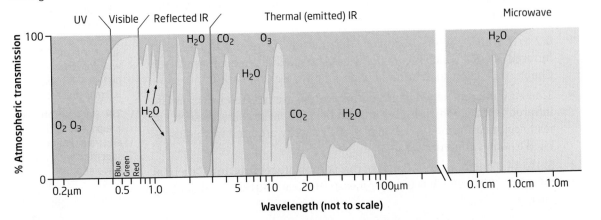

While some light is absorbed by the atmosphere, other light is scattered by it. Scattering occurs due to particles in the atmosphere and is thus always present, but also unpredictable. Particles will absorb energy and redirect it back out in random directions, causing scattering. There are three types of scattering that can occur: The first of these is **Rayleigh scattering**, which occurs when the particles causing the scattering are significantly smaller than the wavelengths being affected. Keeping in mind that the visible-light wavelengths are smaller than one-millionth of a meter, Rayleigh scattering is caused by molecules in the atmosphere. In addition, shorter wavelengths are scattered to a much greater degree than longer wavelengths. This explains why the sky is blue during the day: The shorter blue wavelengths are scattered far more than the longer red wavelengths. However, at sunset or sunrise, sunlight has a longer way to travel to reach Earth, and thus with all of the blue wavelengths already scattered, we see the reds and oranges of longer wavelengths as the Sun goes out of view.

Another type of scattering is **Mie scattering**, when the particles causing the scattering are roughly the same diameter as the wavelengths they're scattering (this would include things like dust or smoke). Last is **nonselective scattering**, where what's causing the scattering is larger than the wavelengths being scattered (such as water droplets or clouds) and scatters all wavelengths equally. Nonselective scattering also explains why we see clouds as being white—the visible colors of blue, green, and red are all scattered the same, thus taking on the color of white (equal levels of red, green, and blue will produce the color white).

Rayleigh scattering scattering of light caused by atmospheric particles smaller than the wavelength being scattered.

Mie scattering scattering of light caused by atmospheric particles the same size as the wavelength being scattered.

nonselective scattering scattering of light caused by atmospheric particles larger than the wavelength being scattered.

What Happens to Energy When It Hits a Target on the Ground?

After absorption and scattering by the atmosphere, the wavelengths that correspond to the windows finally make it to Earth's surface and interact with targets there. One of three things can happen to that energy (on a wavelength-by-wavelength basis)—it can either be transmitted through the target, it can be absorbed by the target, or it can be reflected off the target. **Transmittance** occurs when a wavelength of energy simply passes through a surface to interact with something else later. Think of light passing through a windshield of a car—most of the light will pass through the glass to affect things in the car rather than the windshield itself.

Absorption occurs when the energy is trapped and held by a surface rather than passing through or reflecting off it. Think of walking across a blacktop parking lot during a hot summer day—if you don't have something on your feet, they're going to hurt from the heat in the pavement. Since the blacktop has absorbed energy (and converted it into heat), it's storing that heat during the day. Absorption also explains why we see different colors. For instance, if someone is wearing a bright green shirt, there's something in the dyes of the shirt's material that is absorbing all other colors (that is, the

transmittance when light passes through a target.

absorption when light is trapped and held by a target.

portions of the visible light spectrum) aside from that shade of green, which is then being reflected to your eyes.

Different surfaces across Earth have different properties that cause them to absorb energy wavelengths in various ways. For instance, we see a clear lake as a blue color—it could be dark blue, lighter blue, maybe even some greenish blue, but it's nonetheless a shade of blue. We see it this way because water strongly absorbs all electromagnetic wavelengths aside from the range of 0.4 to 0.5 micrometers (with a little lesser absorption past the edges of that range), which corresponds to the blue portion of the spectrum. Thus, other wavelengths of energy (such as the infrared portion) get absorbed and the blue portion gets reflected.

What remote sensing devices are really measuring is the reflectance of energy from a surface—the energy that rebounds from a target to be collected by a sensor. Like absorption, different items on Earth's surface reflect energy differently. Thus, the total amount of energy that strikes a surface (the **incident energy**) can be computed by adding up the amounts of energy that was transmitted, absorbed, and reflected per wavelength by a particular surface as follows:

incident energy the total amount of energy (per wavelength) that interacts with an object.

$$I = R + A + T$$

where I is the incident energy (the total amount of energy of a particular wavelength) that strikes a surface and is made up of the amount of R (reflection), A (absorption), and T (transmittance) of that particular wavelength. Since remote sensing is focused on the reflection of energy per wavelength, let's change that equation to focus on calculating what fraction of the total incident energy was reflected (and to turn this value into a percentage instead of a fractional amount, we'll multiply by 100):

$$\rho = (R/I) \times 100$$

where ρ is the portion of the total amount of energy composed by reflection (rather than absorption or transmittance) per wavelength. This final value is

Thinking Critically with Geospatial Technology 10.1

How Does Remote Sensing Affect Your Privacy?

The remote sensing processes described in this chapter are systems that measure energy reflection (and turn it into images) without interfering with what's happening below them. For instance, when an aircraft flies overhead to take pictures or a satellite crosses over your house (at more than 500 miles above it), they collect their data and move on. The data collection process doesn't physically affect you or your property in the slightest, and all of this is done without your explicit permission and likely without your knowledge. If you can see your car in your driveway on Google Earth, then it just happened to be parked there when the airplane or satellite collected that image. Does this unobtrusive method of data collection affect your privacy? Is the acquisition of images via remote sensing invasive to your private life? In what ways could this use of geospatial technology intrude on someone's life?

the **spectral reflectance**—the percentage of total energy per wavelength that was reflected off a target—which makes its way toward the sensor and is what is being utilized with remote sensing.

spectral reflectance
the percentage of the total incident energy that was reflected from that surface.

How Can Spectral Reflectance Be Used in Remote Sensing?

All items on Earth's surface reflect energy wavelengths differently. For instance, green grass, bare soil, a parking lot, a sandy beach, and a large lake all reflect portions of the visible, near-infrared, and middle-infrared energy differently (these items emit thermal energy differently as well). In addition, some things reflect energy differently depending on different conditions. As an example of examining these remote sensing concepts, let's look at something simple—measuring reflectance of energy from the leaves on a tree. During late spring and summer, a tree canopy will be fuller, as the leaves stay on the trees and are a healthy green color. The human eye sees healthy leaves as green because the chlorophyll in the leaves is reflecting the green portion of the spectrum and absorbing the red and blue portions. What can't be seen by the human eye (but can be seen by remote sensing instruments) is that healthy leaves also very strongly reflect near-infrared energy. Thus, measurement of near-infrared energy is often used (in part) as an indicator about the relative health of leaves and vegetation.

However, leaves aren't always going to stay green. As autumn progresses, leaves go through a senescence process in which they lose their chlorophyll. As this happens, the leaves reflect less green energy and begin to absorb more red and blue energy, causing the leaves to appear in other colors, such as yellow, orange, and red. As the leaves fall off the trees and turn brown, they have a lot more red reflection, causing their brownish appearance (and there will also be less reflectance of near-infrared energy from the tree).

If a sensor is able to measure all of these energy wavelengths simultaneously for different objects, it would give a person examining the data the capability to tell objects apart by examining their reflectance values in all of these wavelengths. If you were to chart the spectral reflectance values against the wavelengths being measured for each item, you would find that each of these things would have a different line on the chart. This is referred to as an item's **spectral signature** (also called a spectral reflectance curve), as each set of charted measurements will be unique to that item, just like your handwritten signature on a piece of paper is different from other people's signatures. Remote sensing analysts can use these spectral signatures to distinguish items in an image or use them to tell the difference between different types of plants or minerals (see Figure 10.5 on page 304 for an example of various spectral signatures compared to one another).

spectral signature a unique identifier for a particular item, generated by charting the percentage of reflected energy per wavelength against a value for that wavelength.

An application of remotely sensed imagery is its use in assessing the health of green vegetation (such as fields, grass, and leaves on trees). When vegetation is very healthy, it will have a strong reflection of near-infrared energy and

FIGURE 10.5 Examples
of spectral signatures
generated for different
items by charting the
percentage of reflection
on the y-axis and the
wavelengths being
measured on the x-axis.

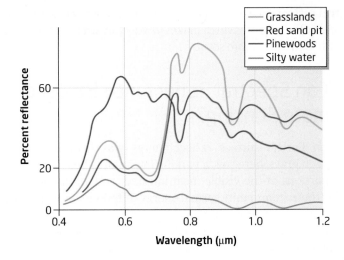

NDVI (Normalized
Difference Vegetation
Index) a method of
measuring the health
of vegetation using
near-infrared and red
energy measurements.

absorption of red energy. As the vegetation becomes more stressed, less near-infrared energy will be reflected and more red energy will be reflected rather than being absorbed. An image can be processed to have a way of measuring the healthiness of vegetated areas. The Normalized Difference Vegetation Index (**NDVI**) is a means of doing so—the process takes a remotely sensed image and creates a new NDVI image from it, containing values related to the relative health of the vegetation being examined. As long as the sensor can measure the red and near-infrared portions of the electromagnetic spectrum, NDVI can be calculated.

NDVI is computed using the measurements for the red and near-infrared bands. The formula for NDVI is:

$$\frac{(NIR - Red)}{(NIR + Red)}$$

NDVI returns a value between −1 and +1, with the higher the value (closer to 1), the healthier the vegetation is at the area being measured. Low values indicate unhealthy vegetation or a lack of biomass. Very low or negative values indicate non-vegetated areas (things like pavement, clouds, or ice). Figure 10.6 shows an example of NDVI being calculated for both a healthy vegetation source (one that reflects a high amount of near-infrared energy and less red energy), resulting in a high NDVI value (0.72) and an unhealthy source (which reflects less near-infrared energy and more red energy), which results in a lower NDVI value (0.14).

NDVI can be calculated by a variety of remote sensing devices, so long as their sensors can measure the red and near-infrared bands (which several of the specific sensors discussed in the next two chapters can do). NDVI gives a quick means of assessing vegetation health, which is used for a variety of environmental applications, including crop monitoring or global vegetative conditions (see Figure 10.7 for an example and *Hands-on Application 10.2: Examining NDVI with NASA ICE* on page 306 to use some remotely sensed imagery to examine the health of a rainforest).

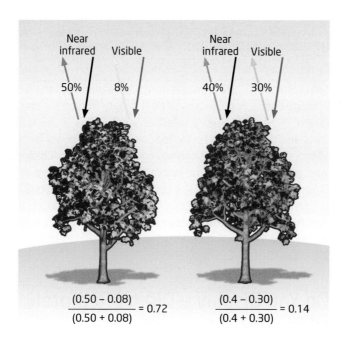

Near infrared   Visible   Near infrared   Visible

50%   8%   40%   30%

$$\frac{(0.50 - 0.08)}{(0.50 + 0.08)} = 0.72$$ $$\frac{(0.4 - 0.30)}{(0.4 + 0.30)} = 0.14$$

FIGURE 10.6 An example of NDVI calculation for healthy (left) and unhealthy (right) vegetation. (Source: Adapted from NASA/Robert Simmon)

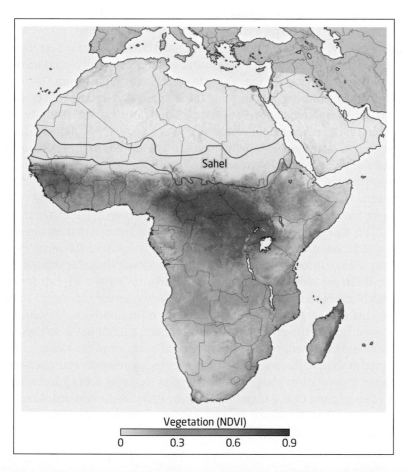

Sahel

Vegetation (NDVI)

0   0.3   0.6   0.9

FIGURE 10.7 An NDVI image of Africa, showing the health (or presence) of vegetation on the continent. (Source: Adapted from NASA/University of Maryland)

Hands-on Application 10.2

Examining NDVI with NASA ICE

NASA (the National Aeronautics and Space Administration) has an online tool called ICE (Image Composite Explorer) that allows users to analyze different applications of remotely sensed data, and one component features an interactive NDVI tool. To use ICE's NDVI exercise, open your Web browser and go to **http://earthobservatory.nasa.gov/Experiments/ ICE/panama/panama_ex2.php**. This section of ICE is part of a larger exercise examining the health of tropical rainforests. The tools will allow you to build NDVI images using the near-infrared and red bands of a Landsat satellite image. To properly see the NDVI imagery, select Channel 4 (this is the near-infrared band) in the red Channel 1 and then select Channel 3 (this is the red band) in the green Channel 2. Click Build and you can view the NDVI image of the area. Change the color table and zoom around the imagery. How can NDVI be used to measure the health of the rainforests?

How Do You Display a Digital Remotely Sensed Image?

spatial resolution the size of the area on the ground being represented by one pixel's worth of energy measurement.

Once this digital data is collected, the next step is to transform it into something that can be viewed on a computer screen, especially since many of these bands of energy are only visible to the sensor and not to the human eye. Sensors can only "see" a certain area of the ground at a time and will record measurements of that particular section. The size of this area on the ground is the sensor's **spatial resolution,** representing the smallest amount of area the sensor can collect information about. For example, if a sensor's spatial resolution was 30 meters, each measurement would consist of a 30 meter × 30 meter section of the ground, and the sensor would measure thousands of these 30 meter × 30 meter sections at once. No matter how many sections are being measured, the smallest area the sensor has information for is a block that is 30 meters by 30 meters. Keep in mind, the spatial resolution of the image is going to constrain the kind of visual information you can obtain from it. For instance, in an image with 4 meter resolution, you may be able to discern that an object is a large vehicle, but no details about it, whereas an image with 0.5 meter resolution may allow you to determine specific features about the vehicle in question.

brightness values the energy measured at a single pixel according to a pre-determined scale. Also referred to as Digital Numbers (DNs).

We'll discuss more about spatial resolution in Chapter 11, but for right now, each one of these blocks of the ground being sensed represents the size of the area on the ground having its energy reflection being measured by the sensor. Due to the effects of the atmosphere, there's not a straight one-to-one relationship between the reflectance of energy and the radiance being measured at the satellite's sensor. The energy measurements at the sensor are converted to a series of pixels, with each pixel receiving a **brightness value,** or BV (also referred to as a Digital Number, or DN), for the amount of radiance being measured at the sensor for that section of the ground. This occurs for each wavelength band being sensed.

Brightness values are scaled to fit a range specified by the sensor. Again, we'll discuss more about these various types of ranges in Chapter 11 (with regard to a sensor's radiometric resolution), but for now, this range of values represents the energy being recorded on a scale from a low number (indicating very little radiance) to a high number (indicating a lot of radiance) and measured as a number of bits. For instance, an 8-bit sensor will scale measurement values between the numbers of 0 (the lowest value) and 255 (the highest value). These values represent the energy being recorded in a range from 0 to 255 (these values are related to the amount of energy being measured). For instance, all values in a band of an image could possibly only be in a range such as 50 to 150, with nothing reaching the maximum value of 255.

Each wavelength band that is being sensed is assigned a BV in this range for each pixel in the image. Brightness values (BVs) can be translated to a grayscale to be viewed on a computer screen. With the range of values associated with **8-bit imagery**, values of 0 represent the color black and values of 255 represent the color white. All other integer values between 0 and 255 are displayed in shades of gray, with lower numbers being darker shades and higher numbers being lighter shades. In this way, any of the wavelength band measurements (including the wavelengths normally invisible to the human eye) can be displayed in grayscale on a computer screen (see **Figure 10.8**). For instance, a near-infrared image would have the brightness value for each pixel representing the near-infrared measurements at the sensor for that area.

If a sensor is measuring the visible portion of the spectrum and treating the entire 0.4 to 0.7 micrometer range as if it was one band, the result will be black-and-white **panchromatic imagery**. However, remote sensing devices are capable of sensing several wavelengths of the electromagnetic spectrum simultaneously. Some common remote sensing devices can sense several different bands at once—sensing 7, 10, or even 36 bands at the same time is currently done on a regular basis by United States government satellite sensors. Imagery created by sensing multiple bands together is referred to as **multispectral imagery**. Some technology can simultaneously sense over 200 bands of the electromagnetic spectrum, producing **hyperspectral imagery**.

8-bit imagery a digital image that carries a range of brightness values from 0 to 255.

panchromatic imagery black and white imagery formed by viewing the entire visible portion of the electromagnetic spectrum.

multispectral imagery remotely sensed imagery comprised of the bands collected by a sensor capable of sensing several bands of energy at once.

hyperspectral imagery remotely sensed imagery comprised of the bands collected by a sensor capable of sensing hundreds of bands of energy at once.

FIGURE 10.8 Pixels of a sample 8-bit image, with their corresponding brightness values and grayscale shading.

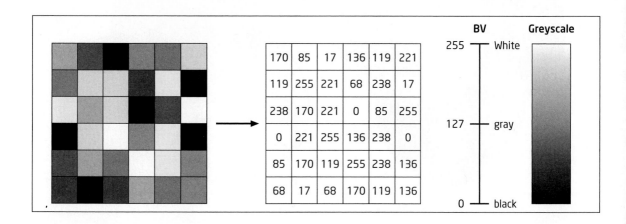

When multiple bands of imagery are available, they can be combined together to view the image in color, rather than grayscale. Displaying imagery in color requires three bands to be displayed simultaneously, each shown in a separate color other than gray. This is simplifying the technical aspects a bit, but a display monitor (like a computer screen) is equipped with devices called **color guns** that can draw a pixel on the screen in varying degrees of red, green, and blue. If an 8-bit image was drawn using the red color gun, the values of 0 to 255 would not be in a range of black, gray, and white, but instead would be in a range of the darkest red, medium red, and very bright red. The same goes for the blue and green color guns—each gun would be able to display an image in its respective color on a range of 0 to 255.

A **color composite** can be generated by displaying the brightness values of one band of imagery in the red gun, a second band of imagery in the green gun, and a third band of imagery in the blue gun. Each pixel would then be drawn using brightness values as the intensity of the displayed color. For instance, a pixel could be displayed using a value of 220 in the red gun, 128 in the green gun, and 75 in the blue gun, and the resulting color shown on the screen would be a combination of these three color intensities. By using the three primary colors of red, green, and blue, and combining them to varying degrees using the range of values from 0 to 255, many other colors can be created (using the 256 numbers in the 0–255 range, with three colors, there are 16,777,216 possible color combinations) to be displayed on the screen (see Figure 10.9 for examples of color formation).

To create colors, different values can be assigned to the red, green, and blue guns. Table 10.1 shows some examples of ways to combine values together to form other colors (keeping in mind that there are millions of possible

color gun equipment used to display a color pixel on a screen through the use of the colors red, green, and blue.

color composite an image formed by placing a band of imagery into each of the three color guns (red, green, and blue) to view a color image rather than a grayscale one.

FIGURE 10.9 Colors that can be formed by adding different colors together (left) and subtracting different colors from each other (right).

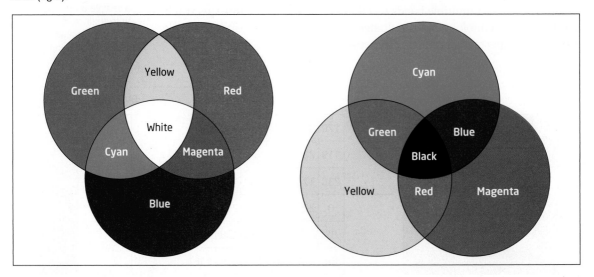

TABLE 10.1 Examples of colors generated through 8-bit combinations of red, green, and blue

Red Value	Green Value	Blue Value	Color Formed
255	255	0	Yellow
255	0	255	Magenta
0	255	255	Cyan
255	128	0	Orange
0	64	128	Dark Blue
255	190	190	Pink
255	255	115	Bright Yellow
115	76	0	Brown
0	197	255	Sky Blue
128	128	128	Gray
255	255	255	White
0	0	0	Black

combinations of these values). See also *Hands-on Application 10.3: Color Tools Online: Color Mixing* for an online tool that can be used to create colors from various 0 to 255 values.

When an image is examined, the color being displayed for each pixel corresponds to particular values of red, green, and blue being shown on the screen. For a remotely sensed image, those numbers are the brightness values of the band being displayed with that color gun. Any of the bands of an image may be displayed with any of the guns, but as there are only three color guns, only three bands can be displayed as a color composite at any time (Figure 10.10 on page 310).

For example, the near-infrared band values could be displayed using the red color gun, the green band values could be displayed using the green color gun, and the red band values could be displayed using the blue color gun. By

Hands-on Application 10.3

Color Tools Online: Color Mixing

The Color Tools Website provides several great resources related to color. The Color Mixer tool available here: **http://www.colortools.net/color_mixer. html** allows you to input values (between 0-255) for red, green, and blue and see the resultant color. On the Website, type a value into the box in front of the "/255" for each of the three colors and check out the result. Try the values listed in Table 10.1 and some of your own to see how different numbers create different colors.

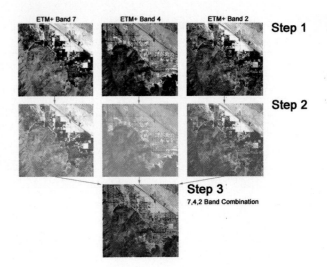

using this arrangement, a brown pixel can be seen on the screen, and after inquiry, it's discovered that the red gun (which has the near-infrared band) has a value of 150, the green gun (which has the green band) has a value of 80, and the blue gun (which has the red band) has a value of 20. These values correspond to the brightness values of their respective bands and are combined together to form that pixel color. If you change up the arrangement of bands and guns, you may have different colors displayed on the screen, but the actual brightness values will not have changed. Table 10.2 shows how the display color will change depending on which band is being displayed with which gun.

Note that while the pixel being displayed on the screen changed color depending on what band was being displayed with a different gun, none of the brightness values of that pixel were altered. Different band combinations can reveal different information about objects in an image due to the varying brightness values of the bands for those items.

Several different types of composites can be formed using the various bands and guns. Whenever a composite has the red band displayed in the red

TABLE 10.2 Different combinations of brightness values for various bands displayed with different color guns from different colors on the screen

Red Value	Green Value	Blue Value	Color Formed
NIR Band (150)	Green Band (80)	Red Band (20)	Brown
Red Band (20)	Green Band (80)	NIR Band (150)	Deep Blue
Green Band (80)	NIR Band (150)	Red Band (20)	Green
Red Band (20)	NIR Band (150)	Green Band (80)	Paler Green

(a)

(b)

FIGURE 10.11 A comparison of the Cleveland area as viewed in (a) a true color composite and (b) a false color composite. (Source: NASA/Purdue Research Foundation)

gun, the green band displayed in the green gun, and the blue band displayed in the blue gun (the way natural colors would be shown), a **true color composite** is formed. A true color composite would look as if you would expect to see a remotely sensed image as if you were looking down at the ground from an airplane (that is, the trees would be green and the water would be blue). Whenever the distribution of bands to guns deviates from this arrangement (some other combination is used besides the true color one), a **false color composite** is generated. For instance, all four of the combinations described in Table 10.2 would result in false color composites as none of them display the red band in the red gun, the green band in the green gun, and the blue band in the blue gun. See **Figure 10.11** for a comparison of true color and false color composites.

With all of these different combinations possible, there is a **standard false color composite** that is often used in many remote sensing studies. In this standard false color arrangement, the near-infrared band is displayed in the red gun, the red band is displayed in the green gun, and the green band is displayed in the blue gun. The false color composite image is similar to the CIR photos discussed back in Chapter 9. Some common effects of this arrangement are the displaying items that would have a strong reflection of near-infrared energy (such as healthy plants, grass, or vegetation) is bright shades of red, while water (which would have very little reflection of the near-infrared, red, and green energy) would be displayed in black. With composites, multispectral imagery can be displayed in color, regardless of what bands are being used and these composites can be used for different types of image analysis (see *Hands-on Application 10.4: Comparing True Color and False Color Composites* on page 312 for an online method of comparing how objects on the ground are shown in true color and false color composites).

true color composite an image arranged by placing the red band in the red color gun, the green band in the green color gun, and the blue band in the blue color gun.

false color composite an image arranged by not placing the red band in the red color gun, the green band in the green color gun, and the blue band in the blue color gun.

standard false color composite an image arranged by placing the near-infrared band in the red color gun, the red band in the green color gun, and the green band in the blue color gun.

Hands-on Application 10.4

Comparing True Color and False Color Composites

An online resource for obtaining remotely sensed imagery is Ohiolink, the statewide online library system. Landsat satellite imagery (which we'll discuss more in Chapter 11) for Ohio can be downloaded from there (if you're in Ohio), and true color and false color composite images of areas around Ohio can be viewed as well. Open your Web browser and go to **http://landsat.ohiolink.edu/GEO/LS7** to get started. The graphic on the page shows Ohio broken up into nine different sectors—these correspond to the Landsat reference system we'll discuss in Chapter 11. For now, select the central block (Path 19/Row 32), which contains imagery of Columbus, Ohio. A new list will appear of all the dates on which Landsat imagery was acquired for the region. Choose the most recent date and other options for viewing will be available.

You can examine a true color composite and a false color composite (among other options) of the area.

Click on the true color composite option to view the satellite imagery. You can choose a new window size and zoom in to the area to see how features such as urban areas, forests, fields, and water appear in true color. Once you've zoomed in and looked around, select the option for near-infrared in the right-hand corner—this will switch to the same view, but in a standard false color composite. How did all of the features you'd been looking at change appearances in the false color composite and why? When you're done, check out some other portions of Ohio along Lake Erie (like Cleveland) to see how other features (such as the lake and lakefront areas) appear in true color and false color composites.

Chapter Wrapup

The basics described in this chapter cover the process of remote sensing and how a digital image can be formed from the measurement of electromagnetic energy reflection. In the next chapter you'll be looking at how these concepts can be applied to perform remote sensing from a satellite platform in space. Remotely sensed imagery is used for numerous applications—Chapters 11 and 12 will examine different types of remote sensing platforms and uses of their imagery, including everything from air pollution to wildfire monitoring to studying water quality.

The lab for this chapter uses a free program called MultiSpec to examine a digital remotely sensed image as well as different band combinations and color composites.

Important note: The references for this chapter are part of the online companion for this book and can be found at http://www.whfreeman.com/shellito1e.

Key Terms

remote sensing (p. 295)
wavelength (p. 297)
electromagnetic spectrum (p. 298)
micrometer (p. 298)
nanometer (p. 298)
visible light spectrum (p. 298)
UV (ultraviolet)(p. 299)
band (p. 299)
blue band (p. 299)
green band (p. 299)
red band (p. 299)
IR (p. 299)
NIR (p. 299)
MIR (p. 299)
TIR (p. 299)
atmospheric windows (p. 300)
Rayleigh scattering (p. 301)
Mie scattering (p. 301)
nonselective scattering (p. 301)

transmittance (p. 301)
absorption (p. 301)
incident energy (p. 302)
spectral reflectance (p. 303)
spectral signature (p. 303)
NDVI (p. 304)
spatial resolution (p. 306)
brightness values (p. 306)
8-bit imagery (p. 307)
panchromatic imagery (p. 307)
multispectral imagery (p. 307)
hyperspectral imagery (p. 307)
color gun (p. 308)
color composite (p. 308)
true color composite (p. 311)
false color composite (p. 311)
standard false color
  composite (p. 311)

Remotely Sensed Imagery and Color Composites

This chapter's lab will introduce you to some of the basics of working with multispectral remotely sensed imagery through the use of the MultiSpec software program.

Objectives

The goals for you to take away from this exercise are:

◉ Familiarizing yourself with the basics of the MultiSpec software program.

◉ Loading various bands into the color guns and examining the results.

◉ Creating and examining different color composites.

◉ Comparing the brightness values of distinct water and environmental features in a remotely sensed satellite image to create basic spectral profiles.

Obtaining Software

The current version of MultiSpec is available for free download at **http://cobweb.ecn.purdue.edu/~biehl/MultiSpec**.

Important note: Software and online resources sometimes change fast. This lab was designed with the most recently available version of the software at the time of writing. However, if the software or Websites have significantly changed between then and now, an updated version of this lab (using the newest versions) is available online at **http://www.whfreeman.com/shellito1e**.

Lab Data

Copy the folder '**Chapter10**'—it contains a Landsat 5 TM satellite image file called 'cle.img' (which is a subset of a Landsat 5 TM scene from 9/11/2005 of northeast Ohio). We will discuss more about Landsat imagery in Chapter 11, but the TM imagery bands refer to the following portions of the electromagnetic (EM) spectrum in micrometers (μm):

◉ Band 1: Blue (0.45 to 0.52 μm)

◉ Band 2: Green (0.52 to 0.60 μm)

◉ Band 3: Red (0.63 to 0.69 μm)

- Band 4: Near infrared (0.76 to 0.90 μm)
- Band 5: Middle infrared (1.55 to 1.75 μm)
- Band 6: Thermal infrared (10.4 to 12.5 μm)
- Band 7: Middle infrared (2.08 to 2.35 μm)

Also, Landsat TM imagery is 30-meter spatial resolution (except for the thermal-infrared band, which is 120 meters). Thus, each pixel you will be examining represents 30 meters square.

Localizing This Lab

Although this lab focuses on a section of a Landsat scene from northeast Ohio, Landsat imagery is available for download via GLOVIS at http://glovis.usgs.gov. This will provide the raw data which will have to be imported into Multi-Spec and processed to use in the program.

Alternatively, NASA has information for obtaining free Landsat data for use in MultiSpec online at http://change.gsfc.nasa.gov/create.html. The page provides step by step instructions for acquiring free data for your own area and getting it into MultiSpec format.

10.1 MultiSpec and Band Combinations

1. Start MultiSpec.

   MultiSpec will start with an empty text box (that will give you updates and reports of processes in the program). You can minimize the text box for now.

2. To get started with a remotely sensed image, select **Open** from the **File** pull-down menu. Alternatively, you can select the **Open** icon from the toolbar:

(Source: Purdue Research Foundation)

3. Navigate to the '**Chapter10**' folder and select '**cle.img**' as the file to open and click **Open**.

4. A new dialog box will appear that allows you to set the display specification for the 'cle' image.

Set Display Specifications for:

cle.img

Area to Display

	Start	End	Interval
Line	1	806	1
Column	1	975	1

Display

Type: 3-Channel Color

Channels:

Red: 4 □ Invert

Green: 3 □ Invert

Blue: 2 □ Invert

Channel Descriptions...

Magnification: 1

Enhancement

Bits of color: 24

Stretch: Linear

Min-max: Clip 2% of Tails

Treat '0' as: Data

Number of display levels: 256

□ Load New Histogram

Cancel   OK

(Source: Purdue Research Foundation)

5. Under Channels, you will see the three color guns available to you (red, green, and blue). Each color gun can hold one band (see the Lab Data section for a listing of which bands correspond to which parts of the electromagnetic spectrum). The number listed next to each color gun represents the band being displayed with that gun.

6. Display **band 4** in the red color gun, **band 3** in the green color gun, and **band 2** in the blue color gun.

7. Accept the other defaults for now and click **OK**.

8. A new window will appear and the 'cle' image will begin loading. This might take a minute or two to completely load.

9. For best results, maximize both MultiSpec as well as the window containing the 'cle' image. Use the bars at the edge of the window to move around the image. You can zoom in or zoom out by using the zoom tools on the toolbar:

The large mountain icon will zoom in and the small mountain icon will zoom out.
10. Zoom around the image, paying attention to some of the city, landscape, and water areas.

Question 10.1 What wavelength bands were placed into which color guns?

Question 10.2 Why are the colors in the image so strange compared to what we're normally used to seeing in other imagery (such as Google Earth or Google Maps)? For example, why is most of the landscape red?

Question 10.3 In this color composite, what colors are the following features in the image being displayed as: water, vegetated areas, and urban areas?

11. Reopen the image, this time using **band 3** in the red gun, **band 2** in the green gun, and **band 1** in the blue gun (referred to as a 3-2-1 combination). Once the image reloads, pan and zoom around the image, examining the same areas you just looked at.

Question 10.4 What kind of composite did we create in this step? How are the bands being displayed in this color composite in relation to their guns?

Question 10.5 Why can we not always use this kind of composite (from Question 10.4) when analyzing satellite imagery?

12. Reopen the image yet again, this time using **band 2** in the red gun, **band 1** in the green gun, and **band 3** in the blue gun (referred to as a 2-1-3 combination). Once it reloads, pan and zoom around the image, examining the same areas you just looked at.

Question 10.6 Once again, what kind of composite was created in this step?

Question 10.7 How are vegetated areas being displayed in this color composite (compared to the arrangement in Question 10.4)?

10.2 Examining Color Composites and Color Formations

1. Reopen the image one more time, returning the 4-3-2 combination (band 4 in the red gun, band 3 in the green gun, and band 2 in the blue gun). You can close the other images, as we'll be working with this one for the rest of the lab.

2. Zoom and move around the image to find and examine Burke Lakefront
    Airport as follows:

(Source: NASA/Purdue Research Foundation)

3. Zoom in to the airfield. From its shape and the pattern of the runways,
    you should be able to clearly identify it in the Landsat image.

(Source: NASA/Purdue Research Foundation)

4. Examine the airfield and its surroundings.

Question 10.8 Why do the areas in between the runways appear red?

5. Open a new image, this time with a 3-4-2 combination. Examine Burke
    Lakefront Airport in this new image and compare it to the one you've
    been working with.

Question 10.9 Why do the areas in between the runways now appear
bright green?

6. Open another new image, this time with a 3-2-4 combination. Examine
    Burke Lakefront Airport in this new image and compare it to the others
    you've been working with.

Question 10.10 Why do the areas in between the runways now appear
blue?

7. At this point, only keep the 4-3-2 image open and close the other two.

10.3 Examining Specific Brightness Values and Spectral Profiles

Regardless of how the pixels are displayed in the image, each pixel in each
band of the image has a specific brightness value set in the 0–255 range. By
examining those pixel values for each band, you can chart a basic 'spectral
profile' of some features in the image.

1. Zoom in to the area around Cleveland's waterfront and identify the
    urban areas (in the image these will mostly be the white or cyan
    regions).

2. From the **Window** pull-down menu, select **New Selection Graph**. Another
    (empty) window (called Selection Graph) will open in MultiSpec.

3. In the image, locate a pixel that's a good example of an urban or
    developed area. The cursor will change to a cross shape, so give the pixel
    one more click.

    Important note: Zoom in so that you are only selecting one pixel with the
    cursor.

4. A chart will appear in the Selection Graph window that will
    graphically show the BVs for each band at that particular pixel
    (see the first page of this lab for the portions of the electromagnetic
    spectrum that match up with each band). The band numbers are on
    the x-axis and the BVs are on the y-axis. The chart can be expanded
    or maximized as necessary to be better able to examine the values.

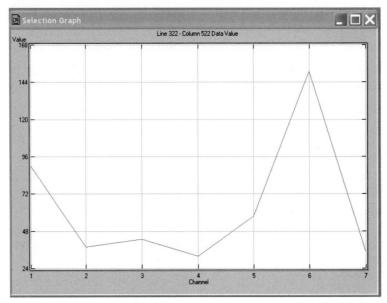

(Source: Purdue Research Foundation)

5. The Selection Graph window now shows the data that can be used to compute a simplified version of a spectral profile for an example of the particular urban land use pixel you selected from the image.

6. For the next question, you'll be required to find a pixel that's a good example of water and another pixel that's a good example of vegetation and translate the data from the chart to a 'spectral profile' for each example. In drawing the 'profiles' from the information on the chart, keep two things in mind:

   a. First, the values at the bottom of the Selection Graph window represent the numbers of the bands being examined. On the answer sheet, the actual wavelengths of the bands are plotted out, so make sure you properly match up each band with its respective wavelength (also note that Band 6 is not being charted).

   b. Second, the values on the x-axis of the Selection Graph window are BVs, not the percentage of reflection as seen in a spectral signature diagram. There are a number of factors involved with transforming BVs into the actual percent reflectance values (A BV and the percent reflectance don't have an exact one-to-one ratio, as there are other factors that affect the measurement at the sensor, such as atmospheric effects), but in this simplified example, you'll just chart the BV for this 'spectral profile.'

7. Examine the image for a good example of a water pixel and a vegetation pixel.

8. Plot a 'spectral profile' diagram for the water and vegetation pixels you chose along the following diagram (remember to plot values as calculated from the BVs).

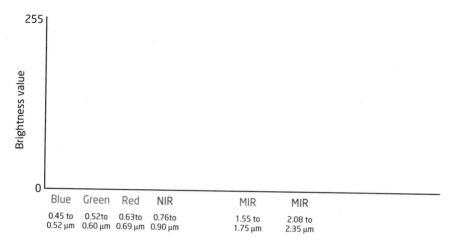

9. Examine your two 'profiles' and answer Questions 10.11 and 10.12.

Question 10.11 What information can you gain from the spectral profile for water about the ability for water to reflect and absorb energy (that is, what types of energy have the greatest amount of reflection and absorption by water)?

Question 10.12 What information can you gain from the spectral profile for vegetation about the ability for vegetation to reflect and absorb energy (that is, what types of energy have the greatest amount of reflection and absorption by vegetation)?

10. Exit MultiSpec by selecting **Exit** from the **File** pull-down menu. There's no need to save any work in this exercise.

Closing Time

Now that you have the basics of working with remotely sensed data and examining the differences in color composites, the next two chapters will build on these skills. The Chapter 11 lab will return to using MultiSpec for more analysis of Landsat satellite imagery.

11

Images from Space

Satellite Remote Sensing, Satellite Orbits, Sensor Resolutions, the Landsat Program, and High-Resolution Satellite Sensors

How many times have you seen something like the following scenario in a television show or movie, or read it in a novel? The heroes are racing against the clock to stop the villains from setting off whatever doomsday weapon they happen to have. In the tense mission-control center, the heroes' computer expert taps a few keys on the keyboard, a spy satellite swings into view, and suddenly the monitors fill up with crystal-clear images from space of the villains' remote hideout. The imagery's crisp enough for the heroes to run some facial recognition software and pick out the international arms dealer selling the weapons of mass destruction (WMDs), as well as read the license plates of the nearby cars. Rest assured, the elite counter-intelligence strike team will show up on cue to save the world, probably watched via live remotely sensed satellite video feed by their teammates at the home base.

Based on this fictional scenario, you might get the impression that remote sensing satellites are magical things that can do anything—rotate into whatever orbital position is required to track people, zoom in (from outer space) to make out the details of someone's license plate, read lips from space, or send live video feeds to the good guys in the field. Certainly, books and films help perpetuate the idea that remote sensing (and anything using satellites) is some sort of utopian technology that can do anything. The goal of this chapter is to examine how satellite remote sensing is really done, as well as its uses and limitations.

The first man-made satellite to achieve orbit around Earth was Sputnik, launched in 1957 by the Union of Soviet Socialist Republics (USSR), and it was quickly followed by the United States the following year with the launch of Explorer I. These initial satellite launches established that satellites carrying equipment (or, in the case of Sputnik II, animals) could be placed into orbit and utilized. As a result, **NASA** (the National Aeronautics and Space Administration) was established in 1958 to head up the United

NASA the National Aeronautics and Space Administration, established in 1958; it is the United States government's space exploration and aerospace development branch.

Corona a United States government satellite remote sensing program utilizing film-based camera equipment, which was in operation from 1960-1972.

States space and aeronautics program. Satellites were used for remote sensing reconnaissance and surveillance beginning with the **Corona** program in 1960.

A Corona satellite had a camera onboard that took pictures of areas on the ground, and then it would eject the film from the satellite to be collected by a United States plane. The film was then developed and interpreted to provide intelligence information about locations around the world (see **Figure 11.1** for an example of Corona imagery). Corona and its counterpart—the USSR's Zenit satellite—provided numerous images from orbit, with Corona itself providing an estimated 800,000 remotely sensed satellite images. Corona ceased operation in 1972 and its imagery was declassified after 1995, although satellite remote sensing has long since continued. A series of satellites is now used by the United States government for acquiring digital imagery from orbit of areas around the world.

Beyond government surveillance, satellites are used today for a wide variety of purposes, constantly collecting data and imagery of the world below. Comparing satellite imaging to Chapter 9's aerial photography methods of remote sensing, several advantages to using a satellite instead of an aircraft are apparent. Satellites are constantly orbiting Earth and taking images—there's

FIGURE 11.1 Corona imagery of Dolon Air Field in the former USSR (taken in 1966). (Source: National Reconnaissance Office)

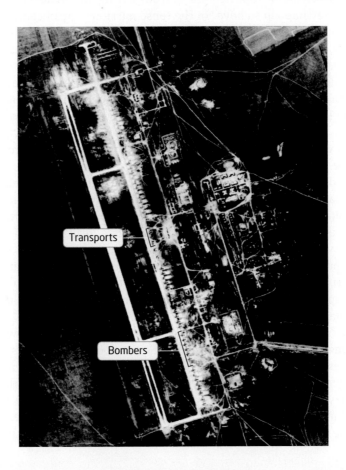

Transports

Bombers

no need to wait for a plane to fly over an area. Satellites can also image a much larger area than a single aerial photograph can. Also, remote sensing satellites provide global coverage and are not restricted to geographic boundaries or constraints the way aircraft are because satellites orbit hundreds of miles above Earth's surface.

How Do Remote Sensing Satellites Collect Data?

Remote sensing satellites are placed in a specific orbital path around Earth. One type of orbit used in remote sensing is **geostationary orbit**, in which satellites rotate at the same speed as Earth. Since a geostationary satellite takes 24 hours to make one orbit, it's always in the same place at the same time. For instance, although it's not a remote sensing system, satellite television utilizes a geostationary orbit—the satellite handling the broadcasting is in orbit over the same area continuously, which is why a satellite reception dish needs to always be pointed at the same part of the sky. Likewise, the WAAS and EGNOS satellites discussed in Chapter 4 are also in geostationary orbit, so they can provide constant coverage to the same area on Earth. There are some remote sensing satellites that use geostationary orbit to always collect information about the same area on Earth's surface (such as the GOES series of satellites used for obtaining weather imagery).

geostationary orbit an orbit in which an object rotates around Earth at the same speed as Earth.

Many remote sensing satellites operate in **near-polar orbit**, a north-to-south path wherein the satellite moves close to the north and south poles while it makes several passes a day about Earth. For instance, the Landsat 7 satellite (a United States satellite we'll discuss in more detail in a few pages) makes slightly more than 14 orbits per day (each orbit takes about 99 minutes)—see Figure 11.2 for Landsat 7's near-polar orbit. Also, check out *Hands-on Application 11.1: Examining Satellite Orbits in Real Time* (page 326) for two methods of tracking satellite orbits in real time.

near-polar orbit an orbital path that carries an object around Earth, passing close to the north and south poles.

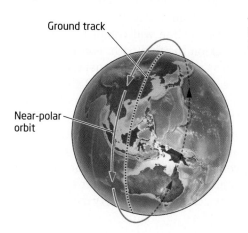

Ground track

Near-polar orbit

FIGURE 11.2 The near-polar orbit of Landsat 7.

Hands-on Application 11.1

Examining Satellite Orbits in Real Time

Think about all the jobs that satellites are servicing—such as communication, satellite TV, weather monitoring, military applications, and GPS—then add remote sensing to that list. When you start considering all these uses, you can see the need for many, many satellites in orbit.

JSatTrak is a very cool program that you can use to view (in real time) the locations of numerous satellites (including remote sensing ones). JSatTrak can be downloaded from **http://www.gano.name/ shawn/JSatTrak/**. When JSatTrak opens, you can access a list of satellites via the Windows pull-down menu and by choosing Satellite Browser. A list of available satellites will appear (many of the remote sensing satellites described in this chapter and in

Chapter 12 can be found by expanding the Weather and Earth Resources option, then selecting. Earth Resources). When you drag the name of a satellite from the list to the Object List window on the right side of the screen, the orbit track and current location of the satellite will be shown on the display window. Check out some satellites like Landsat 5, Landsat 7, Terra, and others. JSatTrak has many other features to explore (such as the 3D Earth Window for looking at satellite tracking).

NASA also provided an online satellite tracking tool called J-Track (also J-Track 3D) that allowed users to similarly track satellite positions in real time. Unfortunately, at the time of this writing, J-Track is currently unavailable, but hopefully will return one day.

While a satellite in continuous orbit passes over Earth, it will be imaging the ground underneath it. However, during an orbit, a satellite can image only a certain size of the ground at one time—it can't see everything at once. The **swath width** is a measurement of how much ground the satellite can image during one pass. For instance, a Landsat satellite's sensor has a swath width of 185 kilometers, meaning that a 185-kilometer-wide area on the ground is imaged as Landsat flies overhead. As satellites orbit, Earth is rotating beneath them, meaning that several days may separate orbital paths (and thus, swaths of the ground being imaged) over nearby geographic areas. For example, the path of a Landsat satellite may pass over the middle of the United States on Day 1 of its orbit, pass over a swath of area hundreds of miles to the east on Day 2, and not return to the path next to the first one until Day 8 (Figure 11.3).

swath width the width of the ground area the satellite is imaging.

After a certain amount of time the satellite will completely pass over all areas on Earth and start again, back on the first path. For instance, the Landsat series of satellites are set up in an orbital path such that they will pass over the same swath on Earth's surface every 16 days. So when Landsat's orbit carries it to the swath where it's collecting imagery of your current location it will be another 16 days before it's able to image your location again. A **Sun-synchronous orbit** occurs when a satellite's orbit takes it over the same area on Earth at the same local time. Because of this, images collected from the satellite will have similar sun-lighting conditions and can be used for comparing changes to an area over time. For instance, Landsat 7 has a Sun-synchronous orbit, allowing it to cross the Equator at a constant time every day.

Sun-synchronous orbit an orbital path set up so that the satellite crosses the same areas at the same local time.

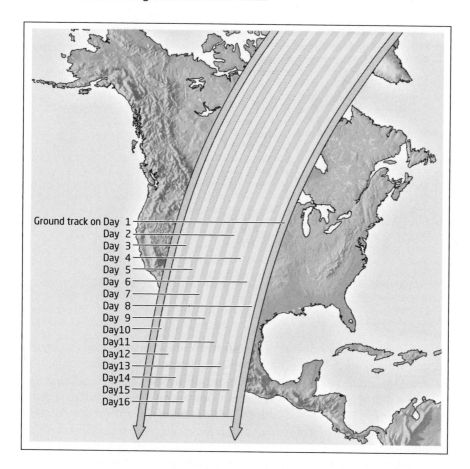

Ground track on Day 1
Day 2
Day 3
Day 4
Day 5
Day 6
Day 7
Day 8
Day 9
Day 10
Day 11
Day 12
Day 13
Day 14
Day 15
Day 16

FIGURE 11.3 An example of the paths covered for one section of the United States during the 16-day revisit cycle of a Landsat satellite.

The sensors onboard the satellites have different ways of actually scanning the ground and collecting information in the swath being imaged. In **along-track** scanning, a linear array is used to scan the ground along the path the satellite is moving in. As the satellite moves, the detectors collect the information from the entire swath width underneath them. This type of detector is referred to as a "pushbroom" sensor. Think of pushing a broom in a straight line across a floor—the broom itself is the sensor and each bristle of the broom represents a part of the array sensing what is on the ground underneath it. A second method is an **across-track** scanner, where a rotating mirror moves back and forth over the swath width of the ground collecting information. This type of detector is referred to as a "whiskbroom" sensor. Think of sweeping a floor with a broom, swinging the broom back and forth across a path on the floor—the broom is the sensor and the bristles represent the detectors collecting the data on the ground below them. Satellite systems usually use one of these types of sensors to obtain the energy reflectance measurements from the ground (see Figure 11.4 on page 328 for examples of the operation of both sensor types).

along-track a scanning method using a linear array to collect data directly on the path the satellite moves on.

across-track a scanning method using a rotating mirror to collect data by moving the device the width of the satellite's swath.

FIGURE 11.4 The
operation of across-track
and along-track scanners.

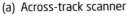

(a) Across-track scanner

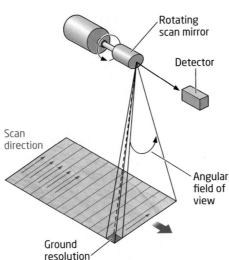

(b) Along-track scanner

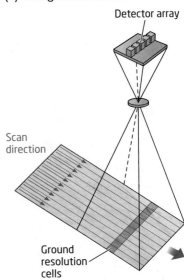

EROS the Earth
Resources Observation
Science Center; located
outside Sioux Falls,
South Dakota, which
serves (among many
other things) as a
downlink station for
satellite imagery.

After a satellite's sensors have collected data, the data needs to be off-loaded from the satellite and sent to Earth. Because data collection is continuous, this information needs to get to people on the surface to process it so it can be utilized. The Earth Resources Observation Science (**EROS**) Center located outside Sioux Falls, South Dakota, is one of the downlink stations in the United States that receives data from numerous remote sensing satellites. Satellite data can be directly sent to a receiver at a station or transmitted to other tracking and data relay satellites, which then send the data to Earth. Once received on the ground, the data can be processed into imagery.

Thinking Critically with Geospatial Technology 11.1

What Effect Does Satellite Remote Sensing Have on Political Borders?

When imagery is acquired by satellite remote sensing for any part of the globe, the satellite sensors effectively ignore political boundaries. After the 2010 earthquake in Haiti, imagery of the country was quickly acquired by satellites and made available for rescue and recovery efforts. When satellite images of nuclear facilities in North Korea or Iran are shown on the news, the satellite's sensors were able to image these places from orbit. Satellites are acquiring global imagery of events ranging from the 2010 Olympics in Vancouver to parades in Beijing, and in all cases the political boundaries of

these places are being ignored by the "eye in the sky" looking down from Earth's orbit. Satellites are not restricted by "no-fly zones" or other restrictions involving a country's airspace and are capable of acquiring imagery of nearly anywhere. Given these types of imaging capabilities, what effect does acquisition of images via satellites have on political borders? There are many areas that can't be approached from the ground without causing international tensions, but these same places can easily be viewed from space. How are political boundaries affected by these types of geospatial technologies?

What Are the Capabilities of a Satellite Sensor?

A satellite sensor has four characteristics that define its capabilities: spatial resolution, radiometric resolution, temporal resolution, and spectral resolution. In Chapter 10 we discussed the concept of resolution in the context of **spatial resolution**, the area on the ground represented by one pixel in a satellite image. A sensor's spatial resolution will affect the amount of detail that can be determined from the imagery. For instance, a sensor with a 1-meter spatial resolution has much finer resolution than a sensor with spatial resolution of 30 meters (see Figure 11.5 for examples of the same area viewed by different spatial resolutions). A satellite's sensor is fixed with one spatial resolution for the imagery it collects; for instance, a 30-meter resolution sensor cannot be "adjusted" to collect 10-meter resolution images.

However, if two bands (one with a higher resolution than the other) are used by the sensor to image the same area, the higher-resolution band can be used to sharpen the resolution of the lower-resolution band. This technique is referred to as **pan-sharpening** because the higher-resolution band used is a panchromatic band. As discussed in Chapter 10, in terms of imagery, a **panchromatic sensor** will be measuring only one large band of wavelengths at once (usually the entire visible portion of the spectrum or the entire visible and part of the near-infrared spectrum). For example, a satellite (like Landsat 7) that senses the blue, green, and red portions of the electromagnetic spectrum at 30-meter resolution could also have a sensor equipped to view the entire visible portion of the spectrum and part of the near-infrared spectrum as a panchromatic band at 15-meter resolution. This panchromatic image can then be fused with the 30-meter resolution color imagery to create a higher-resolution color composite. Many satellites with high spatial-resolution capabilities sense in a panchromatic band that allows for pan-sharpening of the imagery in the other bands.

As mentioned in Chapter 10, a sensor scales the energy measurements into several different ranges (referred to as *quantization levels*) to assign

spatial resolution the ground size represented by one pixel of satellite imagery.

pan-sharpening fusing a higher-resolution panchromatic band with lower-resolution multispectral bands to improve the clarity and detail seen in the image.

panchromatic sensor a sensor that can measure one range of wavelengths.

FIGURE 11.5 The same area on the landscape as viewed by three different sensors, each having different spatial resolutions. (Source: NASA Marshall Space Flight Center)

brightness values to a pixel. The sensor's **radiometric resolution** refers to its ability to measure fine differences in energy. For instance, a sensor with 8-bit radiometric resolution (also mentioned in Chapter 10) places the energy radiance values on a scale of 0 (lowest radiance) to 255 (highest radiance). The wider the range of measurements that can be made, the finer the sensor's radiometric resolution is. The radiometric resolution (and the resultant number of levels) is a measure of the number of the sensor's bits of precision. To calculate the number of levels, take the number of bits and apply it as an exponent to the number 2. For instance, a 6-bit sensor would have 64 levels (2^6) ranging from a value of 0 to a value of 63. An 8-bit sensor would have 256 levels (2^8) ranging from 0 to 255. An 11-bit sensor would have 2048 levels (2^{11}) ranging from 0 to 2047.

The finer a sensor's radiometric resolution, the better it can discriminate between smaller differences in energy measurements. For example, a 2-bit sensor (2^2 or four values) would have to assign every pixel in an image a brightness value of 0, 1, 2, or 3, which would result in most pixels having the same value, and it would be difficult to distinguish between items or to create effective spectral signatures. However, an 11-bit sensor would assign a value anywhere from 0 to 2047, creating a much wider range of value levels and allowing much finer distinctions to be made between items. See Figure 11.6 for examples of differing effects of radiometric resolution on an image when performing remote sensing.

The sensor's **temporal resolution** refers to how often it can return to image the same spot on Earth's surface. For instance, a sensor with 16-day temporal resolution will collect information about the swath containing your house and not return to image it again until 16 days have passed. The finer a sensor's temporal resolution, the fewer days it will take between return times. A way to improve a sensor's temporal resolution is to use **off-nadir viewing** capability, in which the sensor is not fixed to sense what's directly underneath it (the nadir point). A sensor that is capable of viewing places off-nadir can be pointed to image locations away from the current orbital path. Using off-nadir viewing can greatly increase a sensor's temporal resolution, as it can image a target several times during its orbits, even if the satellite isn't directly overhead. This capability is a great help in monitoring conditions that can change quickly or require

FIGURE 11.6 The effect of different levels of radiometric resolution on an image (simulated 2-bit, 4-bit, and actual 8-bit imagery of one band). (Source: NASA)

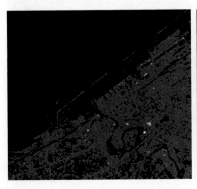  

rapid responses (such as mitigating disasters resulting from hurricanes, torna-does, or widespread fires).

The sensor's last parameter is **spectral resolution**, which refers to the bands and their wavelengths measured by the sensor. For instance, a sensor on the QuickBird satellite can measure four bands of energy simultaneously (wavelengths in the blue, green, red, and near-infrared portions of the spectrum), and the sensor onboard the Landsat 7 satellite can sense in seven different wavelengths at the same time. A sensor with a higher spectral resolution could examine several finer intervals of energy wavelengths (for example, the MODIS sensor onboard the Terra satellite—see Chapter 12—can sense in 36 different bands at once). Being able to sense numerous smaller wavelengths in the near-infrared portion of the spectrum gives much finer spectral resolution than just sensing one wider wavelength of near-infrared energy.

spectral resolution the bands and wavelengths being measured by a sensor.

What Is a Landsat Satellite and What Does It Do?

This chapter and Chapter 10 have made mention of a satellite system called Landsat but haven't really explained what it is, how it operates, or why it's so important. What would become the United States' **Landsat** Program was conceived in the late 1960s as a resource for doing remote sensing of Earth from space. The program's first satellite, Landsat 1, was launched in 1972 (see Figure 11.7 for a graphical timeline of Landsat launches). Landsat 1's sensor was the Landsat **MSS** (Multi-Spectral Scanner) and had a temporal resolution of 18 days. The MSS would continue to be part of the Landsat missions through Landsat 5. Landsat 4 and 5's MSS was capable of sensing four bands (red, green, plus two bands in the near infrared) at 79-meter spatial resolution and 6-bit radiometric resolution. Like the name implies, MSS is an example of a **multispectral sensor**, which can measure multiple wavelengths at once.

Landsat a long-running United States remote sensing project that had its first satellite launched in 1972 and continues today.

MSS the Multi-Spectral Scanner aboard Landsat 1 through 5.

multispectral sensor a sensor that can measure multiple different wavelength bands simultaneously.

FIGURE 11.7 The history of the Landsat program launches, along with upcoming dates.

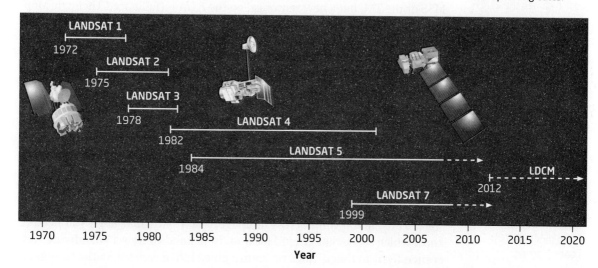

TM the Thematic
Mapper sensor onboard
Landsat 4 and 5.

Landsats 4 and 5 were also equipped with a new remote sensing device called the **TM** (Thematic Mapper). The TM was also a multispectral sensor that allowed for sensing in seven bands in the electromagnetic spectrum simultaneously as follows:

- Band 1: Blue (0.45 to 0.52 micrometers)
- Band 2: Green (0.52 to 0.60 micrometers)
- Band 3: Red (0.63 to 0.69 micrometers)
- Band 4: Near infrared (0.76 to 0.90 micrometers)
- Band 5: Middle infrared (1.55 to 1.75 micrometers)
- Band 6: Thermal infrared (10.4 to 12.5 micrometers)
- Band 7: Middle infrared (2.08 to 2.35 micrometers)

These bands were chosen for TM because they were key to a variety of studies (such as bands 3 and 4 being used for vegetation monitoring and NDVI as discussed in Chapter 10), while band 5 is useful when examining the water content of plants and band 7 is useful in separating various rock and mineral types. The thermal band 6 enabled the monitoring of heat energy in a variety of settings. The TM bands had improved radiometric resolution over MSS of 8 bits while the spatial resolution of bands 1-5 and 7 was improved to 30 meters (band 6 had a spatial resolution of 120 meters) and had a swath width of 185 kilometers. TM also had an improved temporal resolution of 16 days.

Landsat 5 the fifth
Landsat mission,
launched in 1984,
which carries both the
TM and MSS sensors.

Landsat 4 was launched in 1982 and ended its mission in 1993. **Landsat 5** was launched in 1984 and still continues sensing and transmitting viable data to Earth today. Landsat 5 is more than twenty-five years old, so it should raise some curiosity that it's still relied upon so heavily, given the importance of Landsat in remote sensing. Landsat 5 was not likely intended to be the cornerstone of the program—however, Landsat 6 (launched in 1993) did not achieve orbit and was lost, continuing the mission of Landsat 5.

Landsat 7 the seventh
Landsat mission,
launched in 1999,
which carries the ETM+
sensor.

ETM+ the Enhanced
Thematic Mapper sensor
onboard Landsat 7.

Landsat scene a
single image obtained
by a Landsat satellite
sensor.

**Worldwide Reference
System** the global
system of Paths and
Rows that is used to
identify what area
on Earth's surface
is present in which
Landsat scene.

Landsat 7 (Figure 11.8) was launched in 1999, not carrying either the MSS or the TM, but a new sensor called the **ETM+** (Enhanced Thematic Mapper Plus). This new device senses the same 7 bands as TM at the same swath width and radiometric and spatial resolutions (although the thermal band 6's spatial resolution was improved to 60 meters), but also added a new eighth band, a panchromatic one with a spatial resolution of 15 meters. With this improved spatial resolution, the panchromatic band could be used to pan-sharpen the quality of the other imagery. Landsat 7's temporal resolution was still 16 days but could acquire about 250 images per day (see Figure 11.9 for an example of Landsat 7 imagery and *Hands-on Application 11.2: Streaming Landsat Imagery* on page 334 for some examples of recent imagery collected by Landsat).

Each **Landsat scene** measures an area about 170 kilometers long by 183 kilometers wide (for the TM and ETM+ sensors) and is considered a separate image. At this size, it would take several images just to cover one state. Thus, if you want access to imagery of a particular area, you need to know which scene that ground location is in. Landsat uses a **Worldwide Reference System** that divides the entire globe into a series of Paths (columns)

FIGURE 11.8 Landsat 7 in action. (Source: Landsat imagery courtesy of NASA Goddard Space Flight Center and U.S. Geological Survey)

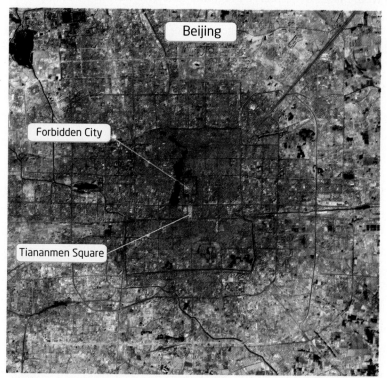

FIGURE 11.9 A Landsat 7 image of Beijing, China, from 2000. (Source: Courtesy of NASA Images acquired by Landsat 7)

Hands-on Application 11.2

Streaming Landsat Imagery

For a dynamic look at streaming Landsat imagery, the EarthNow! Website (sponsored by the USGS and available online at **http://earthnow.usgs.gov**) is a great tool (use of EarthNow! requires you to have Java loaded on your computer). EarthNow! shows recent Landsat imagery (usually from today's date)—a small map on the screen will show the position and orbit of the Landsat satellite that acquired the imagery. The main EarthNow! window will show the imagery collected by Landsat (with attached labels added by those who processed the data). Check out EarthNow! to see where recent Landsat imagery was collected and what the images of the ground that are obtained by Landsat look like.

and Rows. For instance, imagery of Columbus, Ohio, would be from the scene at Path 19, Row 32, while imagery of Toledo would be found in Path 20, Row 31 (Figure 11.10). At an area of 170 kilometers long by 183 kilometers wide, a scene will often be much larger than the area of interest—for example, Path 19, Row 32 contains much more than just Columbus, it encompasses the entire middle section of the state of Ohio, so to examine only Columbus, you would have to work with a subset of the image that matches the city's boundaries. Also, sometimes an area of interest may straddle two Landsat scenes—for instance, Mahoning County in Ohio is entirely located in Path 18, but part of the country is in Row 31 and part of it is in Row 32. In this case, both images would need to be merged together and then have the boundaries of the county used to define the subsetted area.

There's been a growing trend to make Landsat imagery available to the public at a reduced cost. Thanks in part to the activities of groups such as

FIGURE 11.10 The total Landsat coverage for the state of Ohio. It requires nine Landsat scenes to cover the entire state.

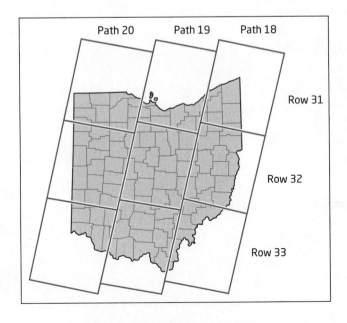

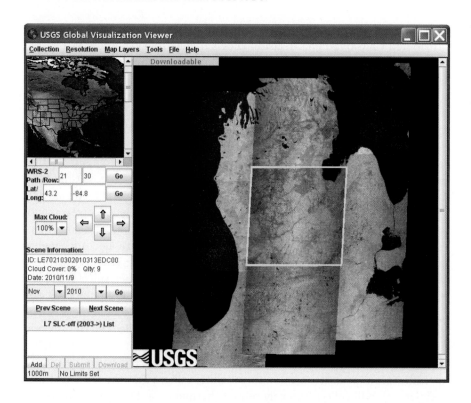

FIGURE 11.11 The USGS Global Visualization Viewer utility used to examine and download a Landsat ETM+ scene (note the size of one scene in relation to the entire state of Michigan). (Source: USGS)

OhioView and AmericaView (see Chapter 15) a lot of Landsat imagery is freely available for distribution over the Internet. The USGS maintains an online utility called **GloVis** (Global Visualization Viewer) that is used to distribute Landsat (and other satellite) imagery across the Internet (see Figure 11.11 and *Hands-on Application 11.3: Viewing Landsat Imagery with GloVis*). In 2008, the USGS announced that the entire Landsat 7 archive of imagery, plus the

GloVis the Global Visualization Viewer set up by the USGS for viewing and downloading satellite imagery.

Hands-on Application 11.3

Viewing Landsat Imagery with GloVis

With GloVis, Landsat imagery from multiple dates (and multiple sensors) can be viewed, examined (for the amount of cloud cover and the available imagery dates), and downloaded for use by a GIS or image processing or viewing software package. To use GloVis, open your Web browser and go to **http://glovis.usgs. gov**—also make sure that you have Java installed on your machine and your browser will allow for a popup window from this Website. Either select a location from the map or enter latitude/longitude coordinates for a place. The Landsat scenes for the area will

appear in a separate window, allowing you to select imagery by a specific path and row, obtain information about the scene (such as the percentage of cloud cover), and available imagery from different sensors.

If the word "downloadable" appears in red in the upper-left-hand corner of the image, the raw band data for that Landsat scene can be obtained directly via GloVis. Take a look at your local area (finding it via Path/Row combination or latitude/longitude) to see what Landsat imagery is available for those coordinates.

available data from Landsats 1 through 5, would be made freely available and distributed via GloVis. This decision allows free worldwide access to a vast library of remote sensing data for everyone, and this availability of the data can only help with furthering research and education in remote sensing.

SLC the Scan Line Corrector in the ETM+ sensor. Its failure in 2003 causes Landsat 7 ETM+ imagery to not contain all data from a scene.

In 2003, Landsat 7's **SLC** (Scan Line Corrector) failed and could not be repaired. Due to the malfunctioning SLC, the ETM+ produces imagery containing missing data. Without a working SLC, an ETM+ image contains only about 75% of the data that should be there. ETM+ data is still used by substituting pixels from other Landsat 7 scenes from the same area but a different date, or by using other methods to fill in the gaps for the missing pixel values. With these types of issues, Landsat 5 TM imagery is still being used by some researchers for a source of remotely sensed Landsat data. However, with Landsat 5 being years beyond its mission life (and having had some mechanical issues in recent years) and Landsat 7 having issues with collecting imagery due to the SLC failure, the question that has to be raised is: What is the future of the Landsat program, as there has been no Landsat launch since 1999?

LDCM the Landsat Data Continuity Mission—the future of the Landsat program, scheduled to be launched in 2012.

OLI the Operational Land Imager, the multispectral sensor that will be onboard LDCM.

The next planned Landsat launch is the Landsat Data Continuity Mission (**LDCM**) scheduled for the end of 2012. The LDCM satellite's instrument will be the next iteration of Landsat sensors (similar to ETM+), called the **OLI** (Operational Land Imager). Unlike ETM+, a thermal band will not be part of OLI, but two new bands, aimed at monitoring clouds and coastal zones, will be developed instead. A separate instrument on the satellite, called TIRS (the Thermal Infra-Red Sensor), will collect remotely sensed thermal data. LDCM is estimated to collect about 400 images per day. There's a strong need for continuing to obtain Landsat imagery for monitoring larger-scale phenomena such as environmental effects, land-use changes, and urban expansion. Long-term gaps in the dates of imagery would have a huge negative impact on these types of studies.

Being able to monitor large-scale conditions on Earth on a regular basis (at 30-meter resolution) with a variety of multispectral bands (including near-infrared, middle infrared, and thermal characteristics) will provide a rich dataset for numerous applications. With Landsat imagery stretching back for more than 30 years providing a constant source of observations, long-term tracking of environmental phenomena (such as ongoing land-use or land-cover changes for an area) can be performed. For instance, the growth of urban areas over time can be observed every 16 days (cloud cover permitting, of course), enabling planners to determine the spatial dimensions of new developments across a region. Figure 11.12 shows an example of utilizing Landsat imagery over time to track the patterns of deforestation in Brazil through examination of clear-cut areas and new urban areas built up in the forest. The color composite images show the rainforest in dark green shades, while the deep-cut patterns of forest clearing are clearly visible and the encroachment of development into the forest can be tracked and measured over time.

Beyond examining land-cover changes over time, the multispectral capabilities of Landsat imagery can be used for everything from measuring crop drought conditions to glacier features to the aftermath of natural disasters. By examining the different band combinations provided by Landsat imagery,

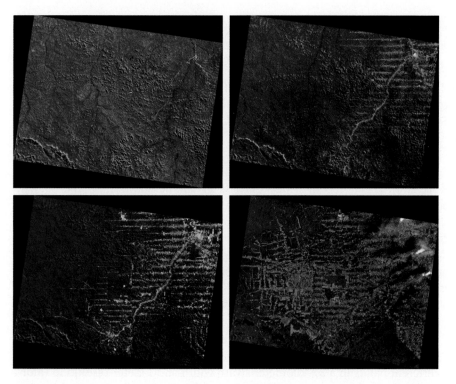

FIGURE 11.12 Multiple dates of Landsat imagery of Rondonia, Brazil, showing the forested areas (in dark green) and the growing patterns of deforestation over time (in lighter shades). (Source: NASA/Goddard Space Flight Center Scientific Visualization Studio)

scientists can find clues to understanding the spread of phytoplankton in coastal waters, enabling environmentalists to try and head off potential algae blooms or red-tide conditions. Threats to animal habitats can be assessed through forest or wetland monitoring via Landsat imagery; Landsat images also allow scientists to track the movements of invasive species that threaten timber industries. *Hands-on Application 11.4: Applications of Landsat Imagery* lets you dig further into the myriad uses of Landsat remote sensing.

NASA also operates a remote sensing project with capabilities related to the Landsat program. **EO-1** (Earth Observing 1) was launched in 2000 as part of NASA's New Millennium Program and placed in orbit so it was traveling

EO-1 a satellite launched in 2000 and set to orbit 1 minute after Landsat 7.

Hands-on Application 11.4

Applications of Landsat Imagery

Landsat imagery has been adopted for numerous environmental monitoring, measuring, and tracking applications. To further investigate these uses, open your Web browser and go to **http://landsat. gsfc.nasa.gov/about/appl_matrix.html**. This Website, maintained by NASA, highlights several different categories of Landsat imagery applications, including Agricultural, Coastal, Geologic, Hydrologic, Mapping, and Environmental Uses. Select one of the categories of interest to you or related to your field and explore some of the case studies and applications presented in that column. How is Landsat imagery being applied to these fields, and how do the long-term consistent uses of multispectral imagery lend themselves to these kinds of applications?

Thinking Critically with Geospatial Technology 11.2

What if There Was No Landsat Data Continuity Mission?

The Landsat Program has been ongoing since 1972, providing global coverage of imagery and a mainstay of the United States remote sensing program. However, it currently relies on the Landsat 5 (launched in 1984, long past its shelf life but still doing the job) and Landsat 7 (launched in 1999 and now impaired due to the SLC error) satellites. With the next stage of the Landsat Program (LDCM) not planned for launch until late 2012, the potential is there for a gap in the acquisition of Landsat imagery if Landsat 5 should fail and Landsat 7's issues should become unsustainable. While this may sound like something of a doomsday scenario, if LDCM is delayed or if disaster should strike and LDCM should fail to achieve orbit (like Landsat 6), and Landsats 5 and 7 would somehow be unavailable for imagery acquisition, then what happens? Would the United States have to purchase imagery from other sources or countries to make up the data gap until another Landsat mission could be launched? Are there other sources that the United States government could use to make up for the lack of Landsat imagery? What impact would a lack of continuous Landsat imagery have on the numerous applications and studies previously described?

ALI the multispectral sensor onboard EO-1.

Hyperion the hyperspectral sensor onboard EO-1.

hyperspectral sensor a sensor that can measure hundreds of different wavelength bands simultaneously.

right behind Landsat 7 (within 1 minute). EO-1 features a multispectral sensor called **ALI** (Advanced Land Imager) that is similar to ETM+ but can image 10 different bands in a 37-kilometer swath width at 30-meter resolution (along with a 10-meter panchromatic band). ALI was also used to test and validate sensor technology planned for use in LDCM. EO-1 also carries **Hyperion**, a sensor that detects 220 bands at 30-meter resolution in a smaller swath width of 7.7 kilometers. Hyperion is an example of a **hyperspectral sensor**, which can measure hundreds of wavelengths simultaneously.

Although Landsat data has been widely used for numerous environmental studies, 30-meter resolution isn't nearly enough to do very detailed, high-resolution studies. At 30-meter resolution, broad land-use change modeling and studies can be performed, but a much finer spatial resolution would be required to pick out individual details in an image. High-resolution sensors have a different purpose than Landsat imagery—with a very fine spatial resolution, crisp details can emerge from a scene. Spy satellites or military surveillance satellites would require this type of finer spatial resolution, but of course the imagery they acquire remains classified. Much of the non-classified and available high-resolution satellite imagery is obtained and distributed commercially rather than from government sources.

What Satellites Have High-Resolution Sensors?

SPOT a satellite program operated by the French Space Agency and the Spot Image Corporation.

There are several remote sensing satellites with high-resolution capabilities. This section describes a few of the more prominent ones. The **SPOT** (Satellite Pour l'Observation de la Terre) program is operated by a French company called

Spot Image, which serves as the distributor of SPOT imagery. The SPOT series of satellites was developed by the French Space Agency, with the first (SPOT 1) being launched in 1986. The most recent satellite in the series is SPOT 5, launched in 2002. SPOT 5 has a revisit cycle of 26 days but has off-nadir viewing capabilities that significantly improve some of its sensors' temporal resolution. The highest spatial resolution of a SPOT 5 image is 2.5 meters, allowing a significant amount of detail to be determined from a single image.

SPOT 5 carries three sensors:

- Vegetation-2: A sensor with a 1-kilometer spatial resolution, used for broad-scale environmental monitoring.

- HRS (High Resolution Stereoscopic): A sensor with 10-meter resolution, used for the development of stereo imagery (see Chapter 15) or digital elevation models (see Chapter 13).

- HRG (High Resolution Geometric): A pair of sensors with multiple resolutions available—20-meter (an infrared band), 10-meter (multispectral sensor with green, red, near-infrared, and one other infrared band), 5-meter (panchromatic), and a 2.5-meter panchromatic Supermode.

Spot Image has announced SPOT 6 and SPOT 7 to be launched in the near future with improved resolution capabilities (such as a 1.5-meter panchromatic spatial resolution). A related program is the **PLEIADES** series of satellites, the first of which is projected to be launched in 2011. PLEIADES will have very high-resolution panchromatic and multispectral capabilities.

Launched in 1999, the GeoEye Company's (formerly SpaceImaging, Inc.) **IKONOS** satellite imagery was one of the benchmarks of commercial high-resolution imagery. The IKONOS satellite has two sensors, a four-band (blue, green, red, and near-infrared) multispectral sensor that has 4-meter spatial resolution and a panchromatic sensor with 1-meter spatial resolution. Both sensors have very high radiometric resolution (11 bits) and due to off-nadir viewing capabilities can achieve temporal resolution of about 3 days. Panchromatic or pan-sharpened 1-meter imagery provides very crisp and detailed image resolution—details of urban environments can be distinguished with little difficulty, although objects smaller than 1 meter square cannot have details made out (see Figure 11.13 on page 340 for an example of IKONOS imagery for the kind of spatial resolution this sensor can provide).

In 2001, another remote sensing company, DigitalGlobe, launched its **QuickBird** satellite, whose sensors provided higher spatial resolution than IKONOS. Like IKONOS, QuickBird featured two sensors—a four-band (blue, green, red, and near-infrared) multispectral sensor with 2.4-meter spatial resolution and a panchromatic sensor with 0.61-meter spatial resolution. Both sensors feature very fine 11-bit radiometric resolution and with off-nadir viewing capability have a temporal resolution between 1 and 3.5 days. 61-centimeter imagery provides extremely high-resolution capabilities, allowing very crisp details to be observed in an image. In 2007, DigitalGlobe launched **WorldView-1**, which featured an improved panchromatic sensor

PLEIADES a high-resolution series of satellites.

IKONOS a satellite launched in 1999 by SpaceImaging, Inc. (now called GeoEye), which features multispectral spatial resolution of 4 meters and panchromatic resolution of 1 meter.

QuickBird a satellite launched in 2001 by the DigitalGlobe company, whose sensors have 2.4-meter multispectral resolution and 0.61-meter panchromatic resolution.

WorldView-1 a satellite launched in 2007 by the Digital Globe company, whose panchromatic sensor has 0.5-meter spatial resolution.

FIGURE 11.13 An IKONOS image of the development of one of the Palm Islands in Dubai, United Arab Emirates. (Source: Satellite Imaging Corporation Copyright © 2007 GeoEye/SIME. All Rights Reserved.)

WorldView-2 a satellite launched in 2009 by the DigitalGlobe company, featuring an 8-band multispectral sensor with 1.84-meter resolution and a panchromatic sensor of 0.46-meter resolution.

GeoEye-1 a satellite launched in 2008 by the GeoEye company, which features a spatial resolution of 41 centimeters with its panchromatic sensor.

with 0.5-meter spatial resolution. DigitalGlobe also launched **WorldView-2** in October 2009 with 0.46-meter panchromatic and 1.84-meter 8-band multispectral spatial resolution.

The highest spatial resolution imagery currently commercially available is from the sensors onboard the **GeoEye-1** satellite, launched in 2008 by the GeoEye company. GeoEye-1 has two sensors, a four-band multispectral (blue, green, red, and near infrared) with 1.65-meter spatial resolution and a

Hands-on Application 11.5

Viewing High-Resolution Satellite Imagery

While the data acquired by satellites like IKONOS and QuickBird is sold commercially, you can view samples of these satellite images at several places online. The Satellite Imaging Corporation's Website provides a means for viewing high-resolution imagery. Open your Web browser and go to **http://www. satimagingcorp.com**. Select the option for Gallery, and then select High-Resolution Sensors. Available imagery includes GeoEye-1, IKONOS, QuickBird, SPOT-5, WorldView-1, and WorldView-2. Examine some examples of each sensor, keeping in mind the specifications of each. You can zoom in on the sample imagery from each satellite to really get a sense of what can be seen via high-spatial resolution satellites. Note that the images available on the Website are just crisp graphics, not the actual satellite data itself.

panchromatic sensor with 0.41-meter resolution. 41-centimeter spatial resolution represents the finest commercial resolution on the market today, currently making GeoEye-1 the satellite with the sensors to beat when it comes to high-resolution imagery.

The GeoEye company has announced plans for the launch of the Geo-Eye-2 satellite in 2012, which would provide 0.25-meter spatial resolution imagery. See *Hands-on Application 11.5: Viewing High-Resolution Satellite Imagery* for examples of all of these high-resolution images online.

Chapter Wrapup

Satellite remote sensing is a powerful tool for everything from covert surveillance to land-use and land-cover monitoring. The next chapter delves into a new set of geospatial satellite tools—the Earth Observing System, a series of orbital environmental satellites that gather data about global processes on an ongoing basis. *Geospatial Lab Application 15.1* returns to using MultiSpec and has you doing more with Landsat imagery and the multispectral capabilities of the TM sensor.

Important note: The references for this chapter are part of the online companion for this book and can be found at **http://www.whfreeman.com/shellito1e.**

Key Terms

NASA (p. 323)
Corona (p. 324)
geostationary orbit (p. 325)
near-polar orbit (p. 325)
swath width (p. 326)
Sun-synchronous orbit (p. 326)
along-track (p. 327)
across-track (p. 327)
EROS (p. 328)
spatial resolution (p. 329)
pan-sharpening (p. 329)
panchromatic sensor (p. 329)
radiometric resolution (p. 330)
temporal resolution (p. 330)
off-nadir viewing (p. 330)
spectral resolution (p. 331)
Landsat (p. 331)
MSS (p. 331)
multispectral sensor (p. 331)
TM (p. 332)
Landsat 5 (p. 332)

Landsat 7 (p. 332)
ETM+ (p. 332)
Landsat scene (p. 332)
Worldwide Reference System (p. 332)
GloVis (p. 335)
SLC (p. 336)
LDCM (p. 336)
OLI (p. 336)
EO-1 (p. 337)
ALI (p. 338)
Hyperion (p. 338)
hyperspectral sensor (p. 338)
SPOT (p. 338)
PLEIADES (p. 339)
IKONOS (p. 339)
QuickBird (p. 339)
WorldView-1 (p. 339)
WorldView-2 (p. 340)
GeoEye-1 (p. 340)

Landsat Imagery

This chapter's lab application builds off the remote sensing basics of the previous chapter and returns to using the MultiSpec program. In this exercise, you'll be starting with a Landsat TM scene and creating a subset of it to work with. During the lab, you'll examine the uses for several Landsat TM band combinations in remote sensing analysis.

Objectives

The goals for this exercise are:

- Further familiarizing yourself and work with satellite imagery in MultiSpec.
- Creating a subset image of a Landsat scene.
- Examining different Landsat bands in composites and comparing the results.
- Examining various landscape features in multiple Landsat bands and comparing them.
- Applying visual image interpretation techniques to Landsat imagery.

Obtaining Software

The current version of MultiSpec is available for free download at **http://cobweb.ecn.purdue.edu/~biehl/MultiSpec**.

Important note: Software and online resources sometimes change fast. This lab was designed with the most recently available version of the software at the time of writing. However, if the software or Websites have significantly changed between then and now, an updated version of this lab (using the newest versions) is available online at **http://www.whfreeman.com/shellito1e**.

Lab Data

Copy the folder '**Chapter11**'—it contains:

- A Landsat 5 TM satellite image (called 'neohio.img') from 9/11/2005 of northeast Ohio, which was constructed from data supplied via OhioView.
- The Landsat TM image bands refer to the following portions of the electromagnetic (EM) spectrum in micrometers (μm):
  - Band 1: Blue (0.45 to 0.52 μm)
  - Band 2: Green (0.52 to 0.60 μm)

- Band 3: Red (0.63 to 0.69 μm)
- Band 4: Near infrared (0.76 to 0.90 μm)
- Band 5: Middle infrared (1.55 to 1.75 μm)
- Band 6: Thermal infrared (10.4 to 12.5 μm)
- Band 7: Middle infrared (2.08 to 2.35 μm)

Also, keep in mind that Landsat TM imagery has a 30-meter spatial resolution (except for the thermal infrared band, which is 120 meters).

Localizing This Lab

Although this lab focuses on a section of a Landsat scene from northeast Ohio, Landsat imagery is available for download via GloVis at http://glovis.usgs. gov. This site will provide the raw data that will have to be imported into MultiSpec and processed to use in the program.

Alternatively, NASA has information for obtaining free Landsat data for use in MultiSpec online at http://change.gsfc.nasa.gov/create.html. This Website provides instructions for acquiring free data for your own area and getting it into MultiSpec format.

11.1 Opening Landsat Scenes in MultiSpec

1. Start MultiSpec.
2. MultiSpec will start with an empty text box (which will give you updates and reports of processes in the program). You can minimize the text box for now.
3. To get started with the Landsat image, select **Open** from the **File** pull-down menu. Alternatively, you can select the **Open** icon from the toolbar:

(Source: Purdue Research Foundation)

4. Navigate to the '**Chapter11**' folder and select '**neohio.img**' as the file to open and click **Open**.
5. A new dialog box will appear to let you set the display specification for the 'neohio' image.

(Source: Purdue Research Foundation)

6. Under Channels, you will see the three color guns available to you (red, green, and blue). Each color gun can hold one band (see the Lab Data section for information on which bands correspond to which parts of the electromagnetic spectrum). The number listed next to each color gun represents the band being displayed with that gun.

7. Display band **4** in the red color gun, band **3** in the green color gun, and band **2** in the blue color gun.

8. Accept the other defaults for now and click **OK**.

9. A new window will appear and the 'neohio' image will load in it.

11.2 Subsetting Images and Examining Landsat Imagery

Right now, you're working with an entire Landsat TM scene, which is an area roughly 170 kilometers long by 183 kilometers wide (as shown in the graphic at the top of page 345—you are using the scene encompassed by path 18, row 31). For this lab, we want to focus only on the area surrounding downtown Cleveland. Thus, you will have to create a subset—in essence, "clipping" out the area that you're interested in and creating a new image from that.

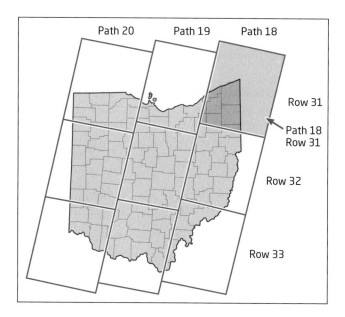

1. Zoom to the part of the 'neohio' image that shows Cleveland (like in the graphic below):

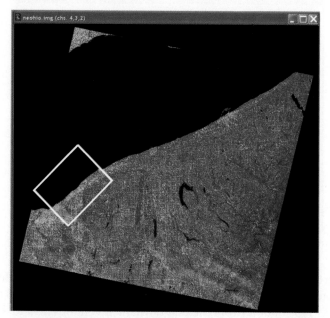

(Source: NASA/Purdue Research Foundation)

2. In the image, you should be able to see many features that make up downtown Cleveland—the waterfront area, much urban development, major roads, and water features.

3. In order to create a new image that shows only Cleveland (a suggested region is shown in the second graphic on page 345), select **Reformat** from the **Processor** pull-down menu, then select **Change Image File Format.**

4. You could draw a box around the area you want to subset using the cursor and the new image that's created will have the boundaries of the box you've drawn on the screen. However, for the sake of consistency in this exercise, use the following values for the Area to Reformat:

   a. Line:
      - Start 4305
      - End 4897
      - Interval 1

   b. Column:
      - Start 641
      - End 1313
      - Interval 1

Leave the other defaults alone and click **OK.**

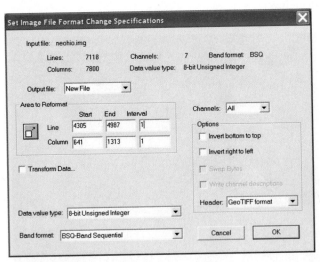

(Source: Purdue Research Foundation)

5. In the Save As dialog box that appears, save this new image in the 'Chapter11' folder (call it '**clevsub.img**'). Choose **Multispectral** for the Save as Type option (from the pull-down menu options next to **Save as Type**). When ready, click **Save.**

6. Back in MultiSpec, minimize the window containing the 'neohio' image.

7. Open the '**clevsub**' image you just created in a new window (in the Open dialog box, you may have to change the Files of Type that it's asking about to "All Files" to be able to select the 'clevsub' image option).

8. Open the 'clevsub' image with a 4-3-2 combination (band 4 in the red gun, band 3 in the green gun, and band 2 in the blue gun).

9. Use the other defaults for the Enhancement options (stretch is Linear and range is Clip 2% of tails).

10. In the Set Histogram Specifications dialog box that opens, select the **Compute new histogram** method, and use the default **Area to Histogram** settings.

11. Click **OK** when ready. The new subset image shows that the Cleveland area is ready to use.

12. Zoom into the downtown Cleveland area and especially areas along the waterfront. Answer Questions 11.1 and 11.2. (You will want to also open Google Earth and compare the Landsat image to its very crisp resolution imagery when answering these questions.)

Question 11.1 What kinds of features on the Cleveland waterfront cannot be distinguished at the 30-meter resolution you're examining?

Question 11.2 Conversely, what specific features on the Cleveland waterfront are apparent at the 30-meter resolution you're examining?

11.3 Examining Landsat Bands and Band Combinations

1. Zoom in the waterfront area, looking at Cleveland Browns Stadium and its immediate surrounding area.

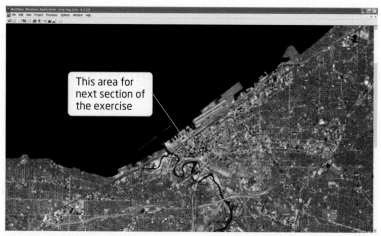

This area for next section of the exercise

(Source: Purdue Research Foundation)

2. Open another version of the 'clevsub' image using the 3-2-1 combination.

3. Arrange the two windows on the screen so that you can examine them together, expanding and zooming in as needed to be able to view the stadium in all three band combinations.

Question 11.3 Which of the two band combinations best brought the stadium and its field to prominence?

Question 11.4 Why did the band combination you chose best help in viewing the stadium and the field? (Hint: You may want to do some brief online research into the nature of the new stadium and its field.)

4. Return to the view of Cleveland's waterfront area. You'll now be examining the water features (such as Lake Erie and the river). Paying careful attention the water features, open four display windows and expand and arrange them side by side to look at differences between them. Create the following image composites (one for each window):

    a. 3,2,1

    b. 7,5,4

    c. 4,2,1

    d. 1,1,1

Question 11.5 Which band combination(s) best let you separate water bodies from land? Why?

5. Return to the view of Cleveland's waterfront area. Focus on the urban features and vegetated features (zoom and pan where necessary to get a good look at urbanization). Open new windows with some new band combinations, noting how things change with each one as follows:

    a. 7,4,2

    b. 1,2,3

    c. 4,3,2

Question 11.6 Which band combination(s) best let you separate urban areas from other forms of land cover (that is, vegetation, trees, etc.)? Why?

6. Open a new display window and examine the full view of the entire 'clevsub' image. Open this window with the following band combination:

    a. 6,6,6

7. Zoom in and move around the Cleveland waterfront area we've been examining.

Question 11.7 Why is the image so "blurry" or "fuzzy" in this combination (compared to all the other band combinations you've looked at in this exercise)?

Closing Time

This exercise wraps up working with Landsat information as well as working with imagery in MultiSpec. Chapter 12 focuses on a whole different set of remote-sensing satellites (part of the Earth Observing System) and its exercise uses a new software program called NASA World Wind to examine this imagery.

Exit MultiSpec by selecting **Exit** from the **File** pull-down menu. There's no need to save any work in this exercise.

12

Studying the Environment from Space

NASA's Earth Observing System Program, Terra, Aqua, Aura, and Viewing and Examining EOS Imagery

"Climate change" and "global warming" have become hot-button issues in today's world, affecting everything from a country's energy policies to its economy to its citizens' choices of homes or vehicles. The study and monitoring of the types of environmental processes that affect the land, oceans, and atmosphere occur on a global scale, and a continuous flow of constant information is needed. For instance, monitoring changes in global sea surface temperature is going to require information captured for the world's oceans on an ongoing basis. In order to acquire the large-scale information of the entire planet in a timely fashion, remote sensing technology can be utilized. After all, satellites are continuously orbiting the planet and collecting data—they provide the ideal mechanism for monitoring Earth's environmental conditions and changes. There are a lot of other remote sensing satellites up in orbit doing their jobs besides Landsat and the high-resolution satellites discussed in Chapter 11. In fact, there's an entire set of science observatories orbiting hundreds of miles over your head dedicated to continuously monitoring things like sea surface temperature, glaciers, trace gases in the atmosphere, humidity levels, and Earth's heat radiation.

The Earth Observing System (**EOS**) is a series of remote sensing satellites operated and maintained by NASA. Each satellite's capabilities are dedicated to an aspect of environmental monitoring or data collection related to Earth's surface, oceans, atmosphere, or biosphere. These EOS satellites often have coarse spatial resolution (for instance, some instruments have a spatial resolution of each pixel representing 1000 meters) but consequently have a much larger swath width (such as the ability to "see" up to 2300 kilometers in one swath). Their instruments are also set up to monitor very specific wavelengths tied to the purpose of that particular sensor. Their revisit times are usually relatively short (like imaging the whole Earth in a couple of days) to provide

EOS NASA's Earth Observing System mission program.

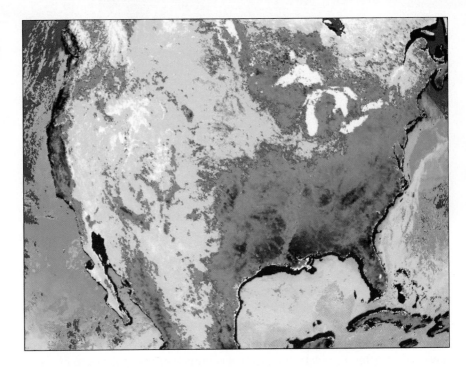

constant data streams of observation data about whatever environmental system they're examining.

There are a lot of different uses and products of EOS data. This chapter can't hope to cover all of them, just demonstrate some sample applications. See Figure 12.1 for an example of the kind of data and imagery that can be collected by EOS—the image was collected by a sensor onboard an EOS satellite over eight days in April, 2000. It shows the health of vegetation (the "greening" of springtime) of North America and the sea surface temperature of its surrounding waters. The overall mission of EOS is to examine global environmental phenomena to advance knowledge and understanding of Earth's systems as a whole. The EOS missions were begun during the 1990s—there are several ongoing EOS missions, with more planned for the future, with three key satellites being Terra, Aqua, and Aura.

What Is Terra and What Can It Do?

Terra the flagship satellite of the EOS program.

Morning Constellation a set of satellites (including Terra, Landsat 7, SAC-C, and EO-1) that pass the Equator in the morning during their orbits.

Terra serves as the "flagship" satellite of the EOS program. It was launched in 1999 as a collaboration between the United States (NASA's Goddard Spaceflight Center, NASA's Jet Propulsion Lab, and the Langley Research Center) and agencies in Canada and Japan, and continues its remote sensing mission today. Terra's orbit is Sun synchronous and it flies in formation as one of four satellites called the **Morning Constellation.** This series of satellites gets this name because Terra crosses the Equator at 10:30 A.M. each day (it was originally named EOS-AM-1 because of this morning crossing). The other satellites in

FIGURE 12.2 A global composite of CERES imagery from June 2010, showing the radiated energy from Earth (blue areas are colder, red areas are warmer). (Source: Kevin Ward, NASA Earth Observations)

this morning constellation are Landsat 7 (which passes at 10:01 A.M.), EO-1 (which passes at 10:02 A.M.), and another satellite called SAC-C.

"Terra" means "Earth" in Latin, which is an appropriate name for the satellite, as its instruments provide numerous measures of (among many other functions) the processes involved with Earth's land and climate. There are five instruments onboard Terra, each with its own purpose and environmental monitoring capabilities. The data and imagery collected by Terra's instruments are processed into a number of specific products that are then distributed for use.

The first of Terra's instruments is **CERES** (Clouds and Earth's Radiant Energy System), a system designed to provide regular energy measurements (such as heat energy) from Earth's surface to the top of the atmosphere. Like its name indicates, CERES is used to study clouds and see what effect they have on Earth's energy. Using CERES to study clouds provides critical information about Earth's climate and temperature. As NASA notes, clouds play a defining role in the heat budget of Earth, but different type of clouds affect Earth in different ways—for instance, low clouds reflect more sunlight (and thus, help cool the planet) while high clouds reflect less but effectively keep more heat on Earth. By combining CERES data with data from other EOS instruments, scientists are able to examine cloud properties and information about Earth's heat and radiation (see Figure 12.2 for an example of using CERES for examining Earth's radiated energy).

CERES the Clouds and Earth's Radiant Energy System instruments onboard Terra and Aqua.

MISR (Multi-angle Imaging SpectroRadiometer) is the second of Terra's instruments. It has nine sensors that examine Earth, but each sensor is set up at a different angle. One of these sensors is at nadir, four of them are set at forward-looking angles, and the remaining four are set at the same angles, but looking backwards. MISR is a multispectral sensor, meaning that objects on the ground get viewed nine times (from different angles) in multiple bands, which means that Terra can effectively cover the entire planet in about nine days. By viewing things at multiple angles, stereo imagery (see Chapter 15) can be constructed, allowing scientists to measure things like cloud heights.

MISR the Multi-angle Imaging SpectroRadiometer instrument onboard Terra.

FIGURE 12.3 The
operation of Terra's MISR
instrument, showing the
Appalachian Mountains,
the Chesapeake Bay, and
Delaware Bay as captured
from looking straight
down (on the right) and
from a variety of other
angles (the images on the
left). (Source: NASA/GSFC/
JPL, MISR Science Team)

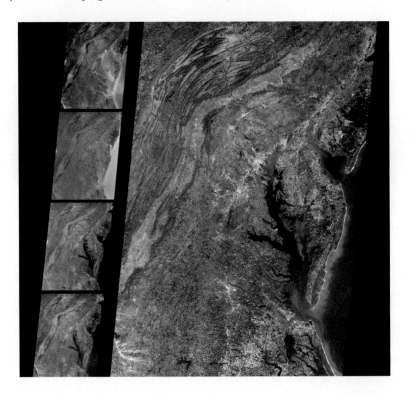

FIGURE 12.3 The operation of Terra's MISR instrument, showing the Appalachian Mountains, the Chesapeake Bay, and Delaware Bay as captured from looking straight down (on the right) and from a variety of other angles (the images on the left). (Source: NASA/GSFC/ JPL, MISR Science Team)

Likewise, the heights of plumes of smoke given off by wildfires or volcanoes can be measured using MISR's stereo capabilities. In addition, by showing the same area nine times in rapid succession, the images can provide a series of very quick temporal snapshots of the same phenomenon (Figure 12.3).

MISR data aids in environmental and climate studies by giving scientists the ability to better examine how sunlight is scattered in different directions, particularly by aerosols and atmospheric particles. For instance, a NASA article by Rosemary Sullivan details how MISR can study air pollution in the form of tiny particulates (caused by things like automobile or factory emissions, or from natural occurrences such as smoke or dust) that can lead to increased health risks. Detailed information on concentrations of aerosols and atmospheric particles can help scientists understand the level of exposure people are having to these types of pollutants. MISR is extremely sensitive to measuring these types of aerosols, and its data can help scientists tell the difference between what's on the ground and what's in the atmosphere. By using MISR (in conjunction with other instruments) to study aerosols, scientists can help determine areas of high aerosol concentrations and provide data to aid in cleaning up air pollution.

Another Terra instrument used for measuring airborne pollutants (but in a different way) is **MOPITT** (Measurements of Pollution in the Troposphere), designed for monitoring pollution levels in the lower atmosphere. MOPITT's sensors collect information to measure the amount of air pollution concentrations across the planet, creating an entire global image about

MOPITT the
Measurements
of Pollution in
the Troposphere
instrument onboard
Terra.

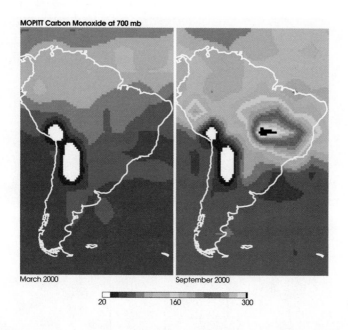

MOPITT Carbon Monoxide at 700 mb

March 2000   September 2000

20   160   300

FIGURE 12.4 MOPITT imagery showing growing carbon monoxide concentration levels in South America. (Source: Images courtesy David Edwards and John Gille, MOPITT Science Team, NCAR)

every three days. Specifically, MOPITT monitors carbon monoxide concentrations in Earth's atmosphere. Carbon monoxide, a global air pollutant, is caused by a number of factors, whether naturally occurring (such as from volcanoes or wildfires) or by human causes (such as vehicle emissions or industrial output). Being able to track long-term carbon monoxide sources can aid in determining areas generating large amounts of pollutants and then monitoring how those pollutants are being circulated through the atmosphere. Figure 12.4 shows an example of this—two MOPITT images of South America. One from March 2000 shows relatively low amounts of carbon monoxide. A second image from September 2000 shows the dimensions of a large mass of carbon monoxide, caused by biomass burning in the Amazon and the movement of carbon monoxide across the ocean to South America from fires in Africa.

Probably the most widely used instrument onboard Terra is **MODIS** (the Moderate Resolution Imaging SpectroRadiometer), a sensor that produces over 40 separate data products related to environmental monitoring. Being able to measure 36 spectral bands (including portions of the visible, infrared, and thermal) makes MODIS an extremely versatile remote sensing instrument, with multiple applications of its imagery. For instance, Figure 12.1 on page 352 was made from MODIS imagery and showed vegetation health and water temperatures. MODIS products are utilized for examining numerous types of environmental features, including (among many others) snow cover, sea surface temperature, sea ice coverage, ocean color, and volcanoes. In addition, MODIS's thermal capabilities can track the heat emitted by wildfires, allowing for mapping of active fires across the globe (see Figure 12.5 on page 356 for an example of MODIS imagery of fires and their aftermath recorded in Wilsons Promontory National Park in Australia).

MODIS the Moderate Resolution Imaging SpectroRadiometer instrument onboard Terra and Aqua.

FIGURE 12.5 MODIS imagery examining the dimensions of a fire and resulting smoke plume at Wilsons Promontory National Park. The other two MODIS images show the park before and after the fire. (Source: NASA/GSFC, MODIS Rapid Response)

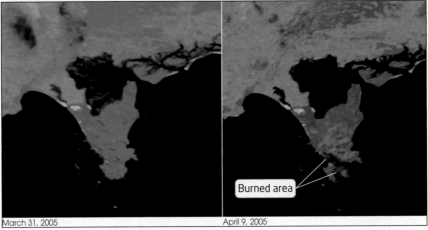

Burned area

March 31, 2005   April 9, 2005

MODIS imagery is very coarse spatial resolution (most images have 1-kilometer resolution). However, the swath width of MODIS is 2330 kilometers, making it possible to examine broad-scale phenomena of large geographic regions. MODIS views most of the globe in one day with complete global coverage being completed on a second day. With this quick revisit time, MODIS allows scientists to look at changes in sea surface temperature or snow cover over time, global leaf-area index studies, and active large-scale fires, all on a regular basis. For instance, the red and near-infrared bands sensed by MODIS can be used to regularly compute global measures of NDVI (see Chapter 10) allowing monitoring of the health of the planet's trees and vegetation. Another application of MODIS is the ability to study the health of phytoplankton in the oceans. As NASA notes, when phytoplankton is unhealthy or under heavy stress, it takes the sunlight it's absorbed and re-emits it as fluorescence, which can then be observed by MODIS. In this way, scientists can track potential harmful algal (phytoplankton) blooms around the world via MODIS. See Figure 12.6 for an example of MODIS imagery of Egypt and the Nile River—new lakes in the desert are visible adjoining the Nile. Also, check out *Hands-on Application 12.1: MODIS Rapid-Fire Online* for examples of current MODIS imagery online.

FIGURE 12.6 MODIS imagery showing new lakes along the Nile River. (Source: Image by Robert Simmon, Reto Stöckli, and Brian Montgomery, NASA GSFC)

The last Terra instrument is **ASTER** (Advanced Spaceborne Thermal Emission and Reflection Radiometer). ASTER has relatively high spatial resolution and is often referred to as a sort of "zoom lens" for Terra's other instruments. With its thermal and infrared measurements, ASTER is also used to obtain information about land surface temperature. In addition, through ASTER's thermal capabilities, natural hazards such as wildfires can be monitored, or phenomena such as volcanoes can be observed and data obtained on their health and lava flows (see Figure 12.7 on page 358 for an example of ASTER

ASTER the Advanced Spaceborne Thermal Emission and Reflection Radiometer instrument onboard Terra.

Hands-on Application 12.1

MODIS Rapid-Fire Online

Recent and archived MODIS data is available online at NASA's MODIS Rapid Response System. Open your Web browser and go to **http://rapidfire.sci. gsfc.nasa.gov/gallery**. Several MODIS images (not the raw data, but processed images) can be viewed. Check out some of the images from recent dates and see what kind of phenomena NASA is tracking. Select the Real-Time link on the Web page for the most recent MODIS acquisition data available. Use the imagery from today's date (or if today's date isn't available yet, use the most recent day). What kinds of features are currently being monitored or examined by MODIS?

In this chapter's Geospatial Lab Application, you'll examine more MODIS imagery using some other tools available from NASA.

FIGURE 12.7 A composite image of the Bezymianny volcano and its lava flow as seen by ASTER's sensors. (Source: NASA/GSFC/MITI/ERSDAC/ JAROS, and U.S./Japan ASTER Science Team, University of Pittsburgh)

Hands-on Application 12.2

ASTER Applications

For a look at a variety of other applications of ASTER, open your Web browser and go to **http:// asterweb.jpl.nasa.gov/gallerymap.asp**—this is the ASTER Web Image Gallery, a collection of imagery detailing how ASTER is used in a wide variety of real-world situations. You can select a geographic location from the map (for instance, select some of the options nearest to where you are now), or choose from the categories on the left-hand menu. Check out how ASTER is used for archeology, geology, and hydrology, or for studying phenomena like volcanoes, glaciers, or natural hazards, or for its use in cities or land-use studies. What are some of the specific uses of ASTER's capabilities for these types of studies?

imagery of the lava flow from a volcano). See *Hands-on Application 12.2: ASTER Applications* for further examples of ASTER imagery.

What Is Aqua and What Does It Do?

Aqua a key EOS satellite whose mission is to monitor Earth's water cycle.

Launched in 2002 as a joint mission between NASA and agencies in Brazil and Japan, **Aqua** is another key EOS satellite. In Latin, "Aqua" means "water," indicating the main purpose of the satellite—to examine multiple facets of Earth's water cycle. Analysis of water in all its forms—solid, liquid, and gaseous—is the key element of all six instruments onboard Aqua.

Terra and Aqua are designed to work in concert with one another—their Sun-synchronous orbits are set up similarly so that while Terra is on a descending path, Aqua is ascending (and vice versa). Because of this setup, Terra crosses the Equator in the morning while Aqua crosses in the afternoon (Aqua was originally called EOS-PM to complement Terra's original EOS-AM-1 name). This connection is further strengthened as both satellites carry a MODIS and a CERES instrument, in essence doubling the data collection performed by these two tools.

Beyond duplicate MODIS and CERES instruments, Aqua carries four others that are unique to its mission of examining Earth's water cycle: AMSU-A, HSB, AIRS, and AMSR-E. The **AMSU-A** (Advanced Microwave Sounding Unit) instrument is used for creating profiles of the temperature in the atmosphere. AMSU-A uses 15 microwave bands and is referred to as a "sounder" because its instruments are examining a three-dimensional atmosphere, similar to the way "soundings" were used by ships to determine water depths. AMSU-A's data provides estimates of not only temperature data but also precipitation and atmospheric water vapor. Similarly, the **HSB** (Humidity Sounder for Brazil) instrument's four microwave bands are used for measuring atmospheric water vapor levels (in other words, humidity) in the atmosphere (but HSB unfortunately stopped operating in 2003). Aqua's fifth instrument is **AIRS** (Advanced Infrared Sounder), whose uses include measuring temperature and humidity levels of the atmosphere as well as information about clouds.

AMSU-A and HSB data is used in close conjunction with AIRS to comprise a sounding system for Aqua. Data products derived from this suite of instruments include three-dimensional maps of atmospheric temperature, cloud types, ozone profiles, carbon dioxide levels, and sea surface temperature. This kind of information can be used for improving weather forecasting as well as studying Earth's atmosphere and climate system.

Figure 12.8 on page 360 shows an example of these three instruments (sometimes referred to as the "AIRS Suite") working together for one purpose—namely, tracking and monitoring a tropical cyclone. The four images show (1) the visible portion of the spectrum, sensed by AIRS, showing the extent and dimension of tropical cyclone Ramasun in 2002, (2) a temperature profile of the cyclone sensed by AIRS, (3) imagery of the surface below the clouds as sensed by AMSU-A, which can see through the cloud cover of the hurricane, and (4) the level of precipitation produced by the hurricane as sensed by HSB. All of the AIRS Suite instruments provide different "snapshots" of the same feature, allowing a wealth of data to be collected about different kinds of weather phenomena.

AMSR-E (the Advanced Microwave Scanning Radiometer for EOS) is the last of the Aqua instruments. It uses 12 microwave bands to cover most of the planet in one day and completes a global dataset in the second day. AMSR-E monitors a variety of environmental factors that affect global climate conditions, including sea ice levels, water vapor, wind speed, and amounts of global rainfall. For example, by assessing the amount of rain across the planet on a near-daily basis, AMSR-E can provide measures of how much precipitation storms can produce as they move across land or oceans.

AMSU-A the Advanced Microwave Sounding Unit instrument onboard Aqua.

HSB the Humidity Sounder for Brazil instrument onboard Aqua.

AIRS the Advanced Infrared Sounder instrument onboard Aqua (used in conjunction with Aqua's HSB and AMSU-A instruments).

AMSR-E the Advanced Microwave Scanning Radiometer for EOS instrument onboard Aqua.

FIGURE 12.8 Imagery of tropical cyclone Ramasun in 2002 as obtained by the AIRS, AMSU-A, and HSB instruments onboard Aqua. (Source: NASA/JPL)

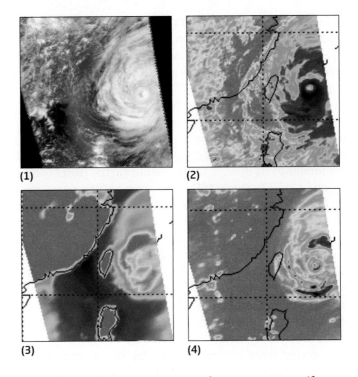

(1)   (2)

(3)   (4)

FIGURE 12.9 An AMSR-E global composite of sea surface temperature from June 2002-September 2003. (Source: NASA/ Goddard Space Flight Center Scientific Visualization Studio)

AMSR-E is also used to measure sea surface temperature (from a millimeter of water on the surface of the ocean) across the globe. As NASA notes, regular sea surface temperatures can be used as an aid in hurricane forecasting or tracking tropical storms. A hurricane requires warm water in the oceans to power up and sustain itself (the water has to be 82°F or greater for a hurricane to form). With regular temperature measurements, conditions for potential hurricanes can be quickly assessed. See Figure 12.9 for an example of AMSR-E sea surface temperature imagery—those places with ocean temperatures of

Hands-on Application 12.3

Aqua Products and the Visible Earth

The Visible Earth is a content-rich online resource set up by NASA for examining applications of the EOS Missions. Graphics and animations are also searchable by geographic region or by event. For instance, if you wanted to view graphics related to water quality, a variety of imagery is available from numerous sensors (and not just limited to the three EOS satellites described in this chapter). Animations of data showing changes over time or short video clips of simulations of satellite orbit or sensors are available as well.

To examine some of the Aqua applications, first open your Web browser and go to **http://visibleearth. nasa.gov**, then the option for Satellites, and then the option for Aqua. All of the Visible Earth categories for Aqua will be displayed. Select a few categories. What types of uses are Aqua's instruments monitoring?

Specific subcategories for Aqua's instruments are also available to examine:

1. AIRS Applications: **http://visibleearth.nasa. gov/view_set.php?sensorID=63**

2. AMSR-E Applications: **http://visibleearth. nasa.gov/view_set.php?sensorID=61**

3. CERES Applications: **http://visibleearth.nasa. gov/view_set.php?sensorID=62**

4. HSB Applications: **http://visibleearth.nasa. gov/view_set.php?sensorID=231**

5. MODIS Applications: **http://visibleearth.nasa. gov/view_set.php?sensorID=64**

After examining how the Aqua instruments are used, be sure to check out a few specific applications (including some of those previously discussed) to see how Aqua instruments are applied to the following topics:

1. How clouds affect Earth's radiation: **http:// visibleearth.nasa.gov/view_rec.php?id=103**

2. How sea surface temperature affects hurricane locations: **http://visibleearth.nasa. gov/view_rec.php?id=13853**

3. How the AIRS Suite is used for studying tropical storms: **http://visibleearth.nasa.gov/ view_rec.php?id=3439**

4. Using AMSR-E to measure extent of sea ice: **http://visibleearth.nasa.gov/view_rec. php?id=3621**

82°F or more are shown in yellow and orange. By knowing where the conditions are right, scientists can forecast the potential for hurricane development. See *Hands-on Application 12.3: Aqua Products and the Visible Earth* for more about AMSR-E and the other Aqua instruments.

What Is Aura and What Does It Do?

The **Aura** EOS satellite was designed as a collaboration between NASA and agencies in Finland, the Netherlands, and the United Kingdom. In Latin, "Aura" means "breeze," which helps describe the satellite's mission—examination of elements in the air, especially the chemistry of Earth's atmosphere. Aura orbits in formation with Aqua and other EOS satellites to form what is referred to as the **A-Train** of satellites. As an aside, the A-Train was to be joined by another satellite, OCO, launched in 2009. Unfortunately, however, OCO did not achieve orbit, leaving the A-Train one satellite short of its projected configuration.

Aura An important EOS satellite dedicated to monitoring Earth's atmospheric chemistry.

A-Train Another term for the Afternoon Constellation.

Afternoon Constellation a set of satellites (including Aqua and Aura) that pass the Equator in the afternoon during their orbits.

HIRDLS the High Resolution Dynamics Limb Sounder instrument onboard Aura.

MLS the Microwave Limb Sounder instrument onboard Aura.

TES the Tropospheric Emission Spectrometer instrument onboard Aura.

OMI the Ozone Monitoring Instrument onboard Aura.

FIGURE 12.10 The configuration of EOS satellites that make up the A-Train.

This organization of satellites is called the A-Train because it serves as the **Afternoon Constellation** of satellites (to complement the Morning Constellation that Terra flies in) and also because two of the key satellites (Aqua and Aura) begin with the letter "A." Currently, Aqua flies lead in the constellation (although the successful 2009 launch of OCO would have put it in the lead position, flying in front of Aqua). CloudSat follows closely after, then CALIPSO about 15 seconds after that, and lastly Aura bringing up the rear (Figure 12.10). A satellite called PARASOL flew between CALIPSO and Aura, although it was recently removed from the formation. Other satellites are currently planned to become part of the A-Train after their launch. The combined data from the A-Train of satellites gives scientists a rich dataset for analysis of climate change questions.

Aura carries four instruments onboard, each utilized in some type of atmospheric observation (such as ozone concentrations or air quality) as it relates to global climate change. The first, **HIRDLS** (High Resolution Dynamics Limb Sounder) measures phenomena such as temperature, water vapor, ozone, and other trace gases to examine qualities such as the transportation of air from one section of the atmosphere to another. As NASA notes, HIRDLS data is also used to examine pollution to see what is naturally occurring (from ozone) and what is human generated. Similarly, Aura's **MLS** (Microwave Limb Sounder) instrument senses microwave emissions in five bands to examine carbon monoxide and ozone in the atmosphere. MLS data can be used as an aid in measuring ozone destruction in the atmosphere. Aura's third instrument, **TES** (Tropospheric Emission Spectrometer), measures things related to pollution, including ozone and carbon monoxide. Since TES is capable of sensing from the land surface up into the atmosphere, its data can be used to assess air-quality levels in urban areas by measuring the levels of pollutants and ozone in cities.

Aura's final instrument is **OMI** (Ozone Monitoring Instrument) which (as the name implies) is dedicated to keeping an eye on changes in ozone across the globe. OMI is a hyperspectral sensor that views sections of the visible-light

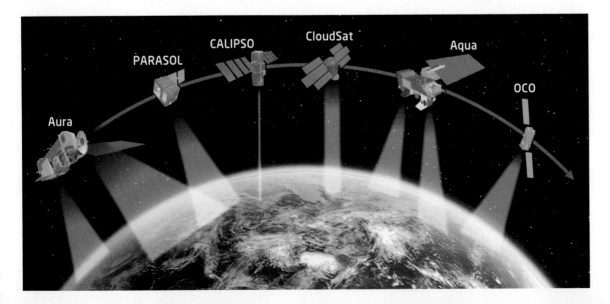

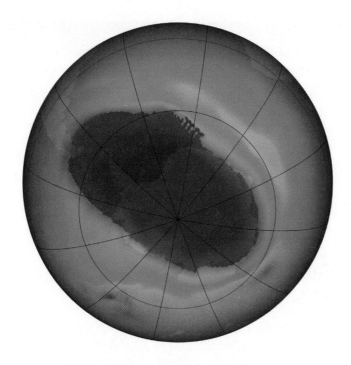

FIGURE 12.11 OMI imagery showing a hole in the ozone layer above Antarctica. (Source: NASA/ Goddard Space Flight Center Scientific Visualization Studio)

spectrum as well as the ultraviolet portion of the electromagnetic spectrum. Back in Chapter 10, we discussed how the atmosphere absorbs a lot of electromagnetic radiation and that ozone is the primary greenhouse gas that absorbs harmful ultraviolet light. OMI data aids scientists in measuring the amount of ultraviolet radiation penetrating to Earth by examining clouds and ozone levels. Figure 12.11 shows an example of imagery from OMI, showing the thinning (or hole) in the ozone layer above Antarctica. See *Hands-on Application 12.4: Aura Products and the Visible Earth* for more examples of Aura's instruments' applications.

Hands-on Application 12.4

Aura Products and the Visible Earth

To examine some of applications of Aura's instruments, the Visible Earth is a great reference. Like in *Hands-on Application 12.3: Aqua Products and the Visible Earth*, open your Web browser and go to **http://visibleearth.nasa.gov**—then select the option for Satellites, and then the option for Aura. All of the Visible Earth categories for Aura will be displayed. What specific effects are being monitored or examined using Aura's instrumentation?

After examining how the Aura instruments are used, be sure to check out a few specific applications

(including some of those previously discussed) to see how Aura is applied to the following topics:

1. Monitoring holes in the ozone layer: **http://visibleearth.nasa.gov/view_rec.php?id=14958**

2. The effects of biomass burning: **http://visibleearth.nasa.gov/view_rec.php?id=15006**

3. Monitoring aerosol and smoke effects from wildfires: **http://visibleearth.nasa.gov/view_rec.php?id=20523**

Thinking Critically with Geospatial Technology 12.1

How Can EOS Data Be Used in Studying and Monitoring Climate Change?

Climate change studies focus on Earth's climate—for example, assessing global temperature changes and the kinds of environmental conditions that influence alterations in Earth's processes. The Earth Observing System satellites provide constant sets of data related to all manner of environmental and climate conditions on our planet. Given the variety of instruments and their uses and data products you've examined in this chapter and their applications, how are EOS satellites' data used for studying climate-change conditions? What are some specific measurements of climate-change-related conditions that EOS data can be used for? How can this data be used in climate change studies?

How Can EOS Applications Be Easily Viewed and Analyzed?

With all of this EOS data and imagery collected on a near-daily basis, you would expect there to be a mechanism to get this information out into the hands of the public for viewing or analysis. Much of the EOS data is compiled into a series of data products that are delivered rather than raw imagery or values. NASA has a number of Web-based platforms in place to distribute these products (as well as EOS imagery), so they can be viewed or analyzed. One of these, **Visible Earth** (used in *Hands-on Applications 12.3* and *12.4* on pages 361 and 363), is a NASA-operated Website from which you can download pictures and animations related to EOS missions (online at: **http://visibleearth.nasa.gov/**). However, the pictures are just that—graphics of the processed data that show images from EOS instruments. The graphics come with detailed descriptions of the scientific event being shown (such as a hole in the ozone layer as measured from OMI or global sea surface temperature as measured by AMSR-E). Visible Earth also features popular detailed composite images such as NASA's **Blue Marble** (showing the entire Earth from space) or the global nighttime city lights.

A second source for accessing and viewing EOS imagery and data is the **NASA Earth Observatory** Website (**http://earthobservatory.nasa.gov/**), a sort of online "magazine" devoted to global environmental issues and how EOS is used with them. The Earth Observatory has articles dating back to 1998, dealing with topics such as global warming, tropical deforestation, and climate and Earth's energy budget. The Website also features an extensive "image of the day" archive (**Figure 12.12**), with each feature showcasing a different environmental occurrence from around the globe. For example, the January 29, 2009, image of the day used MODIS imagery to examine forest and grassland fires in Australia, while the January 27, 2009, image looks at sulfur dioxide emissions in Bulgaria as captured by OMI. Earth Observatory is an excellent source for seeing (and in some

Visible Earth a Website operated by NASA to distribute EOS images and animations of EOS satellites or datasets.

Blue Marble a composite MODIS image of the entire Earth from space.

NASA Earth Observatory a Website operated by NASA that details how EOS is utilized with numerous global environmental issues and concerns.

FIGURE 12.12 Cyclone Oli as viewed by Aqua's MODIS instrument—one of the striking Images of the Day available via Earth Observatory. (Source: NASA Image by Jeff Schmaltz, MODIS Rapid Response Team, Goddard Space Flight Center)

cases interacting with) remotely sensed data and imagery in action and applied to all manner of real-world situations (see *Hands-on Application 12.5: Using the Earth Observatory to Interactively Work With EOS Imagery* on page 366 for more about the interactive components of the Website).

Visible Earth and Earth Observatory are extensive resources for application of imagery, but to access the results of some data products themselves, NASA has set up another Website, NASA Earth Observations (NEO), at http://neo.sci.gsfc.nasa. **NASA NEO** enables users to download processed data from EOS satellites to examine applications centered on five main topics: oceans, atmosphere, energy, land, and life. NEO allows you to download imagery from multiple dates, to view it interactively online, or to download the data in a format compatible with Google Earth. When you view the data in Google Earth format, the EOS imagery "wraps" around the virtual globe in Google Earth, allowing you to interact with the imagery the same as you would with any other Google Earth usage. Working with the data in this interactive format treats the EOS imagery as an overlay on top of Google Earth, allowing you to see how EOS imagery fits with other data layers. In *Geospatial Lab Application 12.1*, you will be making use of NASA NEO and some of its datasets in Google Earth.

A comprehensive EOS imagery tool is **NASA World Wind**, a virtual globe program designed for analysis and manipulation of several types of remote sensing imagery (and is the software used in *Geospatial Lab Application 12.1*). NASA World Wind is available for free download from the Web and includes

NASA NEO the NASA Earth Observations Website, which allows users to view or download processed EOS imagery in a variety of formats, including a version compatible for viewing in Google Earth.

NASA World Wind a virtual globe program from NASA, used for examining various types of remotely sensed imagery.

Hands-on Application 12.5

Using the Earth Observatory to Interactively Work with EOS Imagery

The Earth Observatory also features the ICE (Image Composite Explorer) tool that allows users to analyze applications of remotely sensed data (and was also used back in *Hands-on Application 10.2: Examining NDVI with NASA ICE*). To get started with using EOS data in one of the ICE scenarios, open your Web browser and go to **http://earthobservatory.nasa. gov/Experiments/ICE/Channel_Islands.** This will allow you to examine MODIS products as they relate to conditions in the Channel Islands. The Website gives the background on the Channel Islands and how MODIS is used for monitoring phytoplankton conditions. Within the page is a link to launch ICE, which will allow you to observe imagery related to sea surface temperature, chlorophyll content, and fluorescence. How does MODIS measure these factors and how can EOS imagery and data be used to monitor phytoplankton content in the oceans?

numerous ways of utilizing EOS data. It contains several layers that can be examined, including Landsat true color and false color imagery, Landsat-derived land-cover data and high-resolution orthophotos. These data layers can be used alongside Rapid-Fire MODIS imagery taken from numerous sites around the globe of phenomena like fires, storms, and volcanoes. NASA World Wind also features the Scientific Visualization Studio (SVS), a tool designed to create animations of EOS imagery (such as several days of MODIS imagery of Hurricane Katrina or MOPITT imagery of a year's worth of carbon monoxide pollutants) for observation and analysis. With plenty more features than these added in, NASA World Wind is a very versatile tool that combines the virtual globe environment with EOS imagery and environmental applications.

There area lot of other ongoing EOS ongoing beyond Terra, Aqua, and Aura. For instance, the Jason-1 mission studies properties of the oceans, while the SORCE mission studies energy from the Sun. See *Hands-on*

Hands-on Application 12.6

NASA Eyes on the Earth

NASA has set up a Web application that allows you to view the real-time positions of Terra, Aqua, Aura, and the other EOS satellite missions, as well as viewing 3D data maps of the products and imagery they capture. Open your Web browser and go to **http:// climate.nasa.gov/Eyes/eyes.html** (you may have to download and install a special plugin for all of the functions to work correctly). You can select the EOS satellite mission you wish to view (which includes Terra, Aqua, and Aura, along with other missions such as Acrimsat, Calipso, and Cloudsat), then select either the view of the globe or the view from the satellite itself. Under the "Show Data Map" options, you can select from many different types of data products, including several of the applications discussed in this chapter.

Hands-on Application 12.7

Examining Other Environmental Remote Sensing Applications

NOAA operates two types of remote sensing satellites. The GOES program is a series of geostationary satellites tasked with monitoring weather conditions and are constantly observing weather of the same location. NOAA's POES satellites are polar orbiting and are part of a series of satellites. The NOAA satellites carry the AVHRR (Advanced Very High Resolution Radiometer) sensor that views in six bands (early versions viewed only in five). NOAA's remote sensing program is a critical part of global data collection as it relates to climate. You can examine NOAA satellite images online at the NOAA CoastWatch Website: **http://www.nodc. noaa.gov/dsdt/cw**.

Application 12.6: NASA Eyes on the Earth for a very cool NASA Website filled with interactive information and data about all of the EOS satellites. Also, keep in mind that NASA doesn't have a monopoly on broad-scale environmental remote sensing satellites. For example, NOAA (the National Oceanic and Atmospheric Administration) operates and maintains a series of satellites dedicated to numerous environmental monitoring functions, including weather observations. See *Hands-on Application 12.7: Examining Other Environmental Remote Sensing Applications* for some examples of NOAA satellite imagery as well.

Chapter Wrapup

The satellites of the Earth Observing System provide a constant source of remotely sensed data that can be utilized in numerous studies related to the climate and environment of our planet. Whether used in monitoring wildfires, ozone concentrations, potential algae blooms, or tropical cyclones, the applications described in this chapter just scratch the surface of the uses of the EOS and how these geospatial tools affect our lives. This chapter's lab will have you start working with EOS imagery using NASA World Wind and NEO to get a better feel for some of the applicability of the data.

In the next chapter, we'll dig into a new aspect of geospatial technology, that of modeling and analyzing landscapes and terrain surfaces.

Important note: The references for this chapter are part of the online companion for this book and can be found at **http://www.whfreeman.com/ shellito1e**.

Key Terms

EOS (p. 351)
Terra (p. 352)
Morning Constellation (p. 352)
CERES (p. 353)
MISR (p. 353)
MOPITT (p. 354)
MODIS (p. 355)
ASTER (p. 357)
Aqua (p. 358)
AMSU-A (p. 359)
HSB (p. 359)
AIRS (p. 359)
AMSR-E (p. 359)

Aura (p. 361)
A-Train (p. 361)
Afternoon Constellation (p. 362)
HIRDLS (p. 362)
MLS (p. 362)
TES (p. 362)
OMI (p. 362)
Visible Earth (p. 364)
Blue Marble (p. 364)
NASA Earth Observatory (p. 364)
NASA NEO (p. 365)
NASA World Wind (p. 365)

Earth Observing System Imagery

This chapter's lab will introduce you to some of the basics of examining EOS (Earth Observing System) imagery from the Terra and Aqua satellites. You will be examining data from MOPITT as well as many types of MODIS imagery using online resources along with NASA World Wind and Google Earth.

NASA World Wind is a free program (like Google Earth) created by NASA to incorporate many types of satellite imagery (such as Landsat, Terra, and Aqua) into a virtual globe interface.

Objectives

The goals of this lab are:

* Examining the usage and function of MODIS imagery.
* Utilizing NASA World Wind as a tool for examining remotely sensed imagery.
* Using Terra and Aqua imagery for environmental analysis.
* Examining EOS imagery in Google Earth.

Obtaining Software

* The current version of NASA World Wind (1.4) is available for free download at http://worldwind.arc.nasa.gov/download.html.
* The current version of Google Earth (6.0) is available for free download at http://earth.google.com.

Important note: Software and online resources sometimes change fast. This lab was designed with the most recently available version of the software at the time of writing. However, if the software or Websites have significantly changed between then and now, an updated version of this lab (using the newest versions) is available online at http://www.whfreeman.com/shellito1e.

Lab Data

There is no data to copy in this lab. All data comes as part of the NASA software or will be downloaded from sources on the Internet.

Localizing This Lab

This lab uses EOS data from a variety of locations across the globe, but there are several ways to examine some EOS imagery at a more local level:

◉ In Section 12.2, use a wider variety of dates than those presented to access a greater number of MODIS-monitored events. Look for some that are relatively close to your location or whose MODIS imagery would overlap onto your area.

◉ In Section 12.3, examine one year's worth of MOPITT data as it affects your local area, rather than the globe, and keep track of the carbon monoxide (CO) levels for your region.

◉ In Section 12.4, examine the local effects of some of the Terra and Aqua imagery products on your region, rather than evaluating them on a global scale. Open and examine images from multiple dates in Google Earth for your area, rather than one month's worth of data.

12.1 Using NASA World Wind

1. Start NASA World Wind.

2. World Wind will open with the default setting of NASA's "Big Blue Marble" imagery of Earth. Use the cursor to rotate the globe around to North America and the mouse wheel to zoom in and out of the globe.

3. In World Wind, several options are available for display on the globe, shown at the top of the button bar. If a black triangle appears over the option, that layer is turned "on" and being displayed on the globe; if there is no black triangle, that layer is turned off and is not shown.

12.2 MODIS Rapid Fire in World Wind

1. Click on the option for **Rapid Fire MODIS** to open a new dialog box.

2. The Rapid Fire Modis dialog box will appear.

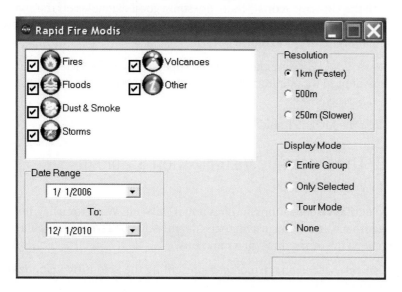

3. Select all six of the phenomena that are being monitored using MODIS.
4. Select **1 km** for the resolution to examine.
5. Use **Entire Group** for the Display Mode.
6. Change the date ranges to those shown here (January 1, 2006, to December 1, 2010).
7. A number of graphics corresponding to one of the six phenomena being monitored by MODIS imagery will appear on the globe.
8. Zoom in to California, where there is a cluster of the "fire" symbols. Double-click on the **fire icon** and a MODIS image will appear overlaid onto the globe. Examine several of the images related to California fires.

Question 12.1 How are the fires being shown in this MODIS imagery?

Question 12.2 How is the extent of the fires being tracked via MODIS? (Hint: What else is visible in the MODIS scene aside from the fires themselves?)

9. Rotate the view around to Western Europe, particularly off the west coast of Portugal. Click on the **green "other"** icon whose label indicates a "phytoplankton bloom." A new MODIS scene will appear. Also, examine some of the other "phytoplankton" images off the coasts of other European countries, such as France or Denmark.

Question 12.3 According to the caption, this is a MODIS image of phytoplankton in the ocean. Why would this be a phenomenon that is important enough to be tracked by MODIS, and what does the imagery show an observer about this phenomenon?

10. Return to North America and examine some of the imagery associated with the **"Storm" icons**—there are some good examples off the coast of Florida (related to Tropical Storm Ernesto) and off the coast of New England (for Tropical Storm Beryl).

Question 12.4 How are the scope and capabilities of the MODIS instrument used for monitoring weather or storm formations such as these Tropical Storms?

11. Close the MODIS Rapid Fire dialog box.

12.3 The Scientific Visualization Studio in World Wind

The Scientific Visualization Studio is a tool in World Wind for using imagery derived from some of the instruments aboard Aqua or Terra to examine Earth phenomena over time through animations.

1. Click on the icon on the toolbar representing the **Scientific Visualization Studio (SVS)**.

2. When started, SVS will perform a short download of current datasets to use. Give it a couple of seconds to do so.

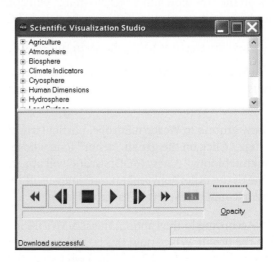

3. SVS will allow you to access various EOS datasets and examine them in a time series. The controls in the SVS dialog box act as DVR-style controls to run the time-series animation forward, backward, stop, or play.

4. Select the **Atmosphere** options, then choose the animation for **AQUA MODIS imagery of Hurricane Katrina**.

5. Click on the **Play** button. It will take a couple minutes to download all of the animation global data.

6. The animation will play, showing several days of imagery from Aqua of Hurricane Katrina building, then making landfall in the Mississippi Gulf Coast/New Orleans region.

Question 12.5 Given what you've seen of MODIS imagery and the information in the SVS, should MODIS be used as the sole instrument to monitor a massive storm formation like Hurricane Katrina? Why or why not?

7. Stop playing the animation, and then from the **Atmosphere** options, select **Global Atmospheric Carbon Monoxide in 2000**. Play the animation (it should replace the Hurricane Katrina imagery and cover the entire globe). Make sure to read the SVS description that accompanies the animation as well. This will show higher levels of carbon monoxide concentration from the MOPITT instrument onboard Terra.

8. Answer Questions 12.6 and 12.7. (You will have to rotate the globe while the animation is playing to answer the questions—also keep in mind the times and dates that are running in the upper-right-hand side of the screen. You may want to pause or rewind to examine certain dates in the past.)

Question 12.6 At what dates are the highest concentrations of carbon monoxide being monitored by MOPITT in Africa and South America?

Question 12.7 Beyond the carbon monoxide levels in Africa and South America (which were likely generated by wildfires), what other geographic areas can you see high concentrations of carbon monoxide developing and spreading from?

9. At this point, you can close World Wind by selecting **Exit** from the **File** pull-down menu.

12.4 Using the NASA Earth Observations (NEO) Web Resources

1. Open your Web browser and go to **http://neo.sci.gsfc.nasa.gov**. This is the Website for NEO (NASA Earth Observations), an online source of downloadable Earth observation satellite imagery. In this portion of the lab, you'll be using EOS imagery in conjunction with Google Earth.

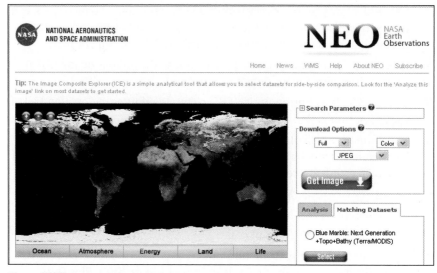

(Source: NASA)

2. Click on the **Atmosphere** tab under the main graphic.

3. Select the option for **Carbon Monoxide (MOPITT)**.

4. This is an image of global carbon monoxide concentration for one month collected using the MOPITT instrument aboard Terra (click on the **Analyze this image** option for more detailed information).

5. Rather than viewing a flat satellite image, NEO gives you the option of examining the imagery draped across a virtual globe.

    Important note: We will be using Google Earth for this portion of the lab, as the imagery can be downloaded in a ".kmz" file (the file format for Google Earth). NASA World Wind has a "kmz" converter tool that will also let you examine the imagery in World Wind if you don't have access to Google Earth.

6. Click on the **Open in Google Earth** option associated with the **October 1, 2010–October 31, 2010**, imagery. The "kmz" file of this dataset will download and Google Earth will automatically open to display it draped around the globe. (If prompted by your Web browser, select **Open using Google Earth**.)

7. Rotate Google Earth to examine the MOPITT imagery draped over the globe.

Question 12.8 Where (geographically) were the highest concentrations of carbon monoxide in October 2010?

8. Back on the NEO Website, select the **Energy** tab, then select **Land/ Surface temperature (Day) MODIS**.

9. Choose the image for **September 1, 2010–October 1, 2010**.

10. Choose to **Open** this composite image in Google Earth.

11. In Google Earth, turn off the MOPITT image (it will be in your Temporary Places listings).

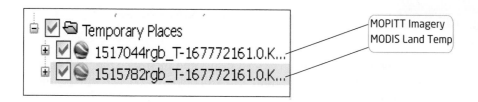

12. Rotate Google Earth to examine the new MODIS image.

Question 12.9 Where (geographically) were the lowest daytime land temperatures in September 2010?

13. Back on the NEO Website, select the **Life** tab, and then select **Vegetation Index NDVI Terra MODIS**. Choose the image for **July 1, 2010, to August 1, 2010**.

14. NDVI is a metric used for measuring vegetation health for a pixel. As described in Chapter 10, the higher the value of NDVI, the healthier the vegetation at that location.

15. Choose to **Open** this composite image in Google Earth.

16. In Google Earth, turn off your other two Terra images.

17. Rotate Google Earth and zoom in to examine the new MODIS composite image.

Question 12.10 Where (geographically) are the areas with the least amount of healthy vegetation on the planet in July 2010?

12.5 EOS Imagery from the National Snow and Ice Data Center

1. NASA NEO is not the only source of EOS data for viewing in a virtual globe format. One example is the National Snow and Ice Data Center (NSIDC). Open your Web browser and go to their Website at http:// nsidc.org/data/virtual_globes.

2. Go to the **Featured Data** section of the Website. Look for the section called Snow.

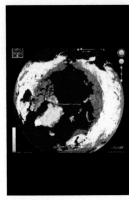

Snow (MODIS Sensor)
View global, monthly average snow cover each month for one year:

NOTE: The following months are missing from this time series: June 2001, March 2002, and December 2003.

<u>Current year snow cover</u> (KML, 673 KB)
<u>2009 snow cover</u> (KML, 670 KB)
2008 snow cover (KML, 670 KB)
2007 snow cover (KML, 670 KB)
2006 snow cover (KML, 670 KB)
2005 snow cover (KML, 670 KB)
2004 snow cover (KML, 670 KB)
2003 snow cover (KML, 670 KB)
2002 snow cover (KML, 670 KB)
2001 snow cover (KML, 670 KB)
2000 snow cover (KML, 670 KB)

Data Source: <u>MODIS/Terra Snow Cover Monthly L3 Global 0.05Deg CMG, Version 5</u>

(Source: NASA)

3. Read about what the imagery is showing, and then open the "KML" file for **2009 snow cover**. The "KML" file will open in Google Earth like the ones from NEO.

4. In Google Earth, turn off your other three NEO images.

5. In Google Earth's Temporary Places, there will be a new expandable item for **NSIDC**. By expanding this item and looking through the options, you'll see another expandable item called **Snow Cover** that shows snow data for each month of 2009 is available as an option that can be clicked on and off. To examine just one month of data, turn off all of the other months. You can also use the slider bar that appears at the top of the Google Earth view to adjust the snow data in Google Earth from month to month.

6. Rotate Google Earth and zoom in to examine the new MODIS image.

Question 12.11 Where (geographically) are the greatest concentrations of snow cover in the Southern Hemisphere in June 2009?

Question 12.12 Where (geographically) are the greatest concentrations of snow cover in the Northern Hemisphere in December 2009?

7. At this point, you can exit Google Earth and your Web browser.

Closing Time

This exercise showed off a number of different ways of viewing and visually analyzing data available from EOS satellites. The next chapter's going to change gears and start looking at the ground itself that those satellites are viewing.

If World Wind is still open, exit it by selecting **Exit** from the **File** pulldown menu. Also, if Google Earth is still open, exit it by selecting **Exit** from the **File** pull-down menu.

13

Digital Landscaping

Digital Topographic Maps, Contours, Digital Terrain Modeling, Digital Elevation Models (DEMs), NED, SRTM, LIDAR, and 3D Views of Landscapes and Terrain

There's one big element missing in geospatial technology that we haven't yet dealt with: the terrain and surface of Earth. All of the maps, imagery, and coordinates (while taking into account numerous critical features of the planet) have been dealing with two-dimensional (2D) images of the surface or of developed or natural features on the land. We haven't yet described how the actual landscape can be modeled with geospatial technology. This chapter delves into how terrain features can be described, modeled, and analyzed using several of the geospatial tools (like GIS and remote sensing) that have been previously described. Whether you are modeling topographic features for construction or recreational opportunities, or interested in seeing what the view from above would be, the landforms on Earth's surface are a necessary component of geospatial analysis (see **Figure 13.1** on page 378 for an example of visualizing Mount Everest with Google Earth).

First and foremost, when we're examining landforms on Earth's surface, each location will have to have an elevation assigned to it. In terms of coordinates, each x/y pair will now have a **z-value** accompanying it that indicates that location's elevation. These elevations have to be measured relative to something—when a point has an elevation of 900 feet, this indicates that it is 900 feet above something. The "something" represents a baseline, or **vertical datum,** that is the zero point for elevation measurements. Some maps of landforms indicate the vertical datum is taken as mean sea level (represented by the National Vertical Datum of 1929). Thus, when this vertical datum is used and an elevation value on a map indicates 1000 feet, that number can be read as 1000 feet above mean sea level. Other geospatial data in North America utilizes the North American Vertical Datum of 1988 (NAVD88). Coastal terrain models may also use mean high water.

z-value the elevation assigned to an x/y coordinate.

vertical datum a baseline used as a starting point in measuring elevation values (either above or below this value).

FIGURE 13.1 An example of Mount Everest modeled as a digital terrain landscape and shown in Google Earth. (Source: © 2009 Google, Image © 2009 DigitalGlobe, Image Copyright 2011 TerraMetrics, Inc. www.terrametrics.com, Image © 2009 Geo Eye)

How Can Terrain Be Represented?

topographic map a map created by the USGS to show landscape and terrain, as well as location of features on the land.

DRG a Digital Raster Graphic—a scanned version of a USGS topographic map.

GeoTIFF a graphical file format that can also carry spatial referencing.

collar the white information-filled border around a topographic map.

A common (and widely available) method of representing terrain features is the **topographic map**. As its name implies, a topographic map is designed to show the topography of the land and the features on it. Topographic maps (commonly called topo maps, for short) are published by the USGS (United States Geological Survey) and are available at a variety of map scales. A common topographic map is a 1:24000 topoquad, also referred to as a 7.5 minute topo map (as it displays an area covering 7.5 minutes of latitude by 7.5 minutes of longitude in size). Topo maps are also available in smaller map scales, such as 1:100000 or 1:2000000. Topo maps are available in a digital format as a **DRG** (Digital Raster Graphic). DRGs are scanned versions of topo maps that have been georeferenced (see Chapter 3), so they'll match up with other geospatial data sources.

Usually, DRGs are available in **GeoTIFF** file format (see Chapter 7 for more information about TIFFs), which allows for high-resolution images while also having a spatial reference. Thus, when a DRG is used as a layer in geospatial technology, it contains information to allow it to match up with other data layers. A topo map (and thus, a DRG) also contains a lot of data (including the scale, the dates of map creation and revision, and similar information) in the white border that surrounds the map. This border is referred to as the map's **collar.** A topo map will either come with all of this information intact, or in some cases, the collar will be removed and you will have only the map itself. See Figure 13.2 for an example of a DRG complete with collar.

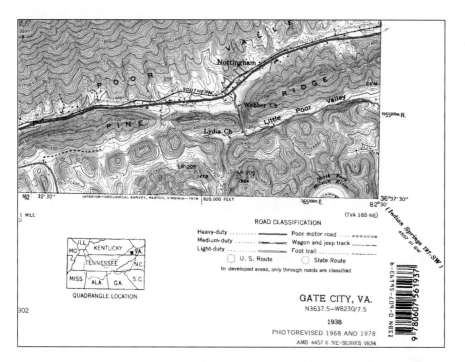

FIGURE 13.2 A portion of a DRG and its map collar. (Source: USGS)

Topo maps (and thus, DRGs) model the landscape through the use of **contour** lines—imaginary lines drawn on the map that represent areas of constant elevation (**Figure 13.3**). The elevations on the surface can be represented by numerous contour lines drawn on the map. However, because elevation is really a continuously variable phenomenon, it's often difficult to draw a line representing every change in elevation (such as from eight feet of elevation to ten feet of elevation) without overloading the map to the point of uselessness because it's filled to the breaking point with contour lines. Thus, contour lines are drawn a certain elevation distance apart (such as drawing a new contour line every 50 feet).

contour an imaginary line drawn on a map to connect points of common elevation.

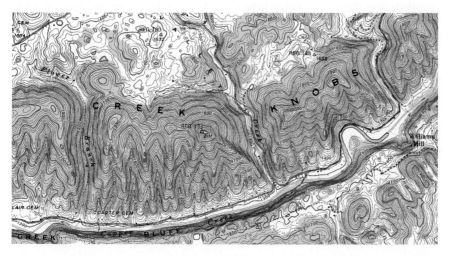

FIGURE 13.3 Contour lines as shown on a DRG (at a 20 foot contour interval). (Source: USGS)

Hands-on Application 13.1

Digital Raster Graphics Online

If you need to look up features on a topographic map, there's no need to wade through a map library filled with topoquads of all sizes and scales. USGS topo maps are made available online as DRGs. The Libre Map Project is a great online resource for 1:24000 DRGs. Open your Web browser and go to **http://libremap.org,** then select the Data tab. From there, select your state. A new menu will appear and let you search for specific features (for instance, selecting Florida for the state will allow you to search for Walt Disney World as a feature, and then download the DRG that contains Disney World).

Use the Libre Map project to search for a prominent feature in your hometown or near your house or school. Then download the DRG in the GeoTIFF file format to your hard drive (and download the TFW file too, to make sure the georeferencing information comes with the map) and view it using a utility like ArcGIS or AEJEE. Look over the DRG to find the feature you're looking for. While you're at it, examine the contour lines near the selected feature and see how they model the nearby elevation and terrain. Lastly, if the DRG comes with a collar, look for the information that tells you what the vertical datum is, when the map was created, and when it was last photo-revised.

---

contour interval the vertical difference between contour lines drawn on a map.

This elevation (or vertical) difference between contour lines is called the **contour interval** and is set up according to the constraints of the map and the area being measured. In general, a wider contour interval is selected when mapping more mountainous terrain (since there are numerous higher elevations) and a narrower contour interval is used when mapping flatter terrain (since there are fewer changes in variation and more details can be mapped). Small-scale maps (see Chapter 7) tend to use a wider contour interval because they cover a larger geographic area and are presenting more generalized terrain information, whereas large-scale maps generally utilize a narrow contour interval for the opposite reason (they show a smaller geographic area and thus can present more detailed terrain information). DRGs (and their contour lines) are frequently used as data sources within geospatial technology (such as overlaying them with other layers in GIS). See *Hands-on Application 13.1: Digital Raster Graphics Online* for a method of obtaining DRGs online.

US Topo a digital topographic map series created by the USGS to allow multiple layers of data to be used on a map in GeoPDF file format.

While DRGs remain available (and many software programs give you the capability of examining "seamless" digital topographic maps), the "next generation" of digital topographic maps is the **US Topo** series. The US Topo maps are delivered in 7.5 minute quad sections, but in GeoPDF file format (see Chapter 7), which allows multiple layers of data to be stored on the same map. These layers can then be turned on and off. US Topo maps include contours, recent transportation features (roads, railroads, airports), transportation names, hydrography, boundaries, and orthophoto images, all as separately accessible layers (see Figure 13.4 for an example of a US Topo map) as well as the full collar of information (which can have its information be displayed or removed as a separate map layer). The USGS makes US Topo maps available for free download (see *Hands-on Application 13.2: US Topo Maps as GeoPDFs* for accessing and utilizing US Topo maps). Although

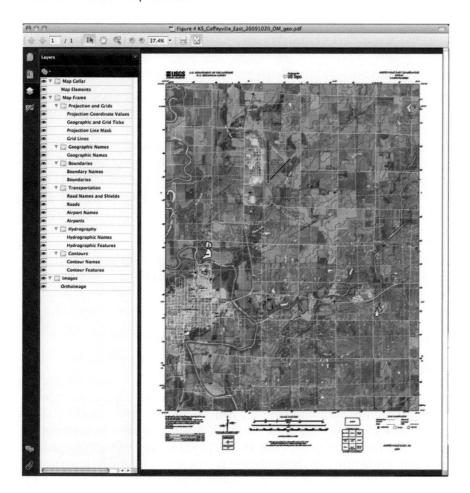

Hands-on Application 13.2

US Topo Maps as GeoPDFs

To see what US Topo maps are available for download in GeoPDF file format, open your Web browser and go to **http://nationalmap.gov/ustopo/index.html**–the main Website for US Topo maps. Click on the link for Coverage to see the current status of which states have been processed into US Topos. Follow the link on the Website to the current download site (at the time of this writing, it was the USGS Store) to access the Map Locator.

At this point, you can type the name of a place to search for available US Topo maps, and then choose to download the maps as a free GeoPDF. For instance, search for Kansas City, then from the available options, download the 7.5 × 7.5 US Topo map for Kansas City. A zip file will be downloaded to your computer containing the GeoPDF. Open the PDF (you may be prompted to install a plug-in to access all of the GeoPDF features) and examine the various layers available within the US Topo map.

Thinking Critically with Geospatial Technology 13.1

If Everything's Digital, Do We Still Need Printed Topo Maps?

A similar question was raised back in Chapter 2: If all of the latest USGS topographic maps are available digitally and in easily accessible formats, is there still a need to have printed copies (at a variety of map scales) on hand? For instance, do surveyors, geologists, botanists, or archeologists (or any other professional who requires topographic information while in the field) need to carry printed topo maps with them? Is there a need for geographers to have printed versions of several quads when seamless digital copies (which can be examined next to one another) can be easily (and freely) obtained? Are there situations where a printed topo map is a necessary item given the readily available digital data?

topographic maps in all of their incarnations can represent terrain with contours, there are more detailed methods of digital terrain modeling available through geospatial technology.

How Can Geospatial Technology Represent Terrain?

DTM a representation of a terrain surface calculated by measuring elevation values at a series of locations.

two-and-a-half-dimensional (2.5D) model a model of the terrain that allows for a single z-value to be assigned to each x/y coordinate location.

three-dimensional (3D) model a model of the terrain that allows for multiple z-values to be assigned to each x/y coordinate location.

TIN Triangulated Irregular Network. A terrain model that allows for non-equally spaced elevation points to be used in the creation of the surface.

In geospatial technology, terrain and landscape features can be represented by more than just contour lines on a map. A Digital Terrain Model (**DTM**) is the name given to a model of the landscape that is used in conjunction with GIS or remotely sensed imagery. The function of a DTM is to accurately represent the features of the landscape and be useful for analysis of the terrain itself. The key to a DTM is to properly represent a z-value for x and y locations. With a z-value, the model can be shown in a perspective view to demonstrate the appearance of the terrain, but this doesn't necessary make it a three-dimensional model. In fact, a DTM is usually best described as a two-and-a-half dimensional model. In a two-dimensional (2D) model, all coordinates are measured with x and y values without a number for z at these coordinates. In a **two-and-a-half-dimensional (2.5D) model**, a single z value can be assigned to each x/y coordinate as a measure of elevation at that location. In a full **three-dimensional (3D) model**, multiple z-values can be assigned to each x/y coordinate. Most DTMs have one elevation value measured for the terrain height at each location, making them 2.5D models (**Figure 13.5**).

An example of a type of DTM that's used in terrain modeling is a **TIN** (Triangulated Irregular Network), in which selected elevation points (those that the system deems the "most important") of the terrain are used in constructing the model. Points are joined together to form non-overlapping triangles, representing the terrain surfaces (see **Figure 13.6**). While TINs are often used in terrain modeling in GIS, there's another very common type of digital terrain model called the DEM.

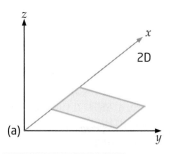 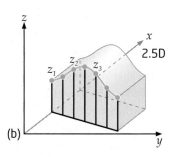 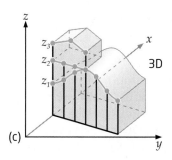

FIGURE 13.5 A comparison of (a) 2D, (b) 2.5D, and (c) 3D models of terrain.

FIGURE 13.6 A TIN representation of the terrain around Gate City, Virginia. (Source: Esri® ArcGIS ArcScene graphical user interface Copyright © Esri. All rights reserved.)

What Is a DEM?

A **DEM** is a Digital Elevation Model, a specific type of model of the terrain and landscape. A DEM is a terrain model produced by the USGS and others. A DEM is based on regularly spaced point data of elevations but can be converted to a raster grid representation (see Chapter 5) for use with other geospatial data, such as satellite imagery or GIS layers. When used in grid format, the elevation values are represented by the grid cell value, while the grid resolution represents the size of the cell being measured. Thus, DEM resolution is measured in a manner similar to the way a remotely sensed image's resolution would be measured (see Chapter 10). If a DEM has 30-meter resolution, then each of the DEM's raster grid cells is set up at 30 meters in size. USGS DEMs have been created using Digital Line Graph (DLG) information. Other methods of DEM creation involve the use of remotely sensed data or stereo imagery (see Chapter 15) and photogrammetry to derive elevation values.

Another source of terrain data is the Shuttle Radar Topography Mission (**SRTM**) a part of a mission of the Space Shuttle Endeavor in the year 2000. For 11 days, Endeavor used a special radar system to map Earth's terrain

DEM Digital Elevation Model—a representation of the terrain surface, created by measuring a set of equally spaced elevation values.

SRTM the Shuttle Radar Topography Mission, flown in the year 2000, which mapped Earth's surface from orbit for the purpose of constructing digital elevation models of the planet.

Hands-on Application 13.3

SRTM Imagery Online

NASA maintains an online gallery of SRTM imagery from the 2000 mission, as well as other products related to SRTM. Open your Web browser and go to **http://www2.jpl.nasa.gov/srtm**. Select the option for Gallery of Images to view image results of SRTM results from around the globe. Check out several of the examples of SRTM imagery that are set up in perspective view—among others, check out the image galleries of North America—United States (Alaska, Utah, and Washington State), Central America (Costa Rica), and Africa (Congo). Note that they're only example graphics of SRTM results, not a way to download the actual SRTM data.

There's also a Multimedia option on the main page, which will allow you to view a series of video files (in Real Player or Quicktime) of SRTM in action. These should help give you a feel for how terrain can be mapped from space.

and topographic features from orbit, resulting in a highly accurate digital elevation model. At the mission's close, roughly 80% of Earth was examined and modeled as 90-meter DEMs (and 30-meter DEM data is available for the United States). See *Hands-on Application 13.3: SRTM Imagery Online* for more about STRM data.

LIDAR Light Detection and Ranging. A process in which a series of laser beams fired at the ground from an aircraft is used for creation of highly accurate DEMs.

An additional remote sensing method of terrain mapping is called **LIDAR** (Light Detection and Ranging). Rather than firing a microwave pulse at a ground target like a RADAR system would, LIDAR uses a laser beam to measure the terrain. In LIDAR, a plane flies over the ground equipped with a special system that fires a series of laser beams (between 2000 and 5000 pulses per second) at the ground. The laser beams are reflected from the ground back to the plane, and based on the distance from the plane to targets on the ground, the elevation of the landscape (as well as objects on the surface of the terrain) can be determined. GPS (see Chapter 4) is used in part to determine where the beams were striking the ground. After the data is collected and processed, the end result of the LIDAR mission is a highly accurate set of x/y locations with a z-value. The elevations of the DEM products derived from LIDAR have a vertical accuracy of 15 centimeters (see Figure 13.7 for an example of a LIDAR-derived DEM).

NED the National Elevation Dataset, which provides digital elevation coverage of the entire United States.

Seamless Server an online utility used by the USGS for the download of different datasets, including elevation data.

The USGS has numerous types of DEMs available, depending on the size of the area being covered and the DEM resolution. The USGS distributes this elevation data for free download via the **NED** (National Elevation Dataset). The purpose of the NED was to have a digital elevation product that covers the entire United States at the best resolution possible and to eliminate any gaps in the data. The NED is designed to be "seamless" in that its data use the same datum, projection, and elevation unit, without having separate tiles of data that the user needs to match up. With the NED, users can select which sections of the national elevation dataset they need and download those areas.

NED data is made freely available online via the USGS and the National Map **Seamless Server**, a very versatile online utility for downloading

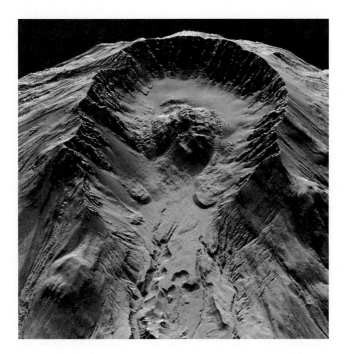

FIGURE 13.7 Mount St. Helens as viewed from LIDAR data. (Source: USGS)

geospatial data (see Chapter 15 for more about the National Map itself). With the National Map Seamless Server, you can specify the geographic region you require data for and select from multiple datasets that cover that region to download. NED data is available at 1 arc second (about 30-meter), 1/3 arc second (about 10-meter), and 1/9 arc second (about 3-meter) resolutions, depending on the region. In addition, digital topo maps and high-resolution orthoimagery are among the many other datasets available for download (see *Hands-on Application 13.4: The National Map Seamless Server* on page 386 for more information on using the Seamless Server).

With terrain data, many different types of terrain analysis can be performed. DEMs are used in geospatial technology for creating viewsheds—maps that show what can be seen or not seen from a particular vantage point. A **viewshed** is used for determining how far a person's visibility is (that is, what they can see) from a location before his or her view is blocked by the terrain. DEMs are also used for a variety of hydrologic applications, such as calculating the accumulation of water in an area or extraction of stream channels or watersheds.

DEMs can be used to derive a new dataset of **slope** information—rather than elevation values, slope represents the change of elevations (and the rate of change) at a location by calculating the rise (vertical distance) over the run (horizontal distance). When a slope surface is created from a DEM, information can be derived not just about heights, but about the steepness or flatness of areas. Similarly, a surface of **slope aspect** can be computed, which will show (for each location) the direction that the slope is facing.

viewshed a data layer that determines what an observer can see and cannot see from a particular location due to terrain.

slope a measurement of the rate of elevation change at a location, found by dividing the vertical height (the rise) by the horizontal length (the run).

slope aspect a determination of the direction that a slope is facing.

Hands-on Application 13.4

The National Map Seamless Server

With the National Map Seamless Server, you can define the geographic area that you want data for, and datasets aren't limited by things like state boundaries, county boundaries, or USGS topoquad size. When you're downloading NED data (for instance), you're selecting the NED information from a "seamless" dataset without breaks in it. Open your Web browser and go to **http://seamless.usgs.gov** and select the option for the Seamless Viewer. A new window will open for the interactive viewer—at this point, examine the options listed under Display. What kinds of data are available beyond just the three types of NED data?

Next, select the option for View and Download United States Data and use the online tools to zoom in an area near you (like your township or your school or workplace's city). Check which kinds of data are available to download. Some other data types (such as the orthoimagery) can be unzipped and viewed using a utility like ArcGIS, AEJEE, or MultiSpec.

Also, from the main Seamless Web page, NED data can be downloaded in a pre-packaged format, enabling you to access NED data for states or counties.

Now, you not only have information about where the steepest slopes are located, but whether they are north-facing, east-facing, and so on. Figure 13.8 shows a comparison of a DEM with the slope and slope aspect maps derived from it.

FIGURE 13.8 A USGS DEM of a portion of the Laurelville/Hocking region of Ohio and the slope and slope aspect grids derived from it. (Source: USGS. Esri® ArcGIS ArcScene graphical user interface Copyright © Esri. All rights reserved.)

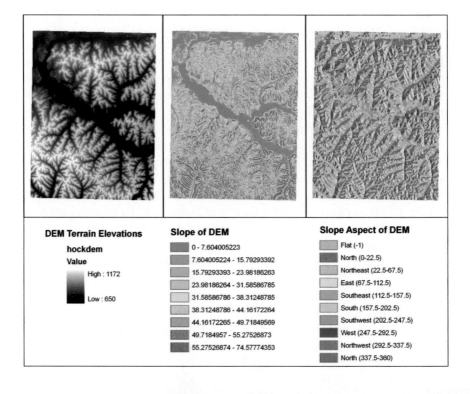

DEM Terrain Elevations
hockdem
Value
High : 1172
Low : 650

Slope of DEM
- 0 - 7.604005223
- 7.604005224 - 15.79293392
- 15.79293393 - 23.98186263
- 23.98186264 - 31.58586785
- 31.58586786 - 38.31248785
- 38.31248786 - 44.16172264
- 44.16172265 - 49.71849569
- 49.7184957 - 55.27526873
- 55.27526874 - 74.57774353

Slope Aspect of DEM
- Flat (-1)
- North (0-22.5)
- Northeast (22.5-67.5)
- East (67.5-112.5)
- Southeast (112.5-157.5)
- South (157.5-202.5)
- Southwest (202.5-247.5)
- West (247.5-292.5)
- Northwest (292.5-337.5)
- North (337.5-360)

How Can You Make Terrain Look More Realistic?

Digital terrain models aren't just limited to flat images (like a DRG) on a computer screen or a paper map—remember, constructs such as DEMs are 2.5D models, and since they have a z-value, that z-value can be visualized in a 3D view. It's more accurate to say these types of visualizations are really **pseudo-3D** views since they're really 2.5D, not really a full 3D model. Setting up a pseudo-3D view of a digital terrain model involves examining it at a **perspective view** (or an oblique view). Points or raster cells are elevated to the height of their z-value and the resultant model is shown from an angular view. In this way, mountain peaks can be seen jutting up from the surface in the same way a meteor crater looks like a depression in the ground.

In many geospatial technology applications, the model then becomes interactive, allowing the user to move or "fly" over the terrain, skimming over the surface and banking past mountains. Beyond viewing the terrain in perspective view, there are numerous ways to make the terrain more realistic-looking, such as artificially altering its appearance or draping imagery over the surface (see Figure 13.9 for an example of a terrain surface shown in perspective view with a digital topographic map overlaid to show off the contour lines in relation to the pseudo-3D model).

Remember that a DEM (or other digital terrain model) is just a map or grid of elevation surfaces. Like any other map, it can be displayed with

pseudo-3D a term often used to describe the perspective view of a terrain model since it is often a 2.5D model, not a full 3D model.

perspective view viewing a digital terrain model at an oblique angle in which it takes on a "three-dimensional" appearance.

FIGURE 13.9 A pseudo-3D view of a digital terrain model enhanced with a digital topographic map displayed over the terrain. (Source: National Geographic Society. Esri® ArcGIS Explorer graphical user interface Copyright © Esri. All rights reserved.)

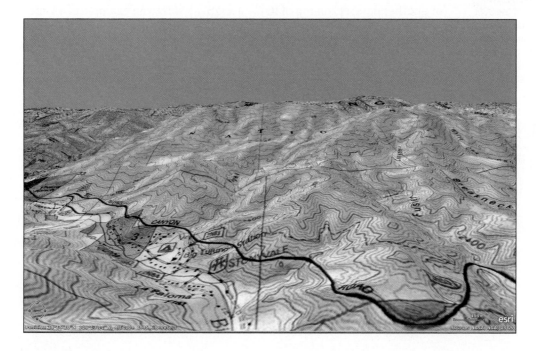

hillshade a shaded relief map of the terrain created by modeling the position of the Sun in the sky relative to the landscape.

Sun altitude the value between 0 and 90 used in constructing a hillshade to model the Sun's elevation above the terrain.

Sun azimuth the value between 0 and 360 used in constructing a hillshade to model the Sun's position in the sky to show the direction of the Sun's rays striking the surface.

draping a process in which an image is given z-values to match the heights in a digital terrain model.

FIGURE 13.10 A DEM of Columbiana County, Ohio, and a hillshade of the DEM made using a Sun elevation of 45° and a Sun azimuth of 315°. (Source: USGS. Esri® ArcGIS graphical user interface Copyright © Esri. All rights reserved.)

various color ramps to show the differences in elevations, but there are several ways to make the terrain model look more realistic (and start to resemble that image of Mount Everest from the beginning of this chapter). The first of these techniques is called a **hillshade**, which models how the terrain would look under different lighting conditions (based on the location of the Sun in the sky). When constructing a hillshade, you can determine where the light source (that is, the Sun) is located with respect to the terrain, thus simulating the effects of shadows being cast on the landscape by the various terrain features.

This is done by setting two parameters for the Sun—the first of these is the **Sun altitude**, a value between 0 and 90, representing the angle of the Sun in the sky between 0 degrees (directly at the surface) and 90 degrees (directly overhead). The second parameter is the **Sun azimuth**, a value between 0 and 360, representing the location of the Sun in relation to where the Sun's rays are coming from. The values represent a circle around the landscape, with 0 being due north, 90 being due east, 180 being due south, and 270 being due west (the values are measured clockwise from due north). By setting values for Sun altitude and Sun azimuth, the terrain can take on various appearances to simulate how the landscape looks under different conditions during the day. A hillshade using a Sun altitude of 45 degrees and a Sun azimuth of 315 degrees is shown in Figure 13.10.

Hillshading provides a good shaded map of what the terrain would look like under various lighting conditions, but there are plenty of features on the landscape (likes roads and land cover) that aren't shown with a hillshade. In order to see these types of features, a process called **draping** is used to essentially show the terrain model with a remotely sensed image (or another dataset) on top of it. Figure 13.11 shows an example of a Landsat TM image draped over a DEM. Draping is achieved by first aligning the image with its corresponding places on the terrain model, then assigning the z-values from those locations on the terrain (in Esri terminology, these

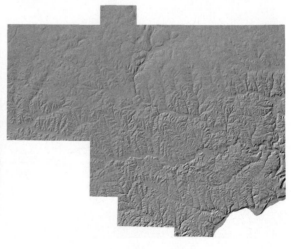

are referred to as **base heights**) to those locations on the image. In essence, locations on the image are assigned a z-value that corresponds with the terrain model.

Draping is a common technique to show remotely sensed imagery on terrain features (similar to how Mount Everest looks in Figure 13.1 on page 378). Virtual globes (like Google Earth and NASA World Wind) can show pseudo-3D landscapes by draping imagery over the terrain models that make up their landscapes. By creating a new draped image in perspective view, one can get new visual information about the appearance of the landscape that's not directly obtainable through examination of contours, DRGs, or non-perspective DEMs (see *Hands-on Application 13.5: Terrain and Imagery Examples in Google Earth* for more information). For example, draping a DRG over a DEM and looking at it in perspective can visually demonstrate how contour lines match up with the elevations that they represent.

Even with hillshading or draping to improve the terrain's appearance, there's no getting around the fact that some sections of terrain have relatively low variability. In these cases, differences in landscape elevations or slopes may be difficult to see when viewing or interacting with a digital terrain model. The solution is to artificially enhance the differences between elevations so that the landscape can be better visualized. **Vertical exaggeration** is the process of artificially altering the terrain model for visualization purposes so that the vertical scale of the digital terrain model is larger than the horizontal

base heights the z-values of a digital terrain model that can then be applied to an image in a draping procedure.

vertical exaggeration a process whereby the z-values are artificially enhanced for terrain visualization purposes.

Hands-on Application 13.5

Terrain and Imagery Examples in Google Earth

Start up Google Earth and "Fly To" Glacier National Park in Montana. Make sure the "Terrain" option is turned on (by default it should be on when Google Earth begins) and use the zoom slider and the other Google Earth navigation tools to change the view so that you're examining the mountains and terrain of Glacier in a perspective view. Remember what you're examining here is imagery that has been draped over a digital terrain model representing the landscape of this section of the country. Fly around the Glacier area (see Figure 13.12 for an example), getting a feel for navigation over draped imagery. You'll be doing more of this (among many other things) in *Geospatial Lab Application 13.1*.

FIGURE 13.12
Examining imagery on a
digital terrain model of
Glacier National Park in
Google Earth. (Source: ©
2009 Google, Image © 2009
Digital Globe)

scale. For instance, if the model's vertical exaggeration were "5x," then the
vertical scale (the z-values) would be five times that of the horizontal scale.
These types of artificial changes can really enhance certain vertical features
(such as making valleys appear deeper or peaks appear higher) for visual purposes. The downside to vertical exaggeration is that it alters the scale of the
data and should be used only for visualizing the data. For a comparison of different vertical exaggerations applied to a DEM, see Figure 13.13.

FIGURE 13.13 Different
levels of vertical
exaggeration for a DEM
(values of none, 2, 5, and
10). (Source: USGS. Esri®
ArcGIS ArcScene graphical user
interface Copyright © Esri. All
rights reserved.)

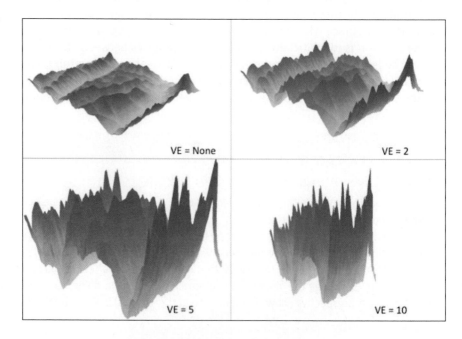

Chapter Wrapup

This chapter explored methods of modeling and visualizing the landforms on Earth's surface. In the next chapter, we'll do a lot more with the 3D aspect of visualization. After all, there's more than just the terrain that can be viewed in 3D—there are plenty of structures, buildings, and natural growth that can be added to a pseudo-3D terrain model to improve the realism of the scene. Chapter 14 picks up where this one leaves off to continue pursuing new avenues of 3D visualization.

This chapter's lab will allow you to implement all of the chapter topics using not only Google Earth but also a new software package (called MICRO-DEM) that allows you to work directly with a DEM and its derivations.

Important note: The references for this chapter are part of the online companion for this book and can be found at **http://www.whfreeman.com/shellito1e**.

Key Terms

z-value (p. 377)
vertical datum (p. 377)
topographic map (p. 378)
DRG (p. 378)
GeoTIFF (p. 378)
collar (p. 378)
contour (p. 379)
contour interval (p. 380)
US Topo (p. 380)
DTM (p. 382)
two-and-a-half-dimensional (2.5D) model (p. 382)
three-dimensional (3D) model (p. 382)
TIN (p. 382)
DEM (p. 383)

SRTM (p. 383)
LIDAR (p. 384)
NED (p. 384)
Seamless Server (p. 384)
viewshed (p. 385)
slope (p. 385)
slope aspect (p. 385)
pseudo-3D (p. 387)
perspective view (p. 387)
hillshade (p. 388)
Sun altitude (p. 388)
Sun azimuth (p. 388)
draping (p. 388)
base heights (p. 389)
vertical exaggeration (p. 389)

Digital Terrain Analysis

This chapter's lab will introduce you to some of the basics of digital terrain modeling—working with DEMs, contours, and draped imagery over the terrain model. You'll be using Google Earth and the MICRODEM software program (developed by Professor Peter Guth of the Oceanography Department of the U.S. Naval Academy) for this lab.

Objectives

The goals of this lab are:

- Examining pseudo-3D terrain and navigating across it in Google Earth.
- Filming an animation of flying over 3D terrain in Google Earth.
- Examining the effects of different levels of vertical exaggeration on the terrain.
- Familiarizing yourself with the MICRODEM operating environment.
- Draping an image over the terrain in Google Earth.
- Deriving hillshades, contours, and slope measurements from a DEM for terrain analysis.
- Examining a DEM in both 2D and perspective view.
- Using a DEM to create a flight path for a video in MICRODEM.

Obtaining Software

- The current version of Google Earth (6.0) is available for free download at http://earth.google.com.
- The current version of MICRODEM (12) is available for free download at http://www.usna.edu/Users/oceano/pguth/website/microdem/microdem.htm.

  Important note: Software and online resources sometimes change fast. This lab was designed with the most recently available version of the software at the time of writing. However, if the software or Websites have significantly changed between then and now, an updated version of this lab (using the newest versions) is available online at http://www.whfreeman.com/shellito1e.

Lab Data

All data used in this lab is either part of the software or comes included as sample data as part of the software installation.

Localizing This Lab

The Google Earth sections of this lab can be performed using areas nearby your location, as this lab features underlying terrain that covers the globe.

The MICRODEM section of the lab can be performed with a DEM of your local area. See *Hands-on Application 13.4: The National Map Seamless Server* for more information on downloading elevation datasets.

13.1 Examining Landscapes and Terrain with Google Earth

1. Start **Google Earth** and for a good example of some variable terrain, **"Fly To"** the Grand Canyon. For more information about the Grand Canyon, check out this Website: http://www.nps.gov/grca/index.htm. By default, Google Earth's "Terrain" options will be turned on for you.

2. Also, select the **Tools** pull-down menu and choose **Options**. In the dialog box that appears, click on the **Navigation** tab, and make sure the box that says "Automatically tilt when zooming" has its radio button filled in. This will ensure that Google Earth will zoom in to areas while tilting to a perspective view.

3. Use the **Zoom Slider** to tilt the view all the way down that it will go, so the view looks like you're standing somewhere in the Grand Canyon (see *Geospatial Lab Application 1.1* for more info on using this tool). Basically, push the **"plus"** button on the Zoom Slider as far down as it'll go and your view will zoom in and tilt down to "eye level."

4. Use your mouse wheel to vertically move the view backwards (do this instead of using the Zoom Slider—otherwise you'll keep tilting backwards as well) so you can see some areas of the Grand Canyon, some of the terrain relief, and the horizon. See the accompanying graphic of Google Earth (below) for an example of examining the Grand Canyon area from this type of perspective view.

(Source: © 2010 Google, Image © DigitalGlobe, Image USDA Farm Service Agency, Image NMRGIS)

5. There are two references to heights or elevation values in the bottom portion of the view:

   ⦿ The heading marked "elev" shows the height of the terrain model (that is, the height above the vertical datum) where the cursor is placed.

   ⦿ The heading marked "Eye alt" shows Google Earth's measurement for how high above the terrain (or the vertical datum) your vantage point is.

   Maneuver your view so your "Eye alt" is at a good level above the terrain so that you can see over the mountains and into the canyons—this will be a good height to start flying over the terrain.

6. Use the mouse and the Move tools to practice flying around the Grand Canyon. You can fly over the terrain, dip in and out of valleys, and skim over the mountaintops. When you've got a good feel for flying over 3D terrain in Google Earth, then move onto the next step.

13.2 Recording Animations in Google Earth

1. On Google Earth's toolbar, select the **Record a Tour** button.

   This option will have Google Earth record your flight and save it as a video. A new set of controls will appear at the bottom of the view:

2. To start recording the video, press the **red** record button.

   Important note: If you have a microphone hooked up to your computer you can get really creative and narrate your flight—your narration or sounds will be saved along with your video.

3. Keep flying around for a short tour (30 to 60 seconds) of the Grand Canyon.

4. When you're done, press the **red** record button again to end the recording of the tour.

5. A new set of controls will appear at the bottom of the screen and Google Earth will begin playing your video.

6. Use the **rewind** and **fast-forward** buttons to skip around in the video, and the **play/pause** button to start or stop. The button with the two arrows will repeat the tour or put it on a loop to keep playing.

7. You can save the tour by pressing the **disk icon** (the Save Tour) button. Give it a name in the dialog box that opens. The saved tour will be added to your Places box (just like all other Google Earth layers).

8. Put together a good tour (perhaps about a minute long) of the Grand Canyon that shows off many of the area's terrain features and save the tour as a Google Earth layer.

Question 13.1 What areas in the Grand Canyon did you select for your tour, and what terrain features did you highlight during the tour?

13.3 Vertical Exaggeration and Measuring Elevation Height Values in Google Earth

Google Earth also allows you to alter the vertical exaggeration of the terrain layer. As discussed on page 389, vertical exaggeration is used for visualization purposes.

1. To look at different levels of vertical exaggeration, select **Options** from the **Tools** pull-down menu. Select the **3D View** tab.

2. You can alter the quality of the terrain (from lowest to highest) for visualization purposes. In the box marked Elevation Exaggeration, you can type a value (between 0.5 and 3) for vertical exaggeration of Google Earth's terrain. Type a value of **1**, then click **Apply** and **OK**, and then re-examine the Grand Canyon.

Question 13.2 Try the following values for vertical exaggeration: 0.5, 1, 2, and 3. How did each value affect the visualization of the terrain? Which was the most useful for a visual representation of the Grand Canyon and why?

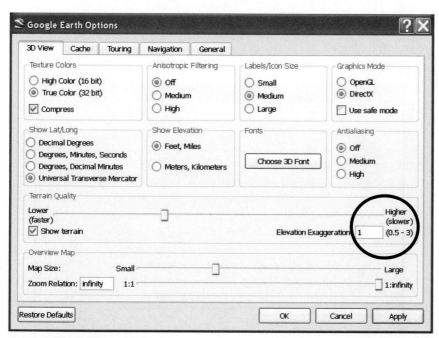

(Source: © 2011 Google)

3. Reset the Elevation Exaggeration to a value of 1 when you're done. From here, we'll examine the elevation values of the terrain surface itself. Although Google Earth will always show the imagery over the terrain model, elevation information of each location is available.

4. In the Layers box, under the **More** option, turn on the **"Parks and Recreation"** layer. Several new symbols will appear around the Grand Canyon, designating specific locales within the park. Wherever you move the cursor on the screen, a new value for elevation is computed in the "elev" option at the bottom of the view. By zooming into one of the new marker spots and placing the cursor on its symbol on the screen, you can determine the elevation value for that location.

Question 13.3 At what elevations are the heights of all the Visitors Centers (labeled with a question mark in a circle as their placemark) within the park?

5. At this point, we'll be moving onto a new software package called MICRODEM, which allows you to work directly with the terrain models and data themselves, not just use the fully processed versions in Google Earth.

Important note: Don't close Google Earth just yet. You'll use it later for another way to take a look at the data you'll use next with MICRODEM.

6. Use Google Earth to "Fly To" Hanging Rock Canyon, California.

13.4 Getting Started with MICRODEM

1. Start **MICRODEM** (the default install is an icon called MICRODEM). It will open with the default screen full of buttons on the toolbar.

2. To examine a DEM, select the **File** pull-down menu and choose **Open DEM**.

3. When MICRODEM installed, it added a new folder called "mapdata" to your computer (the default location is a C:\mapdata folder). Navigate to this folder, select the folder called **DEMs**, and then select the **"HangRockCanyon_DEM_2.tar.gz"** file.

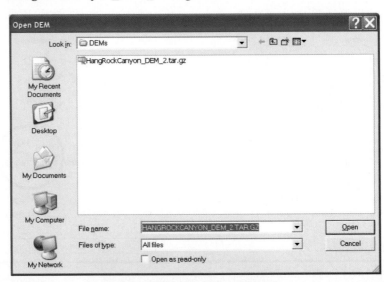

4. Click **Open** to start using this DEM in MICRODEM. (The DEM will open in a new viewer.)

5. Before moving on, use the DEM and the view in Google Earth to orient yourself to where, spatially, Hanging Rock Canyon is and what section of the landscape is being displayed by the DEM.

6. When you have yourself oriented, right-click anywhere on the DEM and a new menu will appear. From the available options, select **Load**, then select **Google Earth overlay**.

7. MICRODEM will then apply a draped image of the DEM over its corresponding area inside of Google Earth (and add a new object called MICRODEM to Google Earth's Temporary Places folder). This will show you where the DEM's parameters match up with the Google Earth imagery.

13.5 Hillshades

1. To change the appearance of the DEM, right-click anywhere on the image and a new menu will appear. From the available options, select **Reflectance Options**.

2. In the Reflectance Map Options dialog box that opens, select **Grays** for the Colors option.

3. You'll also see options for Sun azimuth (default of 335 degrees) and Sun elevation (45 degrees above the surface). The circle in the upper right of the dialog box shows the position of the Sun relative to the DEM with a red dot. This use of the values for azimuth and elevation creates the hillshade effect for viewing the DEM. To create other visualizations of the landscape, you can use other values for Sun azimuth and Sun elevation.

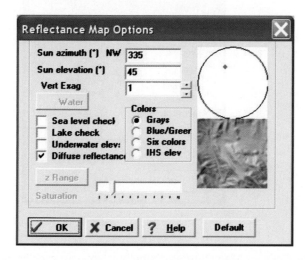

4. Change the Sun azimuth to **90** and the Sun elevation to **10**, and then click **OK**.

Question 13.4 What time of day and Sun conditions do these values for Sun azimuth and Sun elevation simulate? Why?

5. Return to the **Reflectance Map Options**. Test out the effect of other hillshade visualizations based on some other Sun azimuth and Sun elevation values. Answer Question 13.5. When you're done, return to the norm of Sun azimuth = 335 and Sun elevation of 45.

Question 13.5 What values for Sun azimuth and Sun elevation would simulate a "sunset" viewing appearance of the DEM? Why?

13.6 Contour Lines and Contour Interval on a DEM

1. Contour lines can be generated from the DEM and then displayed as a map overlay (among other layers) to examine how the contours match up with the terrain represented by the DEM. To see the overlay options, select the **Manage Overlays** icon from the view's toolbar.

2. In the Map Overlays dialog box, placing a checkmark in one of the options will turn on that particular overlay. Place a checkmark in the **Contours** box and a new button labeled Contours will be added. Press this button to examine the Contour Intervals.

3. In the Contour Map Options dialog box, you can specify the Contour Interval of the contours to be drawn on the DEM.

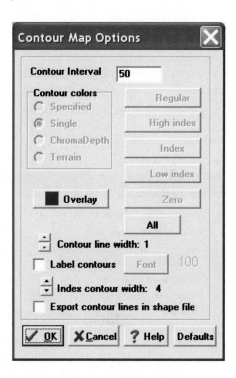

4. Change the Contour Colors option to **Single**.

5. Use the default option of **50** meters for the Contour Interval and then click **OK**. The contours will be drawn over the DEM.

6. Contours drawn with a 50-meter interval will be placed on the DEM as an overlay. Bring up the **Map Overlays** dialog box again, press the **Contours** button, and switch the contour interval to **100** meters. The 50-meter contours will go away and be replaced by new contours drawn at a 100-meter interval.

7. Try out several different contour intervals as follows: 5 meters, 25 meters, 250 meters, and 1 kilometer (in addition to the 50- and 100-meter options). See how the use of those intervals changes how contours are created and drawn.

Question 13.6 From what you've seen of the Hanging Rock Canyon, California region (both in Google Earth via the overlay and from the Hillshade), which of the values for contour intervals (5 meters, 25 meters, 50 meters, 100 meters, 250 meters, and 1 kilometer) best represents the region? That is, if you were drawing a contour map of Hanging Rock Canyon, which would you choose and why?

Question 13.7 Following up on Question 13.6, which of the values for contour intervals (5 meters, 25 meters, 50 meters, 100 meters, 250 meters, and 1 kilometer) worst represents the region? That is, if you were drawing a contour map of Hanging Rock Canyon, which options would you not use and why?

13.7 Slope and a DEM

1. Because a DEM is showing elevations, factors such as the slope (for each location) can be derived. Right-click on the **DEM**, select **Display parameter**, and choose **Slope**.

2. In the Slope Map Options dialog box, select the option for **Trafficability Categories**.

3. Click **OK**.

4. MICRODEM will calculate a measurement for slope (in degrees) for each location on the DEM.

5. Examine the new slope map.

Question 13.8 What areas in Hanging Rock Canyon have the steepest slopes and what physical features are causing these steep slopes? (Refer to some specific areas.)

6. Re-open the DEM again to place it in a new view.

13.8 Three-Dimensional (3D) Visualization of a DEM

1. The last thing we're going to do in this lab is to examine the DEM in a pseudo-3D view and do some more flying. In a program like Google Earth, the terrain data has already been processed for you to work with—in these steps, we're actually going to set up the pseudo-3D view.

2. From the main MICRODEM toolbar, click on the **Oblique Plot** button.

3. The Oblique Plot will enable you to take a section of the DEM and examine it in oblique perspective view. After clicking on Oblique Plot, double-click on a section of the DEM and then drag the mouse to another section. The red box that you are drawing by clicking and dragging represents the area of the DEM that will be shown in perspective view. Double-click again when you've set up the red box for the area you want.

    Important note: Be sure not to select too large an area. MICRODEM may not be able to handle a section that's too big.

4. A new dialog box (called Oblique selection) will occur asking what you want to display in the perspective view. Choose the options for **Reflectance** and **Show overlays on drapes**.

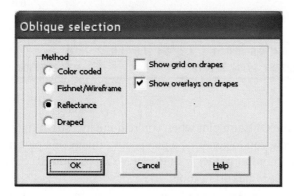

5. Click **OK** in the Oblique selection box to begin the process.
6. The Oblique View of the DEM section will appear. A new toolbar is part of the Oblique View—the controls will allow you to (in order):

    ⦿   Print the image
    ⦿   Save the image (as a graphic)
    ⦿   Edit the image (in a photo manager program)
    ⦿   Copy the image to the clipboard (to use in programs like PowerPoint or Word)
    ⦿   Redraw the image with different parameters
    ⦿   Rotate the image counterclockwise (to see different sides of it)
    ⦿   Rotate the image clockwise
    ⦿   Increase the vertical exaggeration
    ⦿   Decrease the vertical exaggeration

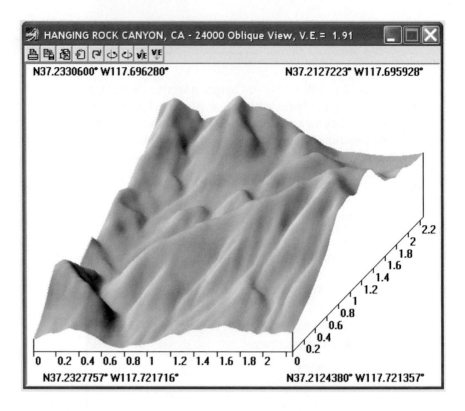

7. Close the **Oblique** plot when you're done investigating it. This tool will allow you to generate a perspective image of a part of the DEM, but in order to interact with the DEM in a pseudo-3D environment, MICRODEM has other tools.

8. In the main MICRODEM toolbar, select the **Flythrough** icon.

9. In the DEM, double-click on a starting point for flying. Drag the mouse in the direction you want to fly, and then double-click the mouse again. A red line will be drawn between the two points. If you want to add a second leg to the flight, drag the line another direction and double-click again. This red line indicates the flight path—once you have the flight path set up, right-click the mouse and choose **End selection**.

10. A new dialog box will appear. Select the **Viewport** tab. These options will allow you to change factors such as how high you will be flying over Earth (the Observer above ground option) or the Observer elevation. For your first flight, just accept the defaults and click **OK**.

11. Two new windows will appear, one showing the flight path as red and black lines and another window that shows a video clip of flying over the DEM.

12. MICRODEM will prompt you to "Continue w/these parameters?" Click **Yes** to watch the video clip.

13. A short clip will play and then a new dialog box will open prompting you to open a movie. Select the file called **"FLYT.MOV"** and click **Open**.

14. A set of controls will appear at the top of the screen as part of the PETMAR Trilobites Movie Player to allow you to play, pause, rewind, set up a loop, or alter the delay between the images shown that make up the animated video.

15. You can stop playing and exit the video at any time and return to change flying options or select a new path to fly over. Play around with some of the flying options, then answer Question 13.9.

Question 13.9 What chosen flight path and elevation parameters enabled the best-appearing flying video for you and why?

Closing Time

MICRODEM is a very powerful program with a lot of functionality and options for working with DEMs and digital terrain analysis, and it has plenty more features to investigate for your own future use. When you're finished, close both MICRODEM and Google Earth.

14

See the World in 3D

3D Geovisualization, 3D Modeling and Design, Prism Maps, Google SketchUp, and Google Earth in 3D

Up until the previous chapter, all topics in this book dealt with geospatial technology in a two-dimensional format, whether as geospatial data, measurements, maps, or imagery. Chapter 13 began to go beyond two dimensions and start on a third, presenting terrain in a perspective and pseudo-3D view. This chapter looks entirely at presenting geospatial information in three dimensions, as well as designing and visualizing 3D data and concepts.

There's no doubt that 3D visualization of data is impressive—as technology has improved and computers have become faster and more powerful, 3D rendering has become more commonplace. Video games and simulators are extremely impressive to watch, as are computer-animated movies. While geospatial technology hasn't reached the level of the newest CGI film (yet), many 3D modeling and visualization techniques are available for creating perspective views, 3D maps, and realistic-looking 3D objects. For instance, Bing Maps and Google Earth both support viewing and interacting with 3D geospatial data. See Figure 14.1 on page 406 for an example of realistic-looking 3D geospatial visualization in Google Earth.

Before proceeding, keep in mind the discussion from Chapter 13 concerning 2.5-dimensional (2.5D) data versus 3-dimensional (3D) data. If only one height value can be assigned to an x/y location, then the data is considered 2.5D, and if multiple z-values can be assigned to an x/y location, then the data is fully 3D (3D data also has volume). For ease of reading and usage, this chapter uses the term "3D" throughout to refer to all geospatial phenomena that incorporate a third dimension into their design or visualization, although technically, some examples will be 2.5D or pseudo-3D.

FIGURE 14.1 Lower Manhattan (New York City) rendered in 3D. (Source: © 2009 Google, Gray Buildings © Sanborn, Image © 2009 DigitalGlobe, Image © 2009 Sanborn)

3D modeling designing and visualizing data that contains a third dimension (a z-axis) in addition to x- and y-values.

z-value the numerical value representing the height of an object.

extrusion the extending of a flat object to have a z-value.

block a flat polygon object that has been extruded.

What Is 3D Modeling?

3D modeling (in geospatial terms) refers to the design of spatial data in a three-dimensional (or pseudo-3D) form. Designing a 3D version of the buildings in a city's downtown or a 3D version of the Eiffel Tower would be examples of 3D modeling of geospatial data. Like all of our spatial data, the 3D models should be georeferenced to take advantage of real-world coordinate systems and measurements. Say for instance you're designing a 3D model of your house—you'd want to start with a georeferenced base to accurately measure the dimensions of your house's footprint. An orthophoto, high-resolution satellite image, or architectural diagram with spatial reference would all be good starting points for a georeferenced base to begin modeling from. By digitizing the house's footprint, you now have a polygon with two dimensions (x and y) to begin with, but to work with a 3D-style object, it has to have a third dimension, or a z-dimension. This **z-value** will be the height of the object above the surface.

In order to create a 3D version of the polygon, it will have to be extruded to reach the height specified by the z-value. **Extrusion** is the process of giving height to an object. If your house is 15 feet high, the footprint polygon can be extruded to a height of 15 feet. Extrusion will change the objects—extruding a polygon will turn it into a **block**. In this case, the building footprint will change to a block object 15 feet high (see Figure 14.2 for the comparison between the two-dimensional polygon footprint and the extruded block).

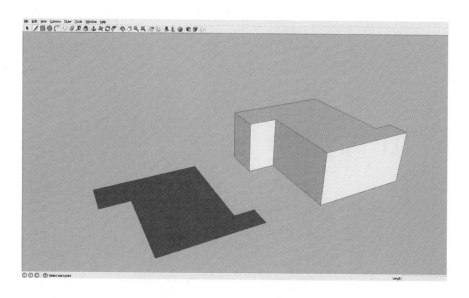

FIGURE 14.2 A flat polygon versus an extruded block. (Source: © 2011 Google SketchUp)

Any object can be extruded to a particular height by assigning a z-value to extrude to. Keep in mind that these items extrude from the ground level up and some objects can't be designed this way. For example, think about what you would do if you were designing a 3D model of a 10-foot-high pedestrian walkway positioned 15 feet over a busy road. If you digitize the bridge's footprint and extrude it to 10 feet, it would look like a wall in the middle of the road rather than a bridge over the road. To get around this, you have to apply an **offset**, or a z-value, where the initial object is placed before extrusion. In this case, the digitized polygon representing the bridge would be offset to a height of 15 feet (placing it well over the road), then extruded to a height of 10 feet (to capture the dimensions of the walkway). **Figure 14.3** illustrates

offset a value applied to objects to move them off the ground level.

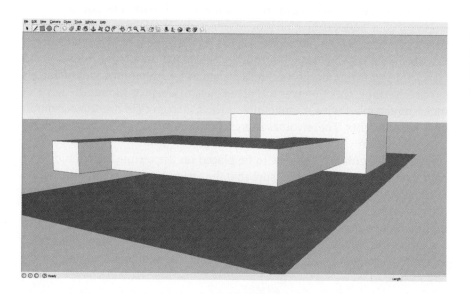

FIGURE 14.3 An offset polygon that has been extruded versus a single extruded block. (Source: © 2011 Google SketchUp)

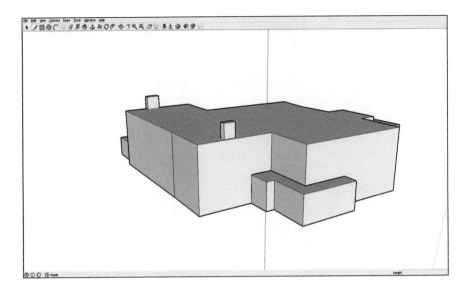

the difference between a regular extruded block and an extruded block that has been offset from the surface.

A building or object being modeled will likely consist of multiple blocks (or other shapes), each representing a different portion of the object. Multiple levels of a house or differently sized sections of a building can each be represented with their own block. Just like a GIS dataset can consist of multiple polygons, a 3D representation can consist of multiple extruded shapes and blocks (see Figure 14.4). Each of these blocks can then be treated as its own object for further modeling.

Another consideration in 3D modeling is that the objects being constructed are not just placed in their correct georeferenced location, but that they also conform correctly to the terrain and real-world elevations. For instance, if your house is in a location 900 feet above sea level, you don't want to start the base of your house at a ground level of zero feet. When you merge the 3D model of your house with other geospatial data, your house would be shown 900 feet underground rather than at its proper location on Earth's surface. Terrain modeling was discussed back in Chapter 13, but the key element to remember with 3D modeling is the concept of (to use Esri terminology) **base heights**. These values represent the elevation of the terrain above a vertical datum.

base heights the
elevation values
assigned to the terrain
upon which the objects
will be placed.

When designing 3D objects to be placed on the terrain (like a model of your house), you want to make sure that the terrain's base heights are applied to the objects you're designing. In this way, the software will understand that the base of your house begins at an elevation of 900 feet, and extruding 15 feet high makes your house's roof top out at a measurement of 915 feet above sea level rather than 15 feet above ground level (zero feet). Many geospatial software packages enable the user to combine 3D modeling techniques with terrain modeling in order to design more realistic 3D visualizations (some specific programs that do this are described later in this chapter).

Once the footprints and elevations are correctly set in place, it's time to start making those gray blocks look like what you're trying to model. A block consists of several **faces**, with each face representing one side of the object. For example, an extruded rectangular block would have six faces—the top, bottom, and four sides. Each face can be "painted" to give it a more realistic appearance. This "painting" can take a variety of forms—simply changing the color of the face (for instance, changing the appearance of a side of the house to a yellow color) can add to its appearance.

face one of the sides of a block object.

More realistic appearances can be achieved by applying a **texture** to a face. A texture is a graphic designed to simulate the appearance of materials (like brick, wood, stone, glass, etc.) on a face. With textures, you don't have to draw individual bricks on the side of a building; just apply a red-brick texture and the face's appearance is changed. Similarly, windows can be designed by applying a translucent texture to a window object drawn onto a face, or different roofing materials can be applied to the faces making up the top of the building (see Figure 14.5 for an example of a 3D building with various textures applied to the block faces).

textures graphics applied to 3D objects to create a more realistic appearance.

Since a texture is just a graphic, it's possible to apply other graphics to a face. For instance, you could take a digital camera picture of your home's particular brick and apply that to a face. You can even take a picture of the entire side of a house or building, resize it, and "paste" that image onto the entire face for an even better appearance. However, it's not just buildings and objects that can be visualized using 3D modeling techniques. 3D maps can also quickly communicate spatial information to the viewer.

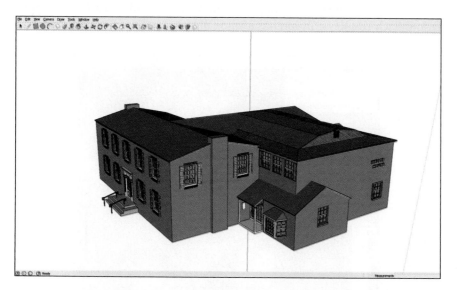

FIGURE 14.5 A version of Youngstown State University's Weller House with other features added, as well as further details, colors, and textures applied to the block faces (compare to the earlier version in Figure 14.4). (Source: © 2011 Google SketchUp)

Thinking Critically with Geospatial Technology 14.1

What's the Advantage of Using 3D Design?

3D visualization has become increasingly common for viewing and using geospatial data. As processors have gotten faster and graphics and animation have improved, rendering large amounts of geospatial data in realistic 3D form is easier than ever. Why is visualization of geospatial data so critical to conveying a message to the user or viewer of the data? Beyond the "cool" factor, what makes 3D visualization so useful for conveying data to the user or viewer as opposed to a simple 2D map? What advantages does 3D visualization carry over regular 2D for presentation of data or communicating spatial information?

How Are 3D Maps Made?

A 3D-style map can be made by applying the previously described 3D modeling concepts. However, this time you're not working with extruding polygons representing building footprints, but instead you're working with the shapes and objects on a map. A regular choropleth or thematic map (see Chapter 7) can be transformed into a 3D-style map by extruding the polygons displayed on the map. The resulting extruded map is referred to as a **prism map**, while the extruded shapes are called raised **prisms**. In a prism map, the shapes on the map are extruded to a relative height that reflects the data value assigned to them. For instance, if a thematic map of the United States showed population by state, then each state's shape would be extruded to a "height" of the number of units equal to that state's population. Thus, the polygon shape of California would be a raised prism extruded to a value of 36.7 million, while neighboring Nevada's shape would be extruded to a value of 2.6 million.

> **prism map** a thematic map that has the map shapes extruded to a value based on the values shown on the map.
>
> **prisms** the extruded shapes on a prism map.

Figure 14.6 shows an example of a prism map dealing with world population. Each country polygon has a value of the population statistics for that

FIGURE 14.6 An example of a prism map showing the 2010 population of each country in the world. (Source: thematicmapping. org. © 2010 Google, Data SIO, NOAA, U.S. Navy, NGA, GEBCO, Image IBCAO, © 2010 CNES Spot Image)

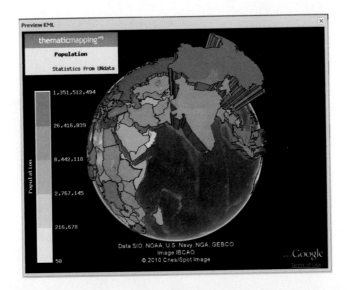

Hands-on Application 14.1

Creating Prism Maps Online

The Thematic Mapping Engine (also used back in Chapter 7) allows you to create prism maps and display them using Google Earth. Open your Web browser and go to **http://thematicmapping. org/engine.** Select an indicator that you want to map (such as population, life expectancy, or mobile phone subscribers) and a year, and then select Prism for the technique. Press Preview to see the results in your browser or Download to open the prism map in Google Earth. Try several of the classification combinations to see how they alter the raised prisms in the data. Lastly, download a prism map of a dataset into Google Earth along with a regular 2D thematic map and examine them to see the difference in visual communication that a prism map offers.

country. Those countries with the highest raised prisms have the largest number of persons (in this example, the map shows China and India with huge populations relative to neighboring countries). The polygon shape of each country is extruded relative to the number of units being measured. See *Hands-on Application 14.1: Creating Prism Maps Online* for an online tool to create prism maps with.

How Can 3D Modeling and Visualization Be Used with Geospatial Technology?

There are numerous 3D design and modeling programs on the market today, and many commercial geospatial programs offer some sort of capability of visualizing or working with 3D data (and there's no way this chapter can get into all of them). For instance, Esri's ArcGIS program contains the 3D Analyst extension, which enables numerous 3D features related to terrain and modeling, but also two other interfaces for working with 3D data—**ArcGlobe** and **ArcScene**. As its name implies, ArcGlobe allows for 3D visualization on a "global" level. If you're examining 3D surfaces and objects in Los Angeles, you can fly north across the globe to view other 3D spatial data in San Francisco, then continue north to look at the landscape in Seattle. ArcGlobe functions similarly to a virtual globe program and works with all types of Esri file formats such as 3D shapefiles. ArcScene contains a full set of 3D-visualization tools for ArcGIS and allows for the same type of 3D modeling, except on a "local" scale. If you were modeling one section of a 3D Los Angeles, that's the extent of the data you'd be working with.

Google Earth also supports 3D visualization. Back in *Geospatial Lab Application 1.1: Introduction to Geospatial Concepts and Google Earth,* you examined a 3D representation of places in South Dakota. Google Earth also contains other 3D objects, such as representations of 3D buildings and structures across the world (such as buildings in New York City, Boston, and other cities

ArcGlobe a 3D visualization tool used to display data in a global environment, which is part of Esri's ArcGIS program.

ArcScene the 3D visualization and design component of Esri's ArcGIS program.

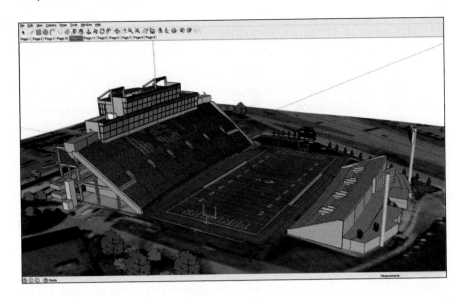

FIGURE 14.7 A version of Youngstown State University's football stadium complex, displayed in Google SketchUp. (Source: © 2011 Google SketchUp)

Google SketchUp a 3D modeling and design software program distributed online by Google, it contains linkages with Google Earth.

around the world). It's easy to create 3D objects, structures, buildings, land-forms, and more through the use of the **Google SketchUp** (GSU) software program and then view them in Google Earth. The regular version of GSU is made freely available for download by Google, while a Pro version with additional features is available for purchase (you'll be downloading and using the free version of GSU in this chapter's lab). GSU is a very versatile (and intuitive to use) program that allows for a full range of 3D design capabilities. Structures created with GSU can be as realistic, detailed, or as basic as you want them to be. See Figure 14.7 for an example of a football stadium in GSU.

Shapes can be molded, extruded, bent, and moved to build any type of structure you can envision. Realistic textures can be applied to faces for a better appearance, and digital photos can also be resized and applied to a face for a more photo-real appearance. GSU also features a process called Photo Match, in which a digital photo of a building can be used as the starting point and you can design the polygon dimensions of the building straight from the photo. From there, the photo can be directly applied to the faces for a realistic-looking 3D design.

components pre-designed (or user-designed) objects in Google SketchUp.

GSU also enables the use of **components** (or pre-made 3D objects) for added realism and detail in designing 3D models. Components allow you to quickly place items such as trees, cars, furniture, landscaping, and similar features into 3D models (see Figure 14.8 for examples of components used in GSU). Like other 3D objects, components are constructed using shapes and textures but are stored for re-use throughout a model. For instance, if you're designing a 3D version of a car dealership, you can quickly add a couple dozen different car components and customize them rather than designing each 3D car model individually. Components are also useful for elements you'll use multiple times in a model. For example, if you're designing a 3D version of a school building, you could design a window once, store it as a component,

FIGURE 14.8 Several sample components displayed in Google SketchUp. (Source: © 2011 Google SketchUp)

and then easily re-use the same model of a window multiple times around the school.

Geospatial Lab Application 14.1 will introduce you to several aspects of using GSU for 3D design, where you'll begin with a building's footprint, use tools to create a shape, design elements of the building (such as the entranceway), as well as design components to use. A file created with GSU (like the one you'll make in the lab) uses the "*.skp" file extension. The **SKP file** consists of information about the polygons that make up the structure, the textures applied, and the geometry of how these objects fit together. However, the SKP format is native to GSU. You (or whoever you share your data with) has to be running GSU to open and examine an SKP file. In addition, you can open only one SKP file at a time, which causes some difficulties if you've designed multiple buildings of a city block (each in its own SKP file) and want to see how they look next to one another.

One way around these issues is to convert the SKP files into another format called a **KMZ**, which can be used in Google Earth. Each building becomes its own KMZ file, and many KMZ files can be opened at once in Google Earth. KMZ is a compressed version of another file format called **KML**, short for Keyhole Markup Language. KML and KMZ are file formats that can be interpreted using Google Earth. By converting an SKP file into a KMZ file, that GSU model can be viewed interactively in Google Earth. Figure 14.9 on page 414 shows an example of this—Youngstown State University's stadium complex (containing both sides of the stadium, stadium lights, the scoreboard, and the goal posts) was converted from an SKP file into a KMZ file. The KMZ file can be opened in Google Earth, which overlays the 3D model of the stadium over the Google Earth imagery, allowing the user to rotate around the stadium, fly across the field, zoom into a bleacher seat, or any of the other usual uses of Google Earth. KMZ has become a standard for creating objects to view in Google Earth—other

SKP file the file format and extension used by an object created in Google SketchUp.

KMZ a file that is the compressed version of a KML file.

KML Keyhole Markup Language—the file format used for Google Earth data.

FIGURE 14.9
Youngstown State
University's Stambaugh
Stadium complex,
converted from an SKP file
in Google SketchUp and
displayed as a "*.kmz" file
(KMZ file) in Google Earth.
(Source: © 2009 Google.
Image NOAA. Image State of
Ohio/OSIP. © 2011 Google
SketchUp)

file formats can be converted to KMZ (such as satellite imagery), which can then be placed or draped across Google Earth (recall how you examined EOS imagery in Google Earth in *Geospatial Lab Application 12.1*).

The big question that should be raised at this point is: How does Google Earth know where to properly place that KMZ file? The Youngstown State University stadium must have had some sort of spatial reference attached to it so that it ends up placed over the imagery of the stadium's location and not at some random point on the globe. The best way of going about this is to know the object's spatial location before any modeling begins—otherwise, you'll be trying to properly georeference the 3D model's location after the fact. GSU works hand in hand with Google Earth to obtain this data. GSU has the capability to take a "snapshot" of imagery (like the area you can see with Google Earth), then import that snapshot (complete with a model of the terrain being shown as well as the spatial reference information for the area in the view) into GSU. Thus, in GSU, you'll have a color overhead image with spatial reference information to use as a starting point for your modeling efforts (see Figure 14.10 for examples). When you've completed your work in GSU, you can convert your SKP file (*.skp) into a KMZ file (*.kmz), which will then contain the information to place the model into its proper spatial location in Google Earth. This is how you'll begin 3D modeling in *Geospatial Lab Application 14.1*—first by obtaining a "snapshot" of the overhead view of the building, and then second by creating the building's footprint from this starting point.

With the popularity of GSU and Google Earth, it's become easy for people to create spatially referenced data in GSU and then place it into Google Earth. To help facilitate the use of the programs (and to foster a geospatial 3D modeling community), Google has created an online **3D Warehouse** as

3D Warehouse
Google's online repository of 3D objects created using Google SketchUp.

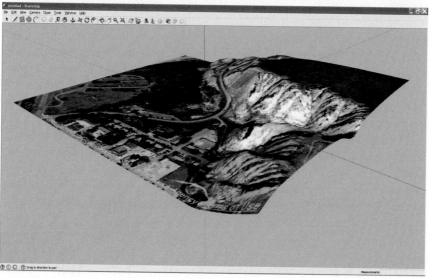

FIGURE 14.10 A section of Theodore Roosevelt National Park in North Dakota in both Google Earth and an image version including terrain in Google SketchUp. (Sources: USDA Farm Service Agency, © 2010 Google; USDA Farm Service Agency, © 2011 Google SketchUp)

a central repository of 3D models from users around the world. In the 3D Warehouse, you can find new components, objects, and a wide variety of 3D buildings, ranging from state capital buildings, to national monuments, to sports arenas from around the world. GSU makes it easy to interact with the 3D Warehouse—the software allows you to upload your model to the warehouse and also to download items from the warehouse into GSU or directly to Google Earth. The 3D Warehouse is a great source of models from users of all levels of ability and contains guidelines on how to model for inclusion into Google Earth (see *Hands-on Application 14.2: Digging into Google's 3D Warehouse* on page 416 for more information about the 3D Warehouse).

Hands-on Application 14.2

Digging into Google's 3D Warehouse

The Google 3D Warehouse is full of all sorts of 3D models from around the world. To start digging through the 3D Warehouse, open your Web browser and go to **http://sketchup.google. com/3dwarehouse**. Take a look at several of the categories—including the "Popular Models," "Featured Collections," and "3D Building Collections." Each one contains several subcategories (such as Cathedrals and Churches of the World or Sculptures and Monuments of the World), which contain numerous 3D models that have been uploaded to the Warehouse. Take a look at some of them, select their particular Warehouse page, and then download the model (you'll see the option for it on its page) into Google Earth to interact with it.

As long as you're looking around the Warehouse, also check out the option for Google Building Maker, a utility that will help you design buildings based on photos and upload them into the Warehouse.

With these types of 3D modeling and design techniques, all manner of 3D features can be created and utilized with other geospatial data. For instance, it's one thing to look at a flat map of a university campus, but it's another thing entirely to have it laid out in a realistic 3D environment that you can interact with. Prospective students are likely to get a better sense of place by viewing a 3D representation of their dorms and where they are in relation to the surrounding urban areas of the campus. The same idea holds true for designing any area in interactive 3D (see *Hands-on Application 14.3: 3D Hawaii* for an example of this).

Future planning efforts can be better implemented with the ability to see the visual impact of numerous types of building plans in relation to their nearby areas (with respect to height, texture, and landscaping). With 3D geospatial data, the locations of items and structures can be seen in relation to one another or to the surrounding terrain. For example, in Google Earth,

Hands-on Application 14.3

3D Hawaii

For an interactive example of 3D with Google Earth, open your Web browser and go to **http://3dhawaii. com**. Note that you'll first need to load the Google Earth plug-in (see also *Hands-on Application 1.5: The Google Earth Plug-In*) in order to work with this Website. (You can download and install the plug-in from **http://www.google.com/earth/explore/ products/plugin.html**.) From the available options, you can select an island (Oahu, Maui, Hawaii Island, Kauai, Molokai, or Lanai) and view many of their building features and landscaping in interactive 3D. The main window will let you pan and zoom around the islands like you would in Google Earth. Check out the variety of features available in interactive 3D—select the option for 3D tour to see animated tours around various areas.

FIGURE 14.11 The Eiffel Tower (and surrounding areas in Paris) as shown in 3D using Google Earth. (Source: © 2010 Google, Image © Institut géographique national, Image © Aerodata International Surveys, Gray Buildings © 2010 CyberCity)

FIGURE 14.12 The U.S. Capitol (and surrounding areas in Washington, D.C.) as shown in 3D using Google Earth. (Source: © 2010 Google, Gray Buildings © District of Columbia (DC GIS) & CyberCity, Gray Buildings © Sanborn)

turning on the "3D Buildings" layer activates a wide range of 3D content tied to spatial locations that you can interact with (see Figures 14.11 and 14.12 for examples of rendered 3D features in a geospatial technology environment and *Hands-on Application 14.4: 3D Buildings in Google Earth* on page 418 for how to interact with this data in Google Earth). There's also Google's Building Maker utility, which allows a user to examine photos of buildings and manipulate shapes to match the photos in order to create a 3D version of the building for Google Earth.

Hands-on Application 14.4

3D Buildings in Google Earth

Figures 14.11 and 14.12 are both images of 3D buildings within Google Earth, but they're far from the only places that have been built and placed in Google Earth. For this application, start up Google Earth and zoom to Boston, Massachusetts, looking at the area around Boston Common and the downtown. From the available options in the Layers Box (if necessary, see Chapter 1 for more information about Google Earth's layout), select the option for 3D Buildings, and place a checkmark in it. Tilt your view down so you're looking at Boston from a perspective view and you'll see various buildings and structures start to appear in 3D. Fly around the city and see how Boston was put together in 3D.

When you're done, try looking at the 3D design of some other urban areas and their 3D buildings, such as New York City, Washington, D.C., or Las Vegas, Nevada. Check some other cities as well to see if their buildings have been designed in 3D for Google Earth.

Chapter Wrapup

The goal of this chapter was to provide an introduction to some of the basic concepts of 3D visualization and how they can be incorporated into geospatial technology. *Geospatial Lab Application 14.1* will get your hands dirty with working with 3D design and visualization using Google SketchUp and Google Earth. However, 3D modeling goes well beyond Google SketchUp—designing, programming, and working with 3D modeling and visualization is a whole industry of its own—video games, movies, television and Web animation, and simulators regularly serve up extremely impressive and ever-more-realistic 3D designs. There's no way to cover all the possible applications, software, companies, and changes going on in this rapidly developing field in just one short chapter.

3D visualization is very cool and just one of the quickly growing aspects of geospatial technology, but there are many new developments (and some other cool data visualizations) yet to come as technology advances. Chapter 15 delves into some of these new and upcoming frontiers in geospatial technology.

Important note: The references for this chapter are part of the online companion for this book and can be found at http://www.whfreeman.com/shellito1e.

Key Terms

3D modeling (p. 406)
z-value (p. 406)
extrusion (p. 406)
block (p. 406)
offset (p. 407)
base heights (p. 408)
face (p. 409)
textures (p. 409)
prism map (p. 410)

prisms (p. 410)
ArcGlobe (p. 411)
ArcScene (p. 411)
Google SketchUp (p. 412)
components (p. 412)
SKP file (p. 413)
KMZ (p. 413)
KML (p. 413)
3D Warehouse (p. 414)

3D Modeling and Visualization

This chapter's lab will introduce you to the concepts of modeling and design using three-dimensional objects. Starting from an aerial image of a building with spatial reference, you will design the building in 3D using Google SketchUp, including its height, textures, and other effects, and place it into its proper location in Google Earth. While there are many methods that can be used for designing a building (including Google's Photo Match and Building Maker tools), this lab shows you how to use a variety of Google SketchUp tools through the context of creating a structure.

Important note: Even though this lab uses a fixed example, it can be personalized by selecting another structure you're more familiar with—such as your home, workplace, or school—and using the same techniques described in this lab to design that structure instead.

Like *Geospatial Lab Application 7.1* there are no questions to answer—the final product you'll create at the end is a 3D representation of a building in Google SketchUp as well as a version of it placed into Google Earth. The answer sheet for this chapter is a checklist of items to help you keep track of your progress in the lab.

Objectives

The goals for this lab are:

- Obtaining imagery to use in a design environment.
- Familiarizing yourself with the use of the Google SketchUp software package, including its operating environment and several of its tools.
- Applying textures to objects.
- Utilizing components in 3D design.
- Transforming your 3D model into the KMZ file format to be placed into Google Earth.

Obtaining Software

- The current version of Google SketchUp (8) is available for free download at http://sketchup.google.com/intl/en/product/gsu.html.
- The current version of Google Earth (6.0) is available for free download at http://earth.google.com.

Important note: Software and online resources sometimes change fast. This lab was designed with the most recently available version of the software at the time of writing. However, if the software or Websites have significantly changed between then and now, an updated version of this lab (using the newest versions) is available online at http://www.whfreeman.com/shellito1e.

Lab Data

There is no data to be copied for this lab. You will, however, use the "Bird's eye" imagery from Bing Maps (see *Hands-on Application 9.4: Oblique Imagery on Bing Maps* on page 280), so start that up in a Web browser.

Localizing This Lab

In this lab you will design a 3D model of a building on Youngstown State University's (YSU's) campus, and then view it in Google Earth. In Section 14.2, you will take a "snapshot" of the building to start with. Rather than examining a campus building at YSU, you can start at this point with a base image of any local building—such as the school building you're in, a nearby landmark or government building, or your house. You can use the same steps in this lab and the GSU tools to design any building or structure you so choose in place of the one used in this lab—the same lab techniques of using Google SketchUp will still apply, you'll just have to adapt them to your own 3D design needs.

14.1 Starting Google SketchUp

1. Start **Google SketchUp (GSU)**. GSU will open in its initial mode. Select the option for **Choose Templates**, and then from the available options select **Google Earth Modeling—Feet and Inches**.

2. There are several different templates available, depending on what type of 3D design you want to do.

3. Next, select the **Start using SketchUp** button.

4. GSU will open with a blank working environment, with three intersecting lines at the center. These represent the axes you will use for 3D design—red and green are the horizontal (x and y) axes, while blue is the vertical (z) axis.

5. You'll see the GSU toolbar at the top of the screen. You'll use many of these tools during the lab—the tools that you'll use often in this lab (from left to right on the toolbar) function as follows:

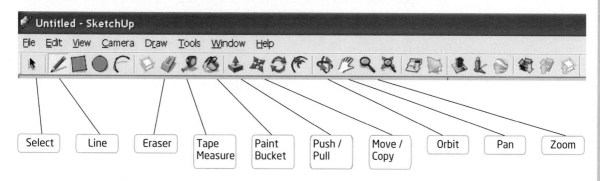

◉ The black cursor is the **Select** tool that lets you choose or select objects.

◉ The pencil is the **Line** tool that lets you draw lines or construction lines.

◉ The pink eraser is the **Eraser** tool that allows you to remove lines and objects.

◉ The extended yellow tape measure is the **Tape Measure** tool that allows you to measure the length or width of objects.

◉ The bucket spilling paint is the **Paint Bucket** tool that allows you to change the color or texture of objects.

◉ The box with the red arrow coming up from it is the **Push/Pull** tool that allows you to alter the shape of polygons.

◉ The four red arrows represent the **Move/Copy** tool that can be used to drag objects around the screen and also to create copies of objects.

◉ The swirling blue arrows represent the **Orbit** tool that allows you to change your view and maneuver around the scene.

◉ The hand is the **Pan** tool that allows you to move around the view.

◉ The magnifying glass is the **Zoom** tool that allows you to zoom in and out of the scene.

◉ There are three other specialized tools (**Add Location, Toggle Terrain, and Preview Model in Google Earth**) that you'll make use of, but the lab will point these out to you when you need to use them.

14.2 Obtaining Google Imagery

GSU directly interfaces with Google Earth (GE), in that you can use Google imagery as a starting point for your design work in GSU and transfer your 3D structures into GE.

1. To begin, start **Google Earth**.

    Important note: This lab will have you design a simplified representation of Lincoln Building (formerly known as Williamson Hall, the original College of Business building) on YSU's campus. All of the measurements of things like heights, window lengths, window measurements, and so on are either just approximations or values generated for this lab and used for this simplified model. To use real values for Lincoln (or your own buildings) you would have to make measurements of the structure or consult blueprints for actual heights and sizes.

2. Lincoln Building is located on YSU's campus at the following latitude/longitude decimal degrees coordinates:

    ◉ Latitude: 41.103978

    ◉ Longitude: −80.647777

3. Zoom GE in to this area and you'll see the top of the building (see the following graphic). Fill the GE view with the entirety of the building and return to GSU. This should give you a good look at the overhead view of the Lincoln Building for reference in the next step. You'll use GE later, so minimize it for now.

(Source: © 2011 Google)

4. Back in GSU, select the **Add Location icon** from the toolbar:

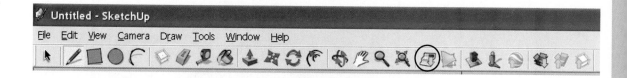

5. Add Location allows you to take a "snapshot" of Google imagery and import it (with information on the coordinates of the view as well as terrain information) into GSU for you to use as a digitizing source.

6. In the Add Location box that appears, type the coordinates of Lincoln Building and click **Search**. Use the zoom and pan tools (which are the same as the Google Earth tools) to expand the view so that you can see all of Lincoln Building in the window (like you just did in Google Earth).

7. When you've got the view set up how you want it, click the **Select Region** button.

(Source: © 2011 Google SketchUp)

8. A new image will appear—a bounding box with blue markers at the corners. Drag the corners of the box so that it stretches all across Lincoln Building (like in the following graphic). This area will be the "snapshot" that GSU takes of the imagery and that you'll use in the modeling process, so be sure to capture the whole area.

(Source: © 2011 Google SketchUp)

9. When everything is ready, click **Grab**.

10. After the importing process is done, the image (in color) should appear in GSU centered on the axes.

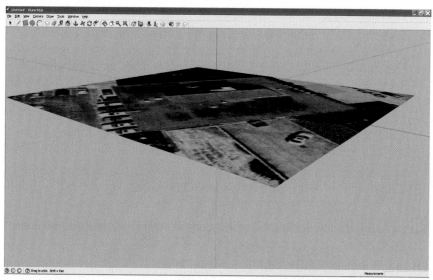

(Source: © 2011 Google SketchUp)

11. In GSU, you can turn the terrain on and off within the snapshot, using the **Toggle Terrain icon** on the toolbar. For now, toggle the terrain off (so the image is perfectly flat).

14.3 Digitizing the Building Footprint

1. You can digitize the dimensions of Lincoln Building's footprint directly from the imported aerial view. The easiest way to do this is to position the view so it's looking directly down on the image. From the **Camera** pull-down menu, select **Standard Views**, then select **Top**.

2. The view will switch to looking at the image directly from above. To zoom in or out so you can see the entire building, select the **Zoom icon** from the toolbar (or if your mouse has a scroll wheel, you can use that to zoom as well). To zoom using the icon, select it, then roll the mouse forward or back to zoom in and out.

3. When you can see the entire base of the building, it's time to start digitizing. Click on the **pencil icon** (the Line tool) on the toolbar to start drawing.

4.   Your cursor will now switch to a pencil. Start at the lower left-hand corner of the building and click the pencil onto the building's corner. A purple point will appear. Now, drag the pencil over to the lower right-hand corner and hover it there for a second. The words "on face in group" will appear. This is GSU's way of telling you just what surface you're drawing on (in this case, the face is the surface of the image).

5.   Click the pencil on the lower right-hand corner, then at the upper right-hand corner, then at the upper left. The guide may change to read "Perpendicular to the edge," indicating you've drawing perpendicular lines. Finally, drag the pencil back to where you started from. When you reach the beginning point, the cursor should turn green and the word "Endpoint" will appear. This indicates you've closed the polygon of the building's footprint in on itself. Click the pencil at the endpoint and the polygon will appear, covering the entire building.

Important note: If you have only lines and not a complete polygon, return to the lower corner and remake the footprint polygon.

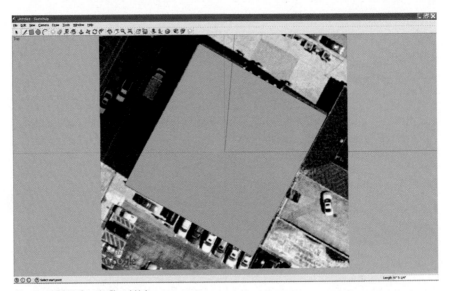

(Source: © 2011 Google SketchUp)

14.4 Extruding the Building Footprint

1.   With the building's footprint successfully digitized, the next step is to give it some height.

2.   From the **Camera** pull-down menu, select **Standard Views** and then select **Iso**. The view will switch to a perspective view.

3.   From the toolbar, select the **Push/Pull** tool.

4. Your cursor will change again. Hover it over the footprint polygon and you'll see it slightly change texture. Click on the footprint and hold down the mouse button, then use the mouse to extrude the polygon by pushing forward.

5. In the lower-right-hand corner of GSU, you'll see the real-world height you're extruding the building to. Lincoln Building is approximately 72 feet high. During the extrusion operation, you can type **72′** and hit the **Enter** key on the keyboard while holding down the mouse button, and the polygon being push/pulled will automatically extrude to that height.

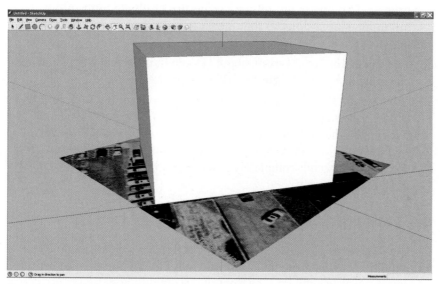

(Source: © 2011 Google SketchUp)

6. You can double-check the measurement by selecting the **Measure Tape icon** from the toolbar.

7. Click the measuring tape at the base of the Lincoln Building block and drag it to the top. The measurement (in feet and inches) will appear. You can use the push/pull tool to readjust the building height if necessary.

14.5 Maneuvering in GSU

1. An easy way to maneuver and examine the 3D objects being created is to use the Orbit and Pan tools available on the toolbar.

2. Use the **Pan** tool (the hand) to drag the current view in any direction.

3. Use the **Orbit** tool (the blue curved arrows) to rotate the view around an object.

4. Practice with the two tools for a couple minutes to get the feel for maneuvering in the GSU environment.

5. Maneuver around so that you can view the front of the building (the part that opens out onto the street).

14.6 Toggle Terrain

1. For this simplified example, we've been assuming the terrain under Lincoln Building is completely flat. When the GE image was imported, it also included data about the terrain built into the image. You would probably want to take terrain into account early in the process if you were modeling a structure or object built into a hill or on a large slope so that measurements and features properly match up.

2. To examine the "real" terrain about Lincoln Building, toggle the terrain on (again using the **Toggle Terrain icon** on the toolbar).

3. Orbit around the building and you'll see that some of the back corners appear to be slightly "floating" over the terrain. To rectify this, orbit underneath the building and the image and use the **Push/Pull** tool to slightly "pull" the building down below the image. This will ensure that the building extends completely into the ground and that none of it will float over the terrain.

14.7 Adding Features to the Building

1. Use Bing Maps' "Bird's eye" feature to get a closer look at Lincoln Building. Viewing from the north, you may be able to see a recessed entranceway leading into the front of Lincoln Building.

2. Set up the entranceway approximately 27 feet (27′) across and beginning approximately 34 feet (34′) from each side of the front of the building. The opening itself is about 9 feet (9′) tall. This is the size of a block you want to draw on the front of the building.

3. In order to find these dimensions, you can use the Tape Measure to create some guide lines (or "construction lines") on the face of the building to use.

4. Select the **Tape Measure icon** and start from the lower left corner (as you face the building). Measure up approximately 9 feet, and then double-click the mouse.

    Important note: A more precise way to use the Tape Measure is to type 9′ (for 9 feet) while dragging the tape in the direction you want to measure. A small mark will appear at the 9′ length, and you can use that as a starting point for more measurements.

5. Next, drag the mouse over the other side of the building and double-click. A dashed line will appear on the face of the building at 9 feet in height. This is a construction line—it's invisible in the modeling process but used for drawing measurements.

6. Starting along one side of the building, measure across 34 feet, double-click the mouse, and then measure down to the base of the building. Do the same on the other side. You should have construction lines set up that block out the entranceway on the face of the building.

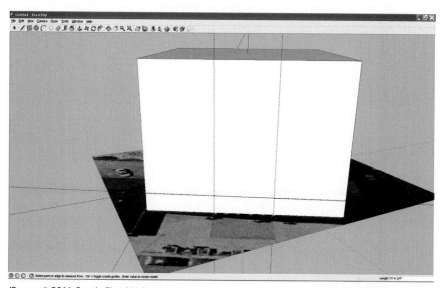

(Source: © 2011 Google SketchUp)

7. If you find you're creating too many construction lines, you can always select them with the cursor icon and delete them by using the **Eraser icon** on the toolbar (in fact, anything you create can be removed by using the eraser).

8. Now, use the pencil to draw the entranceway on the face of the building, tracing over the construction lines.

9. Finally, select the **push/pull** tool and use it to recess the entranceway by approximately 6 feet. Use the mouse to push the block inward, watching the distance bar in the lower right corner.

10. By examining the building with Bing Maps' Bird's eye imagery, you'll see that Lincoln Building has several windows stretched around the top of the building—two sets of four windows on the front and back, and 18 windows on either side. Drawing all 52 of these identical windows individually would be a huge chore, so to reduce the time and effort,

you'll be designing the window only once and reusing it throughout the model.

11. One way of doing this is to set up multiple construction lines across the front of the building. Orbit around to see the top front of the building. Set up construction lines as follows (these are simplifications of the real measurements for purposes of this lab):

    a.   The first window begins 6 feet from the edge of the building.

    b.   The first window begins 3 feet from the top of the building.

    c.   The window is 3 feet wide and 21 feet tall.

    d.   There are five divisions on the window, and their measurements are (from the top):

- 2 feet
- 5 feet
- 7 feet
- 5 feet
- 2 feet

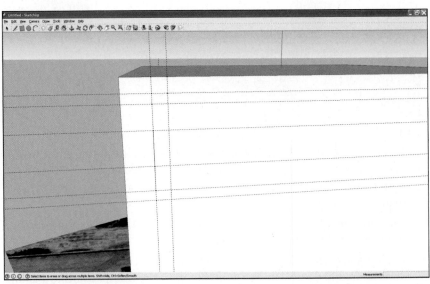

(Source: © 2011 Google SketchUp)

12. Next, draw the outline of the window and the five divisions using the **pencil** tool in conjunction with the construction lines.

13. With the outlines drawn, it's time to make it actually look like a window by adding some textures.

14. Select the **paint bucket icon** from the toolbar.

15. A new window of Materials will open. These are the various textures that can be "painted" on a face. For the purposes of this simplified model, select **Asphalt and Concrete** from the **Materials** window pull-down menu, then select the option for **Concrete Block 8 × 8 Gray**. Back in the model of the window, click the paint bucket on the top and bottom sections of the window. They should change appearance to a "concrete block" finish.

16. Next, select **Translucent** from the **Materials** window pull-down menu and choose the option for **Translucent Glass Blue**. Click on the second and fourth blocks of the window with this new material.

17. Lastly, choose **Colors** from the **Materials** window pull-down menu, select **Color 007** (it's a black color), and then click on the final (third) section of the window. All five sections of the window should now be colored in. You can now close the Materials box.

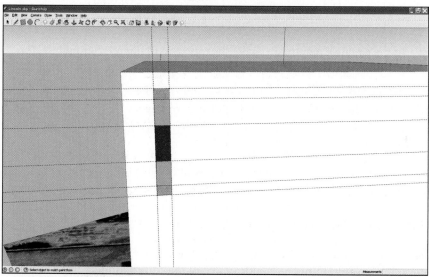

(Source: © 2011 Google SketchUp)

14.8 Working with Components

As noted before, you've made one window, but the model requires a total of 52 of them across the four sides of the building. If you're going to re-use something many times, it's easier to create a Component out of the object.

1. Select the **cursor icon** and drag a box over all sections of the window. They'll become highlighted in blue, signifying they're now selected. Be very careful not to select any of the building faces themselves—you want only the sections of the window. (To de-select items, hold down the **Shift** key and click on the items with the mouse.)

2. Right-click on the selected window and choose **Make Component**.

(Source: © 2011 Google SketchUp)

3. In the Create Component dialog box, type in the name "**Lincwind**" for the new component.

4. Select **Any** for "Glue to" from the drop-down box.

5. Place a checkmark in the "**Cut opening**" option.

6. Place a checkmark in the "**Replace selection with component**" box.

7. Click **Create** when you're done.

8. The selected section of the window will now be a GSU Component and it will be stored for later use in the model.

14.9 Making Multiple Copies

Now that the window is a component, it'll be easier to work with the same item. In this section of the building, there are four windows, each approximately 2.5 feet apart and 3 feet wide. What you're going to do is make multiple copies of the component window, then have GSU automatically space them out for you.

1. Create a construction line 13.5 feet away (the space of two window lengths and the distances between them) from the window's edge, stretching from the top to the bottom of the building. This is where the copy of the window will go.

2. Next, use the cursor to select the window component. It will be outlined in blue when it's selected.

3. Next, choose the **Move/Copy** tool from the toolbar.

4. The cursor will change to the new icon.

   Important note: Hold down the **Ctrl** key on the keyboard—this tells GSU you want to copy the window instead of moving it somewhere. If you don't hold down the **Ctrl** key, GSU will try to move the window somewhere else instead of making a copy.

5. Use the mouse to select and drag the copy of the window over to the right of the new 13.5-foot distant construction line. Make sure that the inference dialog box that appears indicates you're copying it "On the Face" (that is, the copy will be moved to the face of the building instead of into space somewhere).

6. Once the copy of the window is positioned properly, type **/3** on the keyboard. This will tell GSU to make a total of three copies, evenly spaced between the original and the location of the new copy. You will now have four windows on the front left-hand side of the building.

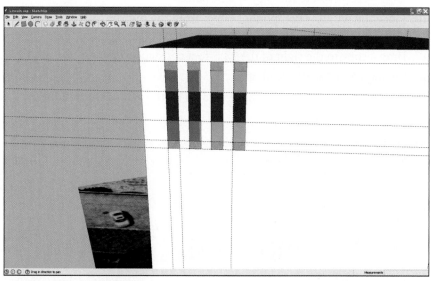

(Source: © 2011 Google SketchUp)

7. Now it's time to put windows on the other side of the front. Create a construction line 6 feet from the right-hand side of the building.

8. To add the component of the window, select **Components** from the **Window** pull-down menu.

9. In the Components dialog box, select the "**house**" icon (under the **Select** tab) to utilize only those components that exist in the current model (the only component should be your "Lincwind" window).

10. Click on the component itself in the window and drag it onto the face of the building in the model, positioning it properly in conjunction with the construction line you just drew.

11. Now, repeat the previous steps (draw a construction line 13.5 feet away and copy the windows, creating a total of three copies properly spaced out from one another).

12. You should now have two sets of four windows each on the front of the building.

13. Rotate to the back of the building. You'll want to create the same layout (two sets of four windows each) using the same measurements on the back face of the building.

14. Next, rotate to one side of the building. Each side of the building has 18 windows on it, so you'll use the same technique—position the component, create a copy at the farthest point away, then have GSU automatically fill in the evenly-spaced copies.

15. Starting at the left-hand side of the building, create a vertical construction line 6 feet from the edge and a horizontal construction line 3 feet from the top. Create another vertical construction line 93.5 feet from the building's edge—this is where the copy will be placed.

16. Place the window component 6 feet from the building's left-hand edge and 3 feet from the top of the building.

17. Drag a copy to the right of the 93.5-foot construction line.

18. Then type /17 on the keyboard.

19. A total of 18 windows should be created on the side of the building.

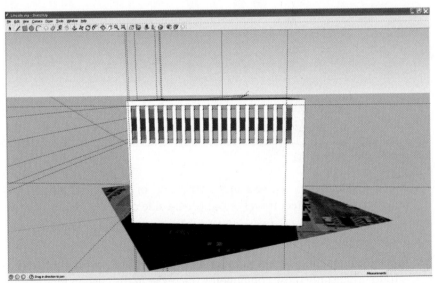

(Source: © 2011 Google SketchUp)

20. Repeat this process on the other side of the building to create an additional 18 windows.

21. At this point, all 52 windows should be created on all four sides of the building.

14.10 Working with Textures

Now, it's time to give the building a textured appearance to make it look like the orange brick in the photos instead of the white block.

1. Select the **paint bucket icon** again, and from the **Materials** pull-down menu, select **Brick and Cladding**. Next, choose the option for **Brick Rough Tan**. Click on each of the sides of the building and each will be "painted" with the new texture.

2. For the roof and the entranceway, select **Colors** from the **Materials** pull-down menu and then select **Color 006**. "Paint" the roof and the entranceway faces with this color.

14.11 Working with 3D Text

The words "LINCOLN BUILDING" are placed on the front of the building, near the lower right-hand side. 3D text can be easily added to a model to create names, signs, or other uses of block lettering.

1. From the **Tools** pull-down menu, select **3D Text**.

2. In the text box, type **LINCOLN BUILDING** (all caps and also hit the enter key between the two words so they appear on two different lines).

3. Choose **Arial** for the Font.

4. Type **8″** (for 8 inches) for the Height of the letters.

5. Type **3″** (for 3 inches) for the Extruded letters.

6. Make sure the **Filled** and **Extruded** checkboxes are marked.

7. When ready, click **Place**.

8. The 3D Text will appear as a selected object—use the cursor to place it on the face of the building near the right-hand side of the entrance.

14.12 Adding Other Features

The purpose of this lab is to get you familiar with using GSU's features and creating an approximation of a building, not modeling it down to its last detail (although GSU is certainly powerful enough to do just that). However, there are other features on the building you may want to add to continue utilizing your GSU skills.

1. Just from looking at the oblique imagery, you can see some other 3D features the building has:

   - The black section of the building stretching up from the entranceway that also contains windows

   - The black pillars in the front of the building that stretch up to the roof.

   - The units on the roof of the building itself

   - The side doorway entrance on the west side of the building

2. Using the techniques described in this lab, you can create representations of these other features using a mixture of extruding items (like drawing a polygon on the roof and extruding it) or textures (drawing in the blocks for the front windows above the entrance and filling them in).

3. When you have the model completed how you want it, there is only one more step to do.

14.13 Placing the Model into Google Earth

1. The final step is to place the finished product into GE as a 3D model. To do this, select the **Preview Model in Google Earth icon** from the toolbar:

2. GSU will export the model to a "*.kmz" file and place it into its proper location in GE. If GE is not already open, this option will automatically start it, then zoom to its proper spot.

3. In GE, use the zoom and rotate tools to examine your 3D model.

(Source: © 2011 Google SketchUp)

14.14 Saving Your Work (and Working On it Later)

1. You can save your work at any time in GSU by selecting **Save** from the **File** pull-down menu.

2. You can reopen your GSU model to work on it later by choosing **Open** from the **File** pull-down menu.

Closing Time

This lab was involved, but served to show off many of the 3D modeling features available within GSU and how to interface GSU together with GE. Exit GSU by selecting **Exit** from the **File** pull-down menu. Make sure you saved the final version of your GSU model and the "*.kmz" version of it as well.

Final Google SketchUp Modeling Checklist

_____   Obtain overhead view of Lincoln Building and place it into GSU

_____   Digitize building footprint

_____   Extrude building to proper height

_____   Set building properly on the terrain

_____   Create entranceway opening on front of building

_____   Create components for windows

_____   Apply windows to all sides of building

_____   Apply appropriate colors and textures to all faces

_____   Create 3D text on front of building

_____   Add other features (front columns, windows, roof unit)

_____   Place 3D model into appropriate place in GE

What's Next for Geospatial Technology?

New Frontiers in Geospatial Technology, Geospatial Technologies in K-12 Education, College and University Educational Programs, How to Easily Get Data, a Geospatial World Online, and Other Developments

It's an understatement to say that geospatial technology is a rapidly changing and advancing field. New Websites, imagery, tools, gadgets, and developments are coming at an incredible rate. GPS, digital maps, and satellite images are part of everyday life. When your phone can provide you with the coordinates of your location, a map of what's nearby, a high-resolution remotely sensed image of your surroundings, directions to where you want to go, and then display it all with a smartphone app, you know that geospatial technologies are part of your life and not going away. Or, as the Geospatial Revolution Project from Penn State University put it: "The location of anything is becoming everything." As noted back in Chapter 1, geospatial technology's being used in a wide variety of fields—as long as there's some sort of location involved, geospatial technologies are somehow involved. Even with all the material covered in the previous 14 chapters of this book, there are a lot of new frontiers to explore in this ever-changing field.

Geospatial technology is increasingly used in our daily lives, whether in our work, travel, or recreation. In fact, observance days have been established to prominently showcase these types of technologies. The National Geographic Society kicks off Geography Awareness Week each November, and since 1998, **GIS Day** has been set up as the Wednesday of the week. GIS Day is sponsored by a number of prominent agencies and groups (including the National Geographic Society, Esri, the USGS, and the Library of Congress, among others). It is aimed at promoting GIS for students, educators, and professionals in the field. Each year on GIS Day, there are several GIS-related events at schools, universities, and workplaces across the world. Similarly,

GIS Day the Wednesday of Geography Awareness Week (observed in November) dedicated to GIS.

Earth Observation Day a day dedicated to remote sensing research and education.

2006 saw the launch of **Earth Observation Day** to showcase remote sensing research and education across the globe.

One of the neatest recent advances is the opportunity for everyday people (without geospatial training) to get involved in this field. With geospatial tools becoming increasingly utilized by everyone, people now have the chance to create their own geospatial content. Whether it's creating new maps, updating existing maps, tying digital pictures to locations in Google Earth, or adding to geocaching databases, people are using geospatial technologies to add to or enhance geospatial data and knowledge. Dr. Michael Goodchild coined the term Volunteered Geographic Information (**VGI**) to give a name to these actions.

VGI Volunteered Geographic Information, a term used to describe user-generated geospatial content and data.

wiki a database available so that anyone can edit it.

A **wiki** is a Web utility that allows users to freely create, manage, update, and change the content of a Website. Possibly the best known one is Wikipedia, an online encyclopedia that anyone can edit. By extending this wiki concept to geospatial data, individuals can contribute their own maps, images, and updates to geospatial content available on the Internet, thereby generating VGI. Websites such as Wikimapia or OpenStreetMap (see Figure 15.1 and *Hands-on Application 15.1: User-Generated Geospatial Content Online* for more information) are dedicated to the VGI principle of users generating or contributing to the geospatial content they contain.

FIGURE 15.1 Detroit, Michigan, and Windsor, Canada, as shown in OpenStreetMap. (Source: http://www.openstreetmap. org/)

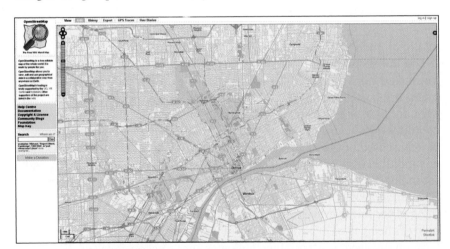

Hands-on Application 15.1

User-Generated Geospatial Content Online

Volunteered Geographic Information (VGI) represents user-generated geospatial content online. Use your Web browser to investigate both of these sites as follows:

1. Wikimapia: **http://wikimapia.org**
2. OpenStreetMap: **http://www.openstreetmap.org**

What types of data and information are available via them, and how can you add your own geospatial content? Search for your local area on both of them to see what's available and what you could potentially add or update.

This chapter explores these types of new frontiers and advances in the geospatial world and who's involved with these types of activities to sponsor and support them. How can you get involved with creating your own maps online or obtaining the newest (and coolest) geospatial devices, tools, and data to use? What kind of educational opportunities are there in higher education, and how is geospatial technology being used at the K–12 education levels? We'll look into all of these topics and some examples of what else is out there in geospatial technology (and, again, there's no way we're going to be able to cover everything or everybody out there—the following sections provide a look at several examples).

Who Is Involved with Geospatial Technology?

There are numerous "power players" in the geospatial world—we've already discussed the role of companies such as Microsoft, Google, Esri, GeoEye, and DigitalGlobe, as well as government agencies including NASA, NOAA, and the USGS in providing geospatial data and infrastructure. Other federal government agencies including the EPA (Environmental Protection Agency) and NGA (the National Geospatial-Intelligence Agency) play important roles in applications of geospatial technology. Many state and local agencies utilize GIS, GPS, or remotely sensed data through city planning, county auditors' offices, or county engineers' offices. Another notable organization in the geospatial field is the Open Geospatial Consortium (**OGC**), a group involved with developing new standards of geospatial data interoperability.

There are also several prominent professional organizations in the geospatial field. For example, the Association of American Geographers (**AAG**) is the national organization for geographers of all fields, whether physical, regional, cultural, or geospatial. Boasting several thousand members, AAG is a key organization for geographers, hosting a large annual national meeting, multiple regional branches, and publishing key geographic journals. Another major professional group is the American Society for Photogrammetry and Remote Sensing (**ASPRS**), a strong national organization dedicated to remote sensing and geospatial research. ASPRS publishes a prominent academic journal in the field and hosts national and regional meetings well attended by geospatial professionals. Also, the National Council for Geographic Education (**NCGE**) is a nationally recognized organization of educators dedicated to promoting geographic education through conferences, research, and journal publications.

There are many ongoing research and educational initiatives dedicated to furthering geospatial technology and its applications, as well as improving and fostering access to geospatial data and tools. For example, **OhioView** is an initiative created in 1996 dedicated to remote sensing research and educational efforts in Ohio. Part of OhioView's efforts led to the free distribution of Landsat imagery (where previously a single Landsat scene used to cost $4000). Today, OhioLINK, the online library system for the state, can be used

OGC the Open Geospatial Consortium, a group involved with developing new standards of geospatial data interoperability.

AAG the Association of American Geographers, the national organization for the field of geography.

ASPRS the American Society for Photogrammetry and Remote Sensing, a professional organization in the geospatial and remote sensing field.

NCGE the National Council for Geographic Education, a professional organization dedicated to geographic education.

OhioView a research and education consortium consisting of the state universities in Ohio along with related partners.

to download Landsat scenes of Ohio via NASA Glenn in Cleveland (and also via OhioView's Website). OhioView serves as a consortium for remote sensing and geospatial research and is composed of members from each of the state universities in Ohio, along with other partners such as the Ohio Aerospace Institute. With all of these statewide affiliations, OhioView serves as a source of many initiatives within the state, including educational efforts.

This type of statewide consortium may have originated with OhioView but it has expanded to a national program called **AmericaView**. Administered in part by the USGS (and set up as a 501(c)(3)), AmericaView aims to spread remote sensing research and educational efforts across all the states of the United States (see Figure 15.2). Many states have developed a **StateView** program similar to OhioView (such as VirginiaView, WisconsinView, TexasView, and so on), with the eventual goal of having a complete set of 50 StateView programs. Each StateView is composed of state universities (and related partners from that state) and offers various geospatial services or data for that state. For example, WisconsinView provides fresh MODIS imagery each day of Wisconsin and the other AmericaView member states, along with other freely available data products (such as high-resolution aerial photography and Landsat imagery). Similarly, TexasView provides access to free remotely sensed images along with links to GIS data of Texas. See *Hands-on Application 15.2: AmericaView and the StateView Programs* to further investigate the resources available through the various StateView programs.

AmericaView
a United States national organization dedicated to remote sensing research and education.

StateView the term used to describe each of the programs affiliated with AmericaView in each state.

FIGURE 15.2 The 2010 membership composition of AmericaView. (Source: Courtesy AmericaView)

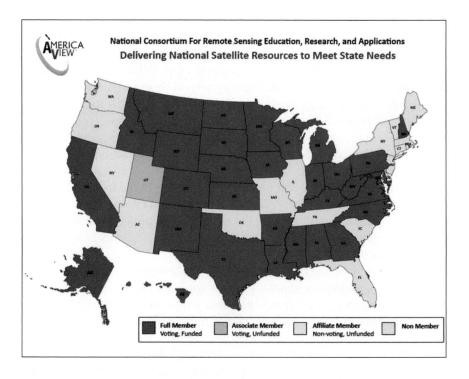

Hands-on Application 15.2

AmericaView and the StateView Programs

To see if your state is part of the AmericaView consortium (and to also see the current version of the program's membership), open your Web browser to the interactive map located at **http://www.americaview.org/membership-map**. What is the status of your home state as a StateView program? If it is a full member, click on the state to bring up the contact info and check out the specific StateView's Website (that is, MichiganView's Website or VirginiaView's Website) to see who that state's members are and what type of data and services they provide. If your state's not (yet) a full member, examine some of your nearby states and see what they offer as a StateView program.

How Is Geospatial Technology Used in K-12 Educational Efforts?

Groups such as NCGE, OhioView, and AmericaView are dedicated to promoting education in this ever-growing field of geospatial technology, but especially in K–12 classrooms. With geospatial technologies becoming increasingly utilized in today's society, introducing concepts early in the grammar-school classroom and using them for illustrating topics such as geography, spatial thinking, or earth science is becoming increasingly common. These efforts are aided by the prevalence of free software programs (such as those used in this book) and plenty of freely available geospatial data. For instance, remotely sensed images from the EOS can help in explaining environmental or climatic conditions such as changing global temperatures, while Landsat imagery can be used to evaluate the health of vegetation or loss of forested lands (and the subsequent effects). NASA and Google have released teacher kits and supplemental lesson plans to accompany their products, and many other examples of geospatial technology being integrated into K–12 curricula can be found on the Internet (see *Hands-on Application 15.3: Educational Resources and Lesson Plans* on page 444 for several examples).

Beyond using Google Earth (and other software) for interactive classroom investigations, many other programs exist that are aimed at promoting geospatial technology at the K–12 level. The National Council for Geographic Education (NCGE) is at the forefront of geographic education, supporting teachers and furthering geographic concepts at the K–12 level. Another organization dedicated to promoting geographic education is the **Geographic Alliance Network**, a national organization of geography educators at the K–12 and higher-education levels. Geographic Alliances are set up in each state as a way of hosting workshops, professional development, meetings, events, involvement with Geography Bees, lesson plans, and more at the statewide level.

Geographic Alliance Network a National Geographic-sponsored organization set up in each state to promote geographic education.

Hands-on Application 15.3

Educational Resources and Lesson Plans

The widespread nature of geospatial technologies and the availability and access to geospatial software have allowed educators to integrate GIS, GPS, and remote sensing into lesson plans in multiple ways. Examine some of the following examples to see how geospatial technology is being incorporated into K-12 curricula. What kinds of activities are teachers using in the classroom that involve geospatial technology?

1. Educaching - GPS lesson plans: **http://www.educaching.com**

2. GIS 2 GPS - GIS and GPS lesson plans: **http://gis2gps.com**

3. Google for Educators - Using Google Earth, Maps, and SketchUp in the classroom: **http://www.google.com/educators/geo.html**

4. Juicy Geography - Lesson plans and ideas related to geospatial technology: **http://www.juicygeography.co.uk**

5. NASA's Teacher's Kit - For use with Landsat 7 imagery: **http://landsat.gsfc.nasa.gov/education/teacherkit**

6. Sciencespot - GPS lesson plans: **http://sciencespot.net/Pages/classgpslsn.html**

7. NASA Teacher's Guide - Using the Image Composite Explorer in the classroom: **http://earthobservatory.nasa.gov/Experiments/ICE/ice_teacher_guide.php**

When you've examined these examples, fire up a search engine and locate some other examples of how geospatial technologies are being applied in the classroom.

SATELLITES Students and Teachers Exploring Local Landscapes to Interpret the Earth from Space—an OhioView initiative involving K-12 teachers and students with geospatial technologies.

Among OhioView's initiatives is **SATELLITES** (Students and Teachers Exploring Local Landscapes to Interpret the Earth from Space), a program designed to instruct K–12 teachers about geospatial technologies and aid them in incorporating it into their classrooms. SATELLITES hosts week-long "Teachers' Institutes" during the summer (in locations in Ohio). These institutes are free for teachers to attend (and teachers can earn free graduate credits, when available, through successful participation). During the week, teachers receive introductory instruction in several of this book's topics (including GIS, GPS, remote sensing, landscapes, and remotely sensed environment or climate data) and hands-on work with the technology. Teachers also receive free equipment to take back to their classrooms, including a GPS receiver and an infrared thermometer. The goal is for the teachers to develop an inquiry-based research project incorporating geospatial technologies and an environmental theme (such as previous SATELLITES institutes that involved the International Polar Year).

An example of a SATELLITES success story is the "Satellite Girls" (a group of four middle school students from Akron, Ohio). Their teacher attended the SATELLITES program in the summer of 2007 and involved a group of his students with data collection and analysis through GLOBE. The girls competed with their project at the state and national levels and were selected to be one of five teams of students to represent the United States at the international GLOBE Learning Expedition in Cape Town, South Africa, in 2008.

FIGURE 15.3 The Website of the GLOBE Program. (Source: The GLOBE Program/The University of Texas at Tyler)

The inquiry-based projects the K–12 students become involved with are part of **GLOBE** (Global Learning and Observations to Benefit the Environment), a program sponsored by NASA and NOAA that incorporates over 20,000 schools from 114 countries across the world dedicated to study of Earth's environment (see Figure 15.3). Using specific data-collection protocols, students involved with GLOBE collect data related to land cover, temperature, water, soils, or the atmosphere. This data is shared with other students around the world via the GLOBE Website, creating rich and extensive datasets for use (for instance, via NASA World Wind). Teachers attending the SATELLITES Institutes learn GLOBE protocols and engage their students through field data collection coupled with geospatial technology applications.

GLOBE Global Learning and Observations to Benefit the Environment-an education program aimed at incorporating user-generated data observations from around the world.

What Types of Educational Opportunities Are Available with Geospatial Technology?

Beyond K–12 involvement, geospatial technology is widely integrated into the curricula of higher education programs around the world. Geospatial technology course offerings are commonly found in college and university geography departments, but also within civil engineering, natural resources, environmental sciences, or geology programs. Bachelor's degrees may offer some sort of concentration in geospatial technology, or at least the option to take several courses as part of the major. A minor in GIS or geospatial technologies is also a common option in geography departments. Graduate programs (both master's and doctorate degrees) in geography often have an option to focus on aspects of geospatial technology as well.

However, with the growth of the geospatial field, it's now possible to get a higher-education degree in geospatial technology (or GIS, or geographic information science, depending on the name of the program). Bachelor's, master's, and even doctorate programs have become available at several universities. Some of these are interdisciplinary programs, sometimes involving elements of various social sciences or computer science and information systems topics such as programming, databases, or Web design, coupled with classes drawn from other related fields. For instance, Youngstown State University offers a bachelor's degree in "Spatial Information Systems" that combines elements from the geospatial side of the geography discipline with computer-related coursework in programming and databases, classes in CAD and professional writing, and an option for the student and advisors to design a set of courses for application (such as environmental science, biology, or archeology).

Certificate programs are also a widespread option for geospatial education. These programs vary from school to school, but typically involve taking a structured set of geospatial classes (perhaps four to seven classes total). These programs tend to be focused squarely on completing the requirements for a set of courses, and are often available at both the undergraduate and graduate levels. For example, the Geospatial Certificate at Youngstown State University involves students taking four required courses—an introductory mapping or geospatial class, an introductory GIS class, an introductory remote sensing course, and either an advanced GIS or advanced remote sensing class. Students take an additional two courses from a pool of options, including field methods, GPS and GIS, a geospatial-related internship, object-oriented programming, and database design. Certificates and degree programs are becoming increasingly available at the community or technical college level as well (see *Hands-on Application 15.4: Degree Programs and Certification for Geospatial Technology* for some further investigation).

With geospatial technology entering so many different fields, students from outside the geography field may seek out a couple of courses or certification to gain a more formal background in geospatial technology to apply to whatever field they're in (such as business, real estate, archeology, geology, or human ecology). Online courses are being offered by a variety of schools to allow students to complete a geospatial certificate or a degree remotely. Companies like Esri and organizations like ASPRS offer their own certification programs, and "virtual universities" offer students at one institution the option to take a geospatial course from a professor at another school. Even more specialized courses are being developed—for example, today you can

Hands-on Application 15.4

Degree Programs and Certification for Geospatial Technology

An online utility is available for examining the available geospatial programs at community colleges or technical colleges throughout the United States—open your Web browser and go to **http://216.69.2.35/flexviewer/index.html**. This map (part of the GeoTech Center Website at **http://www.geotechcenter.org**) shows community and technical colleges involved with geospatial education.

Schools are mapped by categories—whether they offer a degree in an aspect of geospatial technology, a certificate program, or have classes available. Check your local area for what nearby schools (select the option for CC Info on GIS Offerings in the cube menu icon to get the names of the schools) are involved with teaching geospatial courses and what kinds of degree programs they offer.

receive a bachelor's degree from the University of North Dakota in piloting unmanned aerial vehicles (UAVs).

What Other Kinds of Data and Software Are Available Online?

The previous chapters have described numerous datasets and how to obtain them (either for free or for minimal cost), often through the applications in the chapters. To build a database of available geospatial data, you could get your hands on:

- 3D models for Google Earth – see Chapter 14
- Digital Line Graphs (DLGs) – see Chapter 5
- Digital Raster Graphics (DRGs) – see Chapter 13
- Geocaches or other GPS destination locations – see Chapter 4
- Landsat imagery (current or historic) – see Chapter 11
- MODIS images – see Chapter 12
- NAPP aerial photos or orthophotos– see Chapter 9
- National Elevation Data (NED) and DEMs – see Chapter 13
- National Land Cover Database (NLCD) – see Chapter 5
- TIGER/Line files – see Chapter 8
- US Topo maps – see Chapter 13

Beyond these options, there are plenty of other online resources available for obtaining geospatial data, and usually for free. State, county, or municipal Web services will often make their GIS data or remotely sensed imagery available free of charge (as you saw in Chapter 5). Other Web resources used in previous chapters allow you to specify a geographic area and then download multiple datasets that cover that one area. For instance, if you're studying East Liverpool, Ohio, Web resources like the Seamless Server (from Chapter 13) or EarthExplorer (from Chapters 5 and 9) allow you to select many types of data, including remotely sensed imagery, land-cover data, elevation data, and road networks of East Liverpool.

The Seamless Server is only one component of a much larger geospatial data distribution program called the **National Map**. Users of the National Map can access multiple types of datasets, including elevation, hydrography, transportation, land-cover data, boundary files, and more, all in formats readable by GIS software. The National Map is part of larger National Geospatial Program and serves as a source of geospatial data for the United States (see Figure 15.4 and *Hands-on Application 15.5: The National Map Viewer* for more information about the National Map). You can think of the National Map as a "one-stop-shopping" venue for free geospatial data—numerous types of datasets are available for a geographic area that you define, a very useful option that prevents you from having to download data from multiple sources, then clip or alter the data to make it fit the region you're examining.

The United States government has created other online resources, such as the Geospatial One Stop (**GOS**) for obtaining geospatial data. GOS provides users free access to numerous datasets via its portal—orthophotos, parcel information, cadastral data, DEMs, transportation, hydrology datasets, and more are available. There are many data repositories available to easily get the geospatial data you need quickly into your hands. The various State-View programs make Landsat imagery (and sometimes other remotely sensed data) for their state available. For instance, OhioView servers permit the free download of Ohio Landsat data from their multi-year archive. Other online services, such as the GIS Data Depot, contain numerous datasets (including DRGs, DEMs, DOQQs, and hydrologic data).

National Map an online base map of downloadable geospatial data maintained and operated by the USGS and part of the National Geospatial Program.

GOS the Geospatial One-Stop, an online federal government portal used for obtaining geospatial data.

FIGURE 15.4 The online National Map Viewer. (Source: U.S. Geological Survey)

Hands-on Application 15.5

The National Map Viewer

You can access the National Map online at **http://nationalmap.gov/index.html**—then select the option for National Map Viewers. From this new page, choose the option to Open the Viewer. A new Web page will open with the Viewer itself, the tool used in accessing data via the National Map. The National Map can be used to view available data—from the Overlay options on the left side of map, you can select what layers to display. High-resolution imagery can be accessed as an additional layer from the Imagery option on the right side of the map.

The layers you're viewing in the National Map can also be downloaded for use in GIS or other software programs. First, search for an area of interest (for instance, Reno, Nevada), then after the view of the map changes, press the Download Data button in the upper-right corner of the map. New download options will appear that allow you to select the extent of the area you want to download data for (such as counties, congressional districts, or the extent of the view seen on the screen). Next, the data layers available to download will be shown (including US Topo, structures, transportation, boundaries, hydrography, land cover, elevation, and orthoimagery) as well as the formats these layers are available in. Select the layers you want, then choose Add to Cart and follow the remaining steps to download your data (you will have to register with the USGS to receive your data—they may send you an e-mail with a link you can use to retrieve your data).

Sometimes the geospatial data comes available in a format called **SDTS** (the Spatial Data Transfer Standard), a system established by the United States government. The intent with SDTS was to place data in a "neutral" file format rather than a software-specific one. Agencies like the USGS, NOAA, and the Census Bureau utilize SDTS as the format for delivering their geospatial data to you. Software programs will often have a converter that enables the import of data in SDTS format to a form directly readable by the software (for instance, ArcGIS has a utility to import an SDTS DLG into a coverage for you to use).

> **SDTS** the Spatial Data Transfer Standard, a "neutral" file format for geospatial data, allowing it to be imported into various geospatial software programs.

For some other examples of freely available geospatial software online, check out *Hands-on Application 15.6: Some Other Available Geospatial Technology Programs and Tools Online* on page 450.

How Can You Use Geospatial Technology Online?

Online resources are the quickest and easiest way to access geospatial data. Whether you want to acquire archived satellite imagery or get the newest updates for Google Earth, you can do everything via the Internet. The same holds true with distributing your own data and information—setting up a Web interface where your data can be used interactively is the way to go.

Posting parcel databases and DEMs for download will be of great aid to a person with some geospatial knowledge, but other individuals may not know

Hands-on Application 15.6

Some Other Available Geospatial Technology Programs and Tools Online

In the previous chapters, you've used a variety of free software programs, like MultiSpec for examining satellite imagery, MapCruncher for georeferencing a scanned map, and Trimble Planning for looking at GPS satellite conditions. However, there are a lot of other free geospatial tools available out there beyond just the software and programs used in this book. An excellent resource for locating and downloading free geospatial software is the Free Geography Tools Website—open your Web browser and go to **http://freegeographytools.com**. From the Topics list on the right-hand side, choose the type of tool you're interested in, whether it's new additions

for Google Earth, DEM tools, or GPS utilities, and see what kinds of other software programs are available and how you can utilize them.

In addition, this book has made extensive use of Google Earth and other virtual globe programs (like NASA World Wind in Chapter 12 and ArcGIS Explorer in *Geospatial Lab Application 15.1*). Virtual globes are very powerful utilities for examining features around the planet, and are also very cool and fun to use. There are other virtual-globe-style programs available—fire up a search engine and try to locate some other examples of virtual globes or related geospatial programs available online.

what GIS is, or what a shapefile is, or how to georeference imagery, and may be baffled as to how to properly use the data. However, when all of the data is assembled into an online map, allowing a user to find the address of a house for sale, access data about it, determine the land value of the property, and see whether or not the house is on the floodplain, the online utility becomes a lot more useful. In addition, other users may not have access to the specific GIS program you're running, so if all they need is a Web browser to interact with your data, you've removed another obstacle in working with the geospatial data. You've already used some examples of these types of online GIS tools back in the Hands-on Applications in Chapters 5 and 6.

ArcGIS Server an Esri utility for distribution of geospatial data in an interactive mapping format on the Internet.

There are many programs available for taking geospatial data and setting it up online in an easy-to-use interactive format, or to simply create an interactive map and incorporate it into a Website. An example of this is **ArcGIS Server**, Esri software designed to facilitate the distribution of interactive maps and geospatial data across the Internet for viewing and analysis (see **Figure 15.5** for an example of an online mapping and analysis utility created using ArcGIS Server for the city of Philadelphia). Setting up an interactive tool online for managing and analyzing geospatial data is becoming increasingly common for a variety of utilities. Check out *Hands-on Application 15.7: ArcGIS Server Live User Sites* for a way to access and examine several examples of ArcGIS Server in action for working with geospatial data online (rather than being required to have ArcGIS running on your own computer).

Tools like these are part of a larger initiative that Esri refers to as "Mapping for Everyone," which allows users to not just add to existing databases or use special software to distribute geospatial data, but to create their own maps with their own data (or data already processed for them) and publish the maps on their own Website. Like the user-generated content of VGI, this

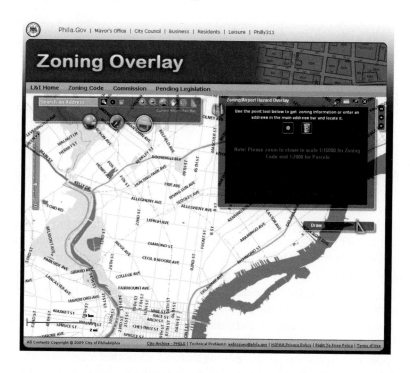

option puts more mapping choices and geospatial data into the hands of end users, allowing you to build, design, and customize your own mapping applications, and then embed them onto a Website or e-mail a link to others.

The basis for "Mapping for Everyone" is an available ArcGIS Javascript (or Flex or Silverlight) **API** (Application Programming Interface) that you can use to develop custom applications. Esri makes several samples (such as base maps or layers, referred to as **Map Services**) available for free as part of their online resources so that creating a simple Web-mapping application is as easy as copying their base code and customizing it. Other companies, such as Google, Microsoft, and Yahoo!, provide APIs that allow you to embed

API Application Programming Interface —the functions and commands used when making requests of software.

Map Services pre-made base maps and layers that can be used in creating Web-mapping applications.

Hands-on Application 15.7

ArcGIS Server Live User Sites

In previous chapters, you've explored several online mapping applications, whether to perform queries, build buffers, check color choices, find a shortest path, or georeference a map. There are more—a lot more—online mapping sites out there. For multiple examples, check out some of the ArcGIS Server Live User Sites—open your Web browser and go to **http://www.esri.com/software/arcgis/arcgisserver/live-user-sites.html** and examine some

of the available maps. Click on a map and a brief description will appear, along with a link to "View Live Site" (that is, interact with that particular map utility itself). Examine several of the Live Sites. All of these mapping sites linked to this Website are examples of using ArcGIS Server to construct Web mapping utilities. What are some of the functions being mapped and how are they being implemented?

Hands-on Application 15.8

ArcGIS Web Mapping

To get started building your own Web map using some available resources, and to see what sort of data is available to start with, open your Web browser and go to **http://www.esri.com/mapping-for-everyone/index.html** and then select one of the options for making a map. What map layers are available for you to utilize in your own maps? By selecting an option, you can choose to get more information about what the data on that map represents. What types of Web applications could you design using this available data?

mashup the combination of two or more map layers into one new application.

cloud a computer structure wherein data, resources, or applications are kept at another location and made available to the user over the Internet.

ArcGIS Online a collection of available Web resources made available by Esri.

their mapping products' (such as Google Earth, Google Maps, Bing Maps, or Yahoo! Maps) functions on a Website or design applications. By combining several map layers together, a user can create a new map **mashup**, which uses different datasets together in one map. Check out *Hands-on Application 15.8: ArcGIS Web Mapping* for some examples of available map layers to get started designing your own Web-mapping applications.

Geospatial technology is increasingly becoming part of a field referred to as **cloud** computing, wherein resources (such as data storage or other geospatial applications, such as GIS) are being utilized at another location but served to a user across the Internet. Several of the previous examples describe uses of cloud computing, where you don't need the analysis tools downloaded and installed on your computer, but are rather accessing them via a Web browser. For instance, geospatial data can be obtained through the cloud structure in the form of aerial or satellite imagery, digital topographic maps, or digital street maps that can be used as a base map with your own applications— but this data is being served to you over the Internet, rather than you having to go out, download it, and set it up on your own computer. For example, Esri makes a number of the base maps just described available through its **ArcGIS Online** resources—when using ArcGIS or ArcGIS Explorer, you can

Thinking Critically with Geospatial Technology 15.1

Who Owns Geospatial Data?

Think about this—you start with a basemap sample (for instance, a world topographic map), obtain data about earthquake locations and other natural hazards, create a mashup, and post it onto your Website. Who owns this geospatial resource or the geospatial dataset you've created? In essence, you've taken other freely available data and created your own unique geospatial resource—but can you really lay claim to this as something you own? Similarly, if you make changes or updates to a resource like Wikimapia, can you claim ownership of a shared online resource? With more and more geospatial data and customization functions becoming available via cloud resources (such as the high-resolution Bing Maps imagery through ArcGIS Online), who really owns the end products created from them (or does anybody really own them)?

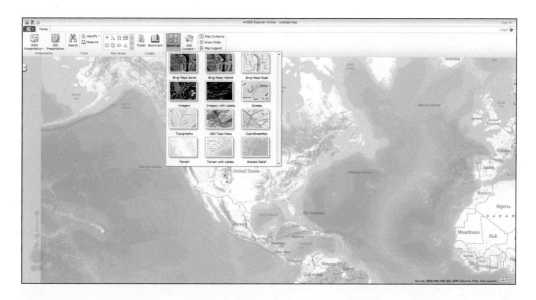

stream these base layers (such as topographic maps, Bing Maps imagery, or street maps) onto your computer through the Internet and use them in your GIS analyses (see Figure 15.6 for an example of the basemap data available through ArcGIS Online).

FIGURE 15.6 The types of base maps available via the Internet through ArcGIS Explorer Online. (Source: USGS, FAO, NPS, EPA, DeLorme, TANA. Esri® ArcGIS ArcExplorer graphical user interface Copyright © Esri)

What Are Some Other Cool Developments in Geospatial Technology?

With computers becoming faster and having the ability to handle geospatial data more quickly and easily, new methods of visualizing data have been developed. It's one thing to view a 2D choropleth map (like in Chapter 7), another thing to create a pseudo-3D representation of the mapped data (like in Chapter 14), and yet another thing altogether to be able to interact with the data in an immersive environment. Examining a realistic geospatial landscape and flying across it (like in Chapter 13) is very neat, but what if you could simulate the full 3D effect of diving through canyons or banking over ridgelines? Immersive "virtual reality" technology allows the user to actually become part of the data, moving with the landscape, being placed inside a structure, or being on the surface of the map. This sense of immersion or interactivity with geospatial data adds immensely to the understanding of the nature of spatial data, both for researchers and students.

Another example of enhanced geospatial visualization is NOAA's Science on a Sphere (**SOS**) device. SOS features a large six-foot sphere that has animated images projected onto its surface, turning the sphere into a global model of Earth. With SOS, viewers can see geospatial data such as ocean currents, sea-surface temperature, land-use information, global fires, and data products derived from satellites in the Earth Observing System (see Chapter 12) on a

SOS Science on a Sphere, a NOAA initiative used in projecting images of Earth onto a large sphere.

FIGURE 15.7 Science on a Sphere in action in Atlanta, Georgia. (Source: NOAA)

simulated planet surface (see Figure 15.7). Think of the global EOS datasets used in *Geospatial Lab Application 12.1* or some of the Google Earth imagery but projected onto an actual spherical surface several feet tall. Museums and science centers around the world are using SOS technology for geospatial education related to Earth and its climate and environment.

Another way of making geospatial data appear more realistic (and adding a sense of immersion) is to view the data in stereo. Stereo imagery shows two images (for instance, two images of the same landscape), rendered apart from each other. When you look at something, both of your eyes are seeing the same thing, just from different points of view, allowing you to see depth and distance (and view the real world in three dimensions). Stereo images use a similar principle, as they are two representations of the same thing, just taken from two different positions of the camera or sensor. Stereo imagery has been used for a long time in remote sensing for measuring heights and depths of areas. Overlapping aerial photography or images acquired from sensors (such as those onboard SPOT 5 or the ASTER or MISR sensors on Terra) can be used for creating stereo data.

A simple way of viewing imagery or data in stereo (which several geospatial software programs have the capability for) is to create a red and blue **anaglyph** of the data—in which the two images are slightly separated, but one is highlighted in red and the other in blue. By wearing a pair of 3D glasses (the ones that have colored gels for lens, with one lens being red and the other being blue), the anaglyph images appear to be raised or elevated off the

anaglyph a new image created by slightly offsetting the features in an existing image slightly from each other and highlighting them in different colors.

FIGURE 15.8 Use of the anaglyph viewing mode of a perspective view of the terrain in southern California (from NASA World Wind). (Source: NASA)

screen in a "3D effect" (see Figure 15.8). This type of stereo effect (where one eye views the red and the other eye views the blue) creates a type of **stereoscopic 3D** that gives the illusion of depth in parts of the image. Some items may appear in the foreground and others in the background. Red and blue anaglyphs are certainly nothing new, but today, when stereoscopic 3D movies have become all the rage (and the top-grossing film of all time being one of them), there are more high-tech ways of using these stereo concepts with geospatial technology.

If you go to a 3D movie today, you're not going to be wearing red and blue lenses, but rather a pair of polarized glasses instead. These same types of 3D glasses can be used with the **GeoWall**, a computer tool used for examining imagery in an immersive three-dimensional environment with stereoscopic 3D viewing techniques. First developed by the members of the GeoWall Consortium in 2001, it has spread to more than 400 systems developed for use in schools and colleges (along with places like the EROS Data Center and NASA's Jet Propulsion Laboratory). The GeoWall can be used for viewing geospatial data in a stereoscopic 3D format.

The GeoWall consists of several components that can be purchased separately to keep the cost of the whole setup affordable (about $10,000 or so). First, two high-power Digital Light Processing (DLP) projectors are used to project the two images needed for stereo viewing. These projectors get hooked into a powerful computer with a dual-output video card (so the computer can send its output into both of the projectors). The images get projected through special filters in front of the projectors' lenses, which polarize the light. The light is projected onto a special screen made of material that will preserve the polarized light (rather than scattering it). Lastly, the user wears a pair of polarized 3D glasses to view the imagery. See Figure 15.9 on page 456 for the components of the GeoWall projectors and their filters.

stereoscopic 3D an effect that simulates the illusion of depth or immersion in an image by utilizing stereo imaging techniques.

GeoWall a powerful computer tool used for displaying data and imagery in stereoscopic 3D.

Wearing the 3D glasses, the viewer perceives the images in a 3D effect as if the items and images are appearing off the screen—similar to watching a stereoscopic 3D movie. Viewing and interacting with imagery like this gives a sense of depth or immersion, as some items will be in the foreground of the image and some will be in the background. Figure 15.10 shows an image of a GeoWall in use from the perspective of an outsider not wearing the polarized glasses (without the glasses, the images would appear blurry and slightly askew from one another). GeoWall viewers position themselves in front of the screen and are able to see the images projected in stereoscopic 3D.

Hands-on Application 15.9

Some Blogs about Geospatial Technologies

Let's face it, geospatial technology is a rapidly changing field, with new innovations and advancements coming all the time, and you'll probably first hear about these things online. Blogs are a great way to keep informed on up-to-date geospatial developments. Check out the following examples of blogs to see what people are buzzing about concerning current or upcoming developments in the geospatial universe:

1. Digital Geography: **http://www. digitalgeography.co.uk**

2. Digital Urban: **http://digitalurban.blogspot. com**

3. Esri blogs: Esri maintains many different blogs about different software programs, development tools, and applications: **http:// www.esri.com/news/blogs/index.html**

4. Google Earth Blog: **http://www.gearthblog. com**

5. Google Lat Long Blog: **http://google-latlong. blogspot.com**

6. Mapperz - The Mapping News Blog: **http:// mapperz.blogspot.com**

7. Official Google SketchUp Blog: **http:// sketchupdate.blogspot.com**

8. Ogle Earth: **http://ogleearth.com**

9. Planet Geospatial: **http://www.planetgs.com**

Once you've examined these examples, fire up a search engine and try to locate some other blogs or forums or pages on a social networking site concerning new developments in geospatial technology.

Using the GeoWall for visualizing geospatial data carries a lot of benefits. For instance, the GeoWall enables the user to see maps draped over a surface model or terrain data displayed in stereoscopic 3D. Certainly, being able to dive inside a meteor crater or view volcanic features, coastlines, or river valleys in a stereoscopic 3D environment allows the viewer to get a better sense of the terrain features. Landscapes, geologic processes, draped satellite imagery, draped topographic maps, stereoscopic photos, and tours of virtual environments are all types of data that can be viewed using a GeoWall setup.

Thinking Critically with Geospatial Technology 15.2

What Do You See as the Next Frontiers for Geospatial Technology?

Right now, you can run Google Earth on your phone, download satellite imagery for free, view landscapes and cities in immersive 3D, create your own geospatial maps and post them on a Website, and earn a doctorate in the geospatial field. If that's what you can do today, what's going to happen tomorrow? Where do you see the geospatial field going? As access to information becomes faster and easier, as computing power advances, and as our lives become more and more tied together in a global web, what's going to be the next stages for geospatial technology? What do you see the geospatial world looking like five years from now?

Chapter Wrapup

So this is it, the end of the last chapter. For 15 chapters now, we've looked at multiple aspects of geospatial technologies, both from the theoretical side and also from working hands-on with several different aspects of the technologies themselves. The geospatial world has changed a lot in the last few years (for example, Google Earth only debuted back in 2005), and it's going to keep changing (see *Hands-on Application 15.9: Some Blogs about Geospatial Technologies* on page 457 for some online blog sources related to geospatial technologies).

GIS, GPS, remote sensing, and all of their applications are only going to continue becoming more important in our lives, and even though many people may still not be familiar with the term "geospatial technology," by this point you've probably got a pretty good idea of what it is and what it can do. There's still one last Geospatial Lab Application. In *Geospatial Lab Application 15.1* you'll be using Esri's ArcGIS Explorer program to examine a variety of geospatial applications as a "summary" of what we've been looking at for 15 chapters.

Important note: The references for this chapter are part of the online companion for this book and can be found at **http://www.whfreeman.com/ shellito1e**.

Key Terms

GIS Day (p. 439)
Earth Observation Day (p. 440)
VGI (p. 440)
wiki (p. 440)
OGC (p. 441)
AAG (p. 441)
ASPRS (p. 441)
NCGE (p. 441)
OhioView (p. 441)
AmericaView (p. 442)
StateView (p. 442)
Geographic Alliance Network
  (p. 443)
SATELLITES (p. 444)

GLOBE (p. 445)
National Map (p. 448)
GOS (p. 448)
SDTS (p. 449)
ArcGIS Server (p. 450)
API (p. 451)
Map Services (p. 451)
mashup (p. 452)
cloud (p. 452)
ArcGIS Online (p. 452)
SOS (p. 453)
anaglyph (p. 454)
stereoscopic 3D (p. 455)
GeoWall (p. 455)

Exploring ArcGIS Explorer

This chapter's lab will introduce you to using Esri's free ArcGIS Explorer software program in the context of many of the concepts introduced throughout the book. In many ways you can think of this lab as a "summary" of several things you've explored in other chapters. However, you'll be using a new geospatial technology program, as ArcGIS Explorer contains numerous GIS tools, as well as access to many different types of GIS and remotely sensed data. You'll also be using "cloud" resources available through ArcGIS Explorer in the form of several types of basemaps, data, and imagery.

Objectives

The goals of this lab are:

- Learning the basics of the ArcGIS Explorer environment.
- Utilizing various imagery and data sources streamed from the cloud.
- Examining digital topographic maps and how terrain is displayed with contours in 3D environments.
- Examining high-resolution Bing Maps imagery and visually identifying features from it.
- Locating an address, creating a buffer around it, and calculating a route between two points on a map.

Obtaining Software

The current version of ArcGIS Explorer Desktop (Build 1500) is available for free download at **http://www.esri.com/software/arcgis/explorer/index.html**.

Important note: Software and online resources sometimes change fast. This lab was designed with the most recently available version of the software at the time of writing. However, if the software or Websites have significantly changed between then and now, an updated version of this lab (using the newest versions) is available online at **http://www.whfreeman.com/shellito1e**.

Lab Data

There is no data to copy in this lab. All data used in this lab comes with the software or is streamed to you via the Internet.

Localizing This Lab

This lab focuses on various features around southern California, but the same techniques can easily be used for a local area. The topographic maps and high-resolution imagery analysis of Sections 15.2 and 15.3 can be adjusted to a local area (as the topographic maps and imagery cover the whole of the United States). Similarly, the addresses and analysis of Section 15.4 can be performed with local landmarks instead of Hollywood ones.

15.1 Getting Started With ArcGIS Explorer

1. Start **ArcGIS Explorer (AGX)**. You'll see that it opens with the default imagery being satellite imagery of the world (acquired by systems like those in described in Chapters 11 and 12).

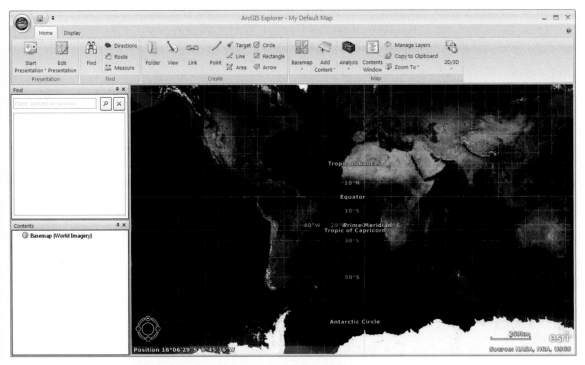

(Source: NASA, NGA, USGS. Esri® ArcGIS ArcExplorer graphical user interface Copyright © Esri)

To start things off, you'll be examining how terrain is modeled and displayed with contour lines on topographic maps (as in Chapter 13). Rather than needing copies of paper topo maps or having to download several quads of digital raster graphics (DRGs) and use them together, ArcGIS Explorer allows you to access an entire set of digital topographic maps (shown at different map scales, depending on what scale you're working at) via the cloud. These topos are also seamlessly blended into one another, so there are no issues with matching map edges against one another.

2. Select the **Find** option.

3. We're going to begin in southern California, just outside of Hollywood. In the AGX Find box, type **Tujunga Valley, California**, and click the **Search** button.

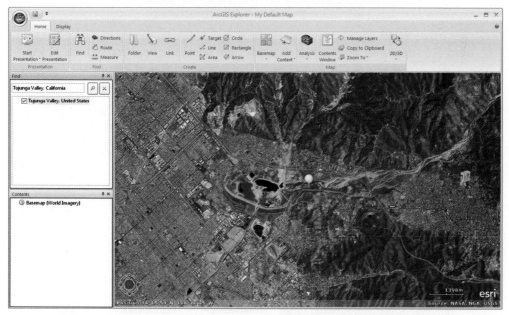

(Source: NASA, NGA, USGS. Esri® ArcGIS ArcExplorer graphical user interface Copyright © Esri)

4. An option for Tujunga Valley, United States, will appear in the Find box, and AGX will zoom in to the valley in southern California.

5. A pushpin will appear to mark Tujunga Valley on the map. Remove the checkmark from the option in the Find box to remove the pushpin. At this point, you can also close the Find box.

6. A set of controls for navigation in AGX is available in the bottom left-hand corner of the view. Scroll your mouse over that area to enlarge the controls and make them visible. You'll see that they are similar to Google Earth's—arrows for navigation and a zoom slider for zooming in and out of a scene. The wheel will spin your view while the "N" button will reset the orientation to the north. You can also use the mouse to pan around and the mouse wheel to zoom in and out.

7. To get your bearings in the area with labels on the various cities and features, select the **Home** tab, then select the **Basemap** option. From the new available options, select **Boundaries and Places**.

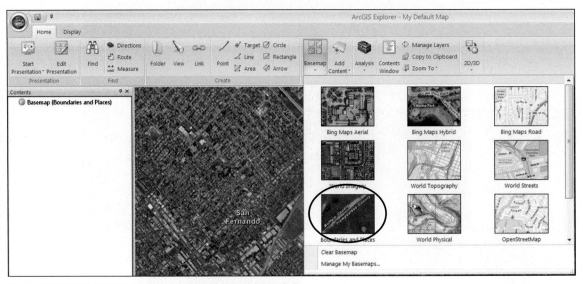

(Source: Esri® ArcGIS ArcExplorer graphical user interface Copyright © Esri)

8. You'll see that labels will now start appearing around the map. By zooming around the image (and zooming out as well), you should be able to make out the cities of San Fernando and Burbank (and Hollywood), with Tujunga Valley separated by two sets of mountains. Zoom in to this region.

15.2 Examining Digital Topo Maps with ArcGIS Explorer

1. To switch to display the topographic maps (that is, stream them in from the cloud), select the **Home** tab, and then select the **Basemap** option. From the available options, select **World Physical**.

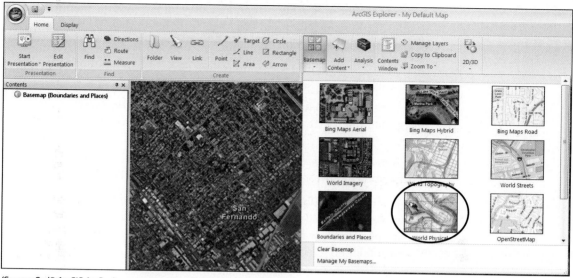

(Source: Esri® ArcGIS ArcExplorer graphical user interface Copyright © Esri)

2. AGX will display the digital versions of the USGS topographic maps draped over the terrain. When you're zoomed out, small-scale digital topos will be displayed and when you're zoomed in, AGX will load large-scale digital topos in their place.

3. Zoom in on Tujunga Valley itself and give AGX a little time to load the larger scale topos for the area.

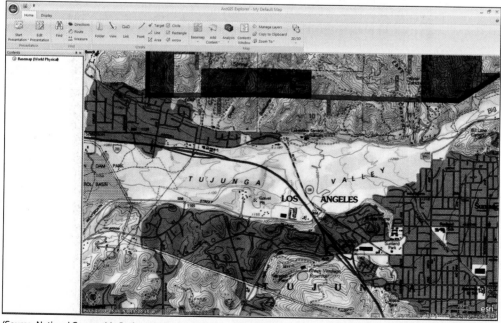

(Source: National Geographic Society. Esri® ArcGIS ArcExplorer graphical user interface Copyright © Esri)

4. Examine the area around Tujunga Valley (along with the mountains to the north of the Valley and Los Angeles to the south) with respect to the contour lines shown on the map.

Question 15.1 How do the contour lines effectively represent the terrain of the valley (as opposed to the mountains)?

Question 15.2 From examining the contour lines on the larger-scale topographic map that appears once you've zoomed in, what is the contour interval of the map?

Question 15.3 From examining the contour lines and heights, what are the highest and lowest elevations of Tujunga Valley itself?

5. A lot of information about the terrain can be gained from the contours on the topographic maps, but AGX also provides a way to examine them in a pseudo-3D way. To set things up in pseudo-3D, first select the **Home** tab, and then select the **2D/3D** option.

6. Select the option for **3D Display**. Give AGX a few seconds to adjust to the 3D view. You'll now see that a new control has been added to the navigation controls—a second zoom slider has appeared on the left-hand side. This second zoom slider is used to tilt the terrain up and down.

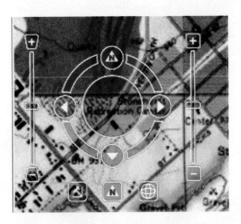

7. Tilt the view down to so that you're looking at the valley (and the surrounding mountains) in perspective view.

8. Next, you'll check the terrain's Vertical Exaggeration. Press the **ArcGIS Explorer Button** (the icon that has the AGX logo—the globe with the arrows—in the upper left-hand corner of the screen) and then click on **ArcGIS Explorer options**.

9. In the new window that opens, select the option for **3D Display Effects** and scroll down to **Vertical Exaggeration**.

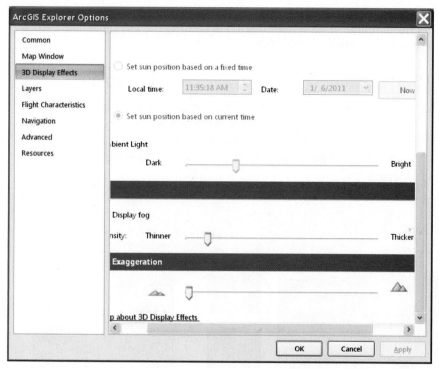

(Source: Esri® ArcGIS ArcExplorer graphical user interface Copyright © Esri)

10. The Vertical Exaggeration option has a slider bar, with the default setting being the slider at the bottom of the scale (the minimum Vertical Exaggeration). You can change the Vertical Exaggeration all the way up to the maximum (click **Apply** after each setting)—try a few different settings—the minimum, about one-fourth of the way across, about one-half the way across, about three-fourths of the way across, and finally at maximum.

Question 15.4 What were the effects of each level of Vertical Exaggeration? Which would be the best level to use for visually analyzing the area and why?

11. Reset the Vertical Exaggeration level to about one-fourth of the way up the slider. This level is good for this visualization part of the exercise, as it will really make the mountains and valleys stand out. Even by looking vertically down on the topos, you should be able to make out some illusion of depth in the image (this is from the topos being draped over AGX's terrain).

12. Use the mouse and controls to fly around the region, zooming and rotating where necessary.

Question 15.5 How does viewing draped imagery in perspective aid in determining the differences in terrain between Tujunga Valley and the nearby mountains? How does this aid in viewing the terrain and landscape features of southern California?

15.3 Examining Imagery with ArcGIS Explorer

1. Back in Chapter 9, you looked at high-resolution digital imagery—this same kind of imagery is available via the cloud resources for AGX. To switch to display the high-resolution imagery, select the **Home** tab, and then select the **Basemap** option. From the available options, select **Bing Maps Aerial**.

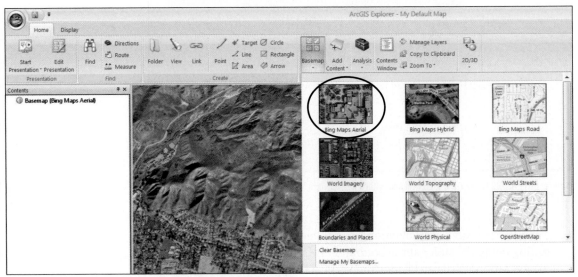

(Source: Esri® ArcGIS ArcExplorer graphical user interface Copyright © Esri)

2. Reposition your view back to Tujunga Valley and carefully look around the region in the valley.

Question 15.6 By using the high-resolution Bing Maps imagery and your visual image interpretation elements, what kinds of land uses are present within Tujunga Valley? For instance, you can make out a highway cutting through the middle of the valley. What is to the immediate east of the highway? What elements of visual image interpretation aid in discerning what this is?

3. To get more information about various features, you can add the labels and roads available through Bing Maps to the imagery. From the **Home** tab, select the **Basemap** option. From the new available options, select **Bing Maps Hybrid**.

4. AGX will switch to a new set of data being streamed to you—not only the images, but labels and other information as well. Examine the new scene of Tujunga Valley and answer Question 15.7 (this will probably also help confirm your answer to Question 15.6).

Question 15.7 What is the name and number of the interstate that goes through the valley?

15.4 Geocoding Addresses and Basic Spatial Analysis with ArcGIS Explorer

1. Back in Chapter 8, you looked at geocoding addresses and routing between them. AGX contains these features as well as several types of GIS spatial analysis functions. To get started, we'll begin at a Hollywood landmark, Grauman's Chinese Theatre (for more information about the theatre, visit http://www.manntheatres.com/chinese).

2. Bring up the **Find** dialog box again (as in Section 15.1) and search for the theater's address: **6925 Hollywood Boulevard, CA 90028**. Click the **Search** button and AGX will change the view so that it centers on the theater, placing a white pushpin in its location along Hollywood Boulevard.

3. You'll now create a buffer around a point representing the theater (as in Chapter 6). From the **Analysis** pull-down menu, select **Buffer point**.

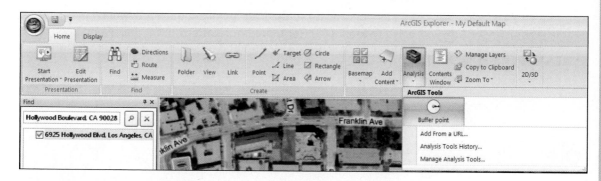

4. In the Buffer point dialog box, you can either specify a location (by coordinates) or interactively select a point to create a buffer around. To select a point representing the theater, press the **select** button next to the **Location to buffer** space. The cursor will change to a

crosshairs—click on the pushpin marking Grauman's Chinese Theatre—AGX will then translate your mouse click into a set of latitude and longitude coordinates representing that point (see Chapter 2 for more about coordinates).

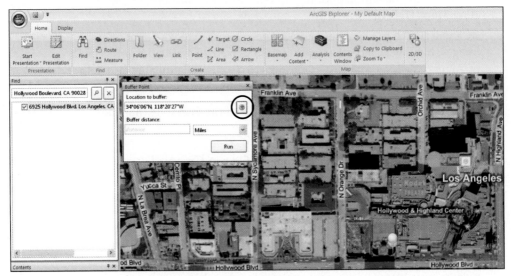

(Source: Esri® ArcGIS ArcExplorer graphical user interface Copyright © Esri)

5. Still in the Buffer Point dialog box, type **1** for the Buffer Distance and select **Miles** as the units to use, then click **Run**. AGX will create the buffer. In the Contents Panel, you'll see a new item added—a one-mile buffer (in red).

6. Zoom out from the theater. You'll see the circular buffer drawn on the screen, representing a one-mile zone of spatial proximity around the theater.

7. Next, let's take a look at the location of another Hollywood landmark—the Hollywood Bowl (for more information on the Hollywood Bowl, visit **http://www.hollywoodbowl.com**)—but in relation to Grauman's Chinese Theatre.

8. Select the **Home** tab, and then click the **Directions** button.

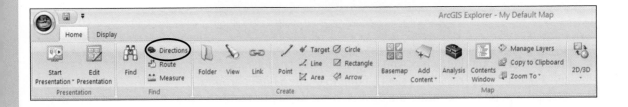

9. In the Directions panel that will open, specify the From Here address as the Chinese Theatre: **6925 Hollywood Boulevard, CA 90028**; and then specify the To Here address as the Hollywood Bowl: **2301 North Highland Avenue, Hollywood, CA 90068**.

    Important note: You can also interactively select points to use as addresses in a manner similar to picking a point to buffer.

10. Select **Miles** to Show Distances in and use **Fastest** as the Route Options. Click **Get Directions** when ready.

11. AGX will compute the directions between the two places and highlight its chosen "fastest path" in green. Examine the route and the two locations, then answer Questions 15.8, 15.9, and 15.10.

Question 15.8 Is the Hollywood Bowl within a one-mile radius of Grauman's Chinese Theatre?

Question 15.9 What is the driving distance between Grauman's Chinese Theatre and the Hollywood Bowl?

Question 15.10 Without transcribing directions, what is the "fastest route" to go from one location to the other?

Closing Time

AGX has a lot of other features beyond these, including several other tools and other types of base data and imagery you can stream off the Internet. Additionally, other geospatial data sources such as rasters, shapefiles, and geodatabases (see Chapter 5), KMZ files (see Chapter 14), and GPS data (see Chapter 4) can be added as separate GIS layers.

At this point you can exit AGX by selecting **Exit ArcGIS Explorer** from the options found in pressing the **Explorer** button in the upper left-hand corner of AGX and say goodbye to labs for the rest of the book.

Glossary

3D modeling designing and visualizing data that contains a third dimension (a z-axis) in addition to x- and y-values.

3D trilateration finding a location on Earth's surface in relation to the positions of three satellites.

3D Warehouse Google's online repository of 3D objects created using Google SketchUp.

8-bit imagery a digital image that carries a range of brightness values from 0 to 255.

AAG the Association of American Geographers, the national organization for the field of geography.

absorption when light is trapped and held by a target.

across-track a scanning method using a rotating mirror to collect data by moving the device the width of the satellite's swath.

address matching another term for geocoding.

address standardization setting up the components of an address in a regular format.

AEJEE the ArcExplorer Java Edition for Educators—a free GIS program produced by Esri that can display data, create maps, and perform basic analysis.

aerial photography the acquisition of imagery of the ground taken from an airborne platform.

affine transformation a linear mathematical process by which data can be altered to align with another data source.

Afternoon Constellation a set of satellites (including Aqua and Aura) that pass the Equator in the afternoon during their orbits.

AIRS the Advanced Infrared Sounder instrument onboard Aqua (used in conjunction with Aqua's HSB and AMSU-A instruments).

algorithm a set of steps used in a process (for example, the steps used in computing a shortest path).

ALI the multispectral sensor onboard EO-1.

almanac data concerning the status of a GPS satellite, which is included in the information being transmitted by the satellite.

along-track a scanning method using a linear array to collect data directly on the path the satellite moves on.

AmericaView a United States national organization dedicated to remote sensing research and education.

AMSR-E the Advanced Microwave Scanning Radiometer for EOS instrument onboard Aqua.

AMSU-A the Advanced Microwave Sounding Unit instrument onboard Aqua.

anaglyph a new image created by slightly offsetting the features in an existing image slightly from each other and highlighting them in different colors.

AND the Boolean operation that corresponds with an Intersection operation.

API Application Programming Interface—the functions and commands used when making requests of software.

Aqua a key EOS satellite whose mission is to monitor Earth's water cycle.

Arc/Info an Esri GIS product that required command-line entry for tools and capabilities.

ArcGIS Esri's current primary GIS software package.

ArcGIS Explorer a virtual Earth tool that can be downloaded for free from Esri.

ArcGIS Online a collection of available Web resources made available by Esri.

ArcGIS Server an Esri utility for distribution of geospatial data in an interactive mapping format on the Internet.

ArcGlobe a 3D visualization tool used to display data in a global environment, which is part of Esri's ArcGIS program.

ArcMap the component of ArcGIS used for viewing and analyzing data.

ArcScene the 3D visualization and design component of Esri's ArcGIS program.

ArcView an Esri GIS product developed with a Windows interface.

ASPRS the American Society for Photogrammetry and Remote Sensing, a professional organization in the geospatial and remote sensing field.

ASTER the Advanced Spaceborne Thermal Emission and Reflection Radiometer instrument onboard Terra.

atmospheric windows those wavelengths of electromagnetic energy in which most of the energy passes through Earth's atmosphere.

A-Train another term for the Afternoon Constellation.

attribute table a spreadsheet-style form where the rows consist of individual objects and the columns are the attributes associated with those objects.

attributes the non-spatial data that can be associated with a spatial location.

Aura an important EOS satellite dedicated to monitoring Earth's atmospheric chemistry.

band a narrow range of wavelengths being measured by a remote sensing device.

base heights the elevation values assigned to the terrain upon which the objects will be placed.

batch geocoding matching a group of addresses together at once.

block a flat polygon object that has been extruded.

blue band the range of wavelengths between 0.4 and 0.5 micrometers.

Blue Marble a composite MODIS image of the entire Earth from space.

Boolean operator one of the four connectors (AND, OR, NOT, XOR) used in building a compound query.

brightness values the energy measured at a single pixel according to a pre-determined scale. Also referred to as Digital Numbers (DNs).

buffer a polygon of spatial proximity built around a feature.

C/A code the digital code broadcast on the L1 frequency, which is accessible by all GPS receivers.

cartography the art and science of creating and designing maps.

Catalog the component of ArcGIS used for managing data (which contains the functionality of the previous ArcCatalog).

CERES the Clouds and Earth's Radiant Energy System instruments onboard Terra and Aqua.

CGIS the Canadian Geographic Information System—a large land inventory system used in Canada and the first time the name "GIS" was utilized for this type of system.

channels the number of satellite signals a GPS unit can receive.

choropleth map a type of thematic map in which data is displayed according to one of several different classifications.

CIR photo color infrared photo—a photo where infrared reflection is shown in shades of red, red reflection is shown in shades of green, and green reflection is shown in shades of blue.

cloud a computer structure wherein data, resources, or applications are kept at another location and made available to the user over the Internet.

CMYK a color scheme based on using the colors cyan, magenta, yellow, and black.

collar the white information-filled border around a topographic map.

color composite an image formed by placing a band of imagery into each of the three color guns (red, green, and blue) to view a color image rather than a grayscale one.

color gun equipment used to display a color pixel on a screen through the use of the colors red, green, and blue.

color ramp a range of colors that are applied to the thematic data on a map.

Compass China's GNSS, currently in development.

components pre-designed (or user-designed) objects in Google SketchUp.

compound query a query that contains more than one operator.

connectivity the linkages between edges and junctions of a network.

constellation the full complement of satellites comprising a GNSS.

continuous field view the conceptualization of the world that all items vary across Earth's surface as constant fields, and values are available at all locations along the field.

contour interval the vertical difference between contour lines drawn on a map.

contour an imaginary line drawn on a map to connect points of common elevation.

control points point locations where the coordinates are known—these are used in aligning the unreferenced image to the source.

control segment one of the three segments of GPS, consisting of the control stations that monitor the signals from the GPS satellites.

Corona a United States government satellite remote sensing program utilizing film-based camera equipment, which was in operation from 1960–1972.

CORS the Continuously Operating Reference Stations—a system operated by the National Geodetic Survey to provide a ground-based method of obtaining more accurate GPS positioning.

coverage a data layer represented by a group of files in a workspace consisting of a directory structure filled with files and also associated files in an INFO directory.

data classification various methods used for grouping together (and displaying) values on a choropleth map.

datum a reference surface of Earth.

datum transformation changing measurements from one datum to measurements in another datum.

decimal degrees (DD) the fractional decimal equivalent to coordinates found using degrees, minutes, and seconds.

degrees, minutes, and seconds (DMS) the measurement system used in GCS.

DEM Digital Elevation Model—a representation of the terrain surface, created by measuring a set of equally spaced elevation values.

DGPS differential GPS—a method using a ground-based correction in addition to the satellite signals in position determination.

digitizing the creation of vector objects through sketching or tracing representations from a map or image source.

Dijkstra's Algorithm an algorithm used in calculating the shortest path between an origin node and other destination nodes in a network.

discrete object view the conceptualization of the world that all items can be represented with a series of objects.

dissolve the ability of the GIS to combine polygons with the same features together.

DLG a Digital Line Graph—the features (such as roads, rivers, or boundaries) digitized from USGS maps.

DOQ a Digital Orthophoto Quad. Orthophotos that cover an area of 3.75 minutes of latitude by 3.75 minutes of longitude, or one-fourth of a 7.5 minute USGS quad.

DPI Dots Per Inch—a measure of how coarse (lower values) or sharp (higher values) an image or map resolution will be when exported to a graphical format.

draping a process in which an image is given z-values to match the heights in a digital terrain model.

DRG a Digital Raster Graphic—a scanned version of a USGS topographic map.

DTM a representation of a terrain surface calculated by measuring elevation values at a series of locations.

dual frequency a GPS receiver that can pick up both the L1 and L2 frequency.

Earth Observation Day a day dedicated to remote sensing research and education.

easting a measurement of so many units east (or west) of some principal meridian.

edge a term used for the linkages of a network.

EGNOS a Satellite Based Augmentation System that covers Europe.

electromagnetic spectrum the light energy wavelengths and the properties associated with them.

ellipsoid a model of the rounded shape of Earth.

EO-1 a satellite launched in 2000 and set to orbit 1 minute after Landsat 7.

EOS NASA's Earth Observing System mission program.

ephemeris data referring to the GPS satellite's position in orbit.

Equal Interval a data classification method that selects class break levels by taking the total span of values (from highest to lowest) and dividing by the number of desired classes.

Equator the line of latitude that runs around the center of Earth and serves as the 0 degree line to make latitude measurements from.

EROS the Earth Resources Observation Science Center; located outside Sioux Falls, South Dakota, which serves (among many other things) as a downlink station for satellite imagery.

Esri the Environmental Systems Research Institute, a key developer and leader of Geographic Information Systems products.

ETM+ the Enhanced Thematic Mapper sensor onboard Landsat 7.

Exclusive Or the operation wherein the chosen features are all of those that meet the first criteria as well as all of those that meet the second criteria, except for the features that have both criteria in common in the query.

extrusion the extending of a flat object to have a z-value.

face one of the sides of a block object.

false color composite an image arranged by not placing the red band in the red color gun, the green band in the green color gun, and the blue band in the blue color gun.

false easting a measurement made east (or west) of an imaginary meridian set up for a particular zone.

false northing a measurement made north (or south) of an imaginary line such as used in measuring UTM northings in the southern hemisphere.

fields the columns of an attribute table.

fonts various styles of lettering used on maps.

Galileo the European Union's GNSS, currently in development.

geocaching using GPS to find locations of hidden objects based on previously obtaining a set of coordinates.

geocoding the process of using the text of an address to plot a point at that location on a map.

geodatabase a single file that can contain multiple datasets, each as its own feature class.

geodesy the science of measuring Earth's shape.

GeoEye-1 a satellite launched in 2008 by the GeoEye company, which features a spatial resolution of 41 centimeters with its panchromatic sensor.

Geographic Alliance Network a National Geographic-sponsored organization set up in each state to promote geographic education.

geographic coordinate system (GCS) a set of global latitude and longitude measurements used as a reference system for finding locations.

Geographic Information System (GIS) a computer-based set of hardware and software used to capture, analyze, manipulate, and visualize spatial information.

geographic scale the real-world size or extent of an area.

geoid a model of Earth using mean sea level as a base.

GeoPDF a format that allows for maps to be exported to a PDF format, yet contain geographic information or multiple layers.

geoprocessing the term that describes when an action is taken to a dataset that results in a new dataset being created.

georeferencing a process whereby spatial referencing is given to data without it.

geospatial data items that are tied to a specific real-world location.

geospatial technology a number of different high-tech systems that acquire, analyze, manage, store, or visualize various types of location-based data.

geostationary orbit an orbit in which an object rotates around Earth at the same speed as Earth.

GeoTIFF a graphical file format that can also carry spatial referencing.

GeoWall a powerful computer tool used for displaying data and imagery in stereoscopic 3D.

GIS Day the Wednesday of Geography Awareness Week (observed in November) dedicated to GIS.

GIS Model a representation of the factors used for explaining the processes that underlie an event or for predicting results.

Global Positioning System (GPS) acquiring real-time location information from a series of satellites in Earth's orbit.

GLOBE Global Learning and Observations to Benefit the Environment—an education program aimed at incorporating user-generated data observations from around the world.

GLONASS the former USSR's (now Russia's) GNSS, currently rebuilding to a constellation of satellites.

GloVis the Global Visualization Viewer set up by the USGS for viewing and downloading satellite imagery.

GNSS the Global Navigation Satellite System, an overall term for the technologies using signals from satellites for finding locations on Earth's surface.

Google Earth a freely available virtual globe program first released in 2005 by Google.

Google SketchUp a 3D modeling and design software program distributed online by Google, it contains linkages with Google Earth.

GOS the Geospatial One-Stop, an online federal government portal used for obtaining geospatial data.

GPS the Global Positioning System, a technology using signals broadcast from satellites for navigation and position determination on Earth.

graduated symbols the use of different sized symbology to convey thematic information on a map.

great circle distance the shortest distance between two points on a spherical surface.

green band the range of wavelengths between 0.5 and 0.6 micrometers.

grid cell a square unit, representing some real-world size, which contains a single value.

hillshade a shaded relief map of the terrain created by modeling the position of the Sun in the sky relative to the landscape.

HIRDLS the High Resolution Dynamics Limb Sounder instrument onboard Aura.

HSB the Humidity Sounder for Brazil instrument onboard Aqua.

Hyperion the hyperspectral sensor onboard EO-1.

hyperspectral imagery remotely sensed imagery comprised of the bands collected by a sensor capable of sensing hundreds of bands of energy at once.

hyperspectral sensor a sensor that can measure hundreds of different wavelength bands simultaneously.

identity a type of GIS overlay that retains all features from the first layer along with the features it has in common with a second layer.

IKONOS a satellite launched in 1999 by SpaceImaging, Inc. (now called GeoEye), which features multispectral spatial resolution of 4 meters and panchromatic resolution of 1 meter.

incident energy the total amount of energy (per wavelength) that interacts with an object.

International Date Line a line of longitude that follows a similar path as the 180th meridian (but changes away from a straight line to accommodate geography).

intersect a type of GIS overlay that retains the features that are common to two layers.

Intersection the operation wherein the chosen features are those that meet both criteria in the query.

interval data a type of numerical data in which the difference between numbers is significant, but there is no fixed non-arbitrary zero point associated with the data.

IR (infrared) the portion of the electromagnetic spectrum with wavelengths between 0.7 and 100 micrometers.

join a method of linking two (or more) tables together.

JPEG the Joint Photographic Experts Group image, or graphic file format.

junction a term used for the nodes (or places where edges come together) in a network.

key the field that two tables have to have in common with each other in order for the tables to be joined.

KML Keyhole Markup Language—the file format used for Google Earth data.

KMZ a file that is the compressed version of a KML file.

label text placed on a map to identify features.

Landsat a long-running United States remote sensing project that had its first satellite launched in 1972 and continues today.

Landsat 5 the fifth Landsat mission, launched in 1984, which carries both the TM and MSS sensors.

Landsat 7 the seventh Landsat mission, launched in 1999, which carries the ETM+ sensor.

Landsat Scene a single image obtained by a landsat satellite sensor.

large-scale map a map with a higher value for its representative fraction. Such maps will usually show a small amount of geographic area.

latitude imaginary lines on a globe north and south of the Equator that serve as a basis of measurement in GCS.

layout the assemblage and placement of various map elements used in constructing a map.

LDCM the Landsat Data Continuity Mission—the future of the Landsat program, scheduled to be launched in 2012.

legend a graphical device used on a map as an explanation of what the various map symbols and color represent.

LIDAR Light Detection and Ranging. A process in which a series of laser beams fired at the ground from an aircraft is used for creation of highly accurate DEMs.

line segment a single edge of a network that corresponds to one portion of a street (for instance, the edge between two junctions).

linear interpolation a method used in geocoding to place an address location among a range of addresses.

lines one-dimensional vector objects.

longitude imaginary lines on a globe east and west of the Prime Meridian that serve as a basis of measurement in GCS.

map a representation of geographic data.

Map Algebra combining datasets together using simple mathematical operators.

map projection the translation of locations on a three-dimensional (3D) Earth to a two-dimensional (2D) surface.

map scale a metric used to determine the relationship between measurements made on a map and their real-world equivalents.

Map Services pre-made basemaps and layers that can be used in creating Web mapping applications.

map template a pre-made arrangement of items in a map layout.

mashup the combination of two or more map layers into one new application.

MCE Multi-Criteria Evaluation—the use of several factors (weighted and combined) to determine the suitability of a site.

metadata descriptive information about geospatial data.

micrometer a unit of measurement equal to one-millionth of a meter. Abbreviated μm.

Mie scattering scattering of light caused by atmospheric particles the same size as the wavelength being scattered.

MIR (middle infrared) the portion of the electromagnetic spectrum with wavelengths between 1.3 and 3.0 micrometers.

MISR the Multi-angle Imaging SpectroRadiometer instrument onboard Terra.

MLS the Microwave Limb Sounder instrument onboard Aura.

MODIS the Moderate Resolution Imaging SpectroRadiometer instrument onboard Terra and Aqua.

MOPITT the Measurements of Pollution in the Troposphere instrument onboard Terra.

Morning Constellation a set of satellites (including Terra, Landsat 7, EO-1, and SAC-C) that pass the Equator in the morning during their orbits.

MSAS a Satellite Based Augmentation System that covers Japan and nearby regions.

MSS the Multi-Spectral Scanner aboard Landsat 1 through 5.

multipath effect an error caused by a delay in the signal due to reflecting from surfaces before reaching the receiver.

multispectral imagery remotely sensed imagery comprised of the bands collected by a sensor capable of sensing several bands of energy at once.

multispectral sensor a sensor that can measure multiple different wavelength bands simultaneously.

NAD27 the North American Datum of 1927.

NAD83 the North American Datum of 1983.

nadir the location under the camera in aerial photography.

nanometer a unit of measurement equal to one-billionth of a meter. Abbreviated nm.

NASA the National Aeronautics and Space Administration, established in 1958; it is the United States government's space exploration and aerospace development branch.

NASA Earth Observatory a Website operated by NASA that details how EOS is utilized with numerous global environmental issues and concerns.

NASA NEO the NASA Earth Observations Website, which allows users to view or download processed EOS imagery in a variety of formats, including a version compatible for viewing in Google Earth.

NASA World Wind a virtual globe program from NASA, used for examining various types of remotely sensed imagery.

National Map an online base map of downloadable geospatial data maintained and operated by the USGS and part of the National Geospatial Program.

Natural Breaks a data classification method that selects class break levels by searching for spaces in the data values.

NAVSTAR GPS the United States Global Positioning System.

NCGE the National Council for Geographic Education, a professional organization dedicated to geographic education.

NDGPS the National Differential GPS—it consists of ground-based DGPS beacons around the United States.

NDVI (Normalized Difference Vegetation Index) a method of measuring the health of vegetation using near-infrared and red energy measurements.

near-polar orbit an orbital path that carries an object around Earth, passing close to the north and south poles.

NED the National Elevation Dataset, which provides digital elevation coverage of the entire United States.

Negation the operation wherein the chosen features are those that meet all of the first criteria and none of the second criteria (including where the two criteria overlap) in the query.

network a series of junctions and edges connected together for modeling concepts such as streets.

NIR (near infrared) the portion of the electromagnetic spectrum with wavelengths between 0.7 and 1.3 micrometers.

NLCD the National Land Cover Database is a raster-based GIS dataset that maps the land cover types for the entire United States at 30-meter resolution.

nominal data a type of data that is a unique identifier of some kind. If numerical, the differences between numbers are not significant.

nonselective scattering scattering of light caused by atmospheric particles larger than the wavelength being scattered.

non-spatial data data that is not directly linked to a spatial location (such as tabular data).

normalized altering count data values so that they are at the same level of representing the data (such as using them as a percentage).

north arrow a graphical device on a map used to show the orientation of the map.

northing a measurement of so many units north (or south) of a baseline.

NOT the Boolean operator that corresponds with a Negation operation.

oblique photo an aerial photo taken at an angle.

off-nadir viewing the capability of a satellite to observe areas other than the ground directly underneath it.

offset a value applied to objects to move them off the ground level.

OGC the Open Geospatial Consortium, a group involved with developing new standards of geospatial data interoperability.

OhioView a research and education consortium consisting of the state universities in Ohio along with related partners.

OLI the Operational Land Imager, the multispectral sensor that will be onboard LDCM.

OMI the Ozone Monitoring Instrument onboard Aura.

OR the Boolean operator that corresponds with Union operation.

ordinal data a type of data that refers solely to a ranking of some kind.

orthophoto an aerial photo with uniform scale.

orthorectification a process used on aerial photos to remove the effects of relief displacement and give the image uniform scale.

overlay the combining of two or more layers in the GIS.

P code the digital code broadcast on the L1 and L2 frequencies, which is accessible by the military.

panchromatic black-and-white aerial imagery.

panchromatic imagery black-and-white imagery formed by viewing the entire visible portion of the electromagnetic spectrum.

panchromatic sensor a sensor that can measure one range of wavelengths.

pan-sharpening fusing a higher-resolution panchromatic band with lower-resolution multispectral bands to improve the clarity and detail seen in the image.

parsing breaking an address up into its component parts.

pattern the arrangement of objects in an image. An element of image interpretation.

PDOP the Position Dilution of Precision—it describes the amount of error due to the geometric position of the GPS satellites.

perspective view viewing a digital terrain model at an oblique angle in which it takes on a "three-dimensional" appearance.

photo scale the representation used to determine how many units of measurement in the real world are equivalent to one unit of measurement on an aerial photo.

photogrammetry the process of making measurements using aerial photos.

PLEIADES a high-resolution series of satellites.

points zero-dimensional vector objects.

polygons two-dimensional vector objects.

Prime Meridian the line of longitude that runs through Greenwich, England, and serves as the 0 degree line of longitude to base measurements from.

principal point the center point of an aerial photo.

prism map a thematic map that has the map shapes extruded to a value based on the values shown on the map.

prisms the extruded shapes on a prism map.

pseudo-3D a term often used to describe the perspective view of a terrain model since it is often a 2.5D model, not a full 3D model.

pseudorange the calculated distance between a GPS satellite and a GPS receiver.

Quantile a data classification method that attempts to place an equal number of data values in each class.

query the conditions used to retrieve data from a database.

QuickBird a satellite launched in 2001 by the DigitalGlobe company, whose sensors have 2.4-meter multispectral resolution and 0.61-meter panchromatic resolution.

radiometric resolution a sensor's ability to determine fine differences in a band of energy measurements.

raster data model a conceptualization of representing spatial data with a series of equally spaced and sized grid cells.

ratio data a type of numerical data in which the difference between numbers is significant, but there is a fixed non-arbitrary zero point associated with the data.

Rayleigh scattering scattering of light caused by atmospheric particles smaller than the wavelength being scattered.

records the rows of an attribute table.

red band the range of wavelengths between 0.6 and 0.7 micrometers.

reference database the base network data used as a source for geocoding.

reference map a map that serves to show the location of features, rather than thematic information.

relational operator one of the six connectors (=, < >, <, >, >=, or =<) used to build a query.

relief displacement the effect seen in aerial imagery where tall items appear to "bend" outward from the photo's center toward the edges.

remote sensing the process of collecting information related to the reflected or emitted electromagnetic energy from a target by a device a considerable distance away from the target onboard an airborne or spacecraft platform.

reproject changing a dataset from one map projection (or measurement system) to another.

RF representative fraction—a value indicating how many units of measurement in the real world are equivalent to how many of the same units of measurement on a map.

RGB a color scheme based on using the three primary colors red, green, and blue.

Root Mean Square Error (RMSE) an error measure used in determining the accuracy of the overall transformation of the unreferenced data.

rotated when the unreferenced image is turned during the transformation.

satellite imagery digital images of Earth acquired by sensors onboard orbiting spaceborne platforms.

SATELLITES Students and Teachers Exploring Local Landscapes to Interpret the Earth from Space—an OhioView initiative involving K–12 teachers and students with geospatial technologies.

SBAS Satellite Based Augmentation System—a method of using correction information sent from an additional satellite in the GPS position determination.

scale bar a graphical device used on a map to represent map scale.

scaled when the scale of the unreferenced image is altered during the transformation.

SDTS the Spatial Data Transfer Standard, a "neutral" file format for geospatial data, allowing it to be imported into various geospatial software programs.

Seamless Server an online utility used by the USGS for the download of different datasets, including elevation data.

Selective Availability the intentional alteration of the timing and position information transmitted by a GPS satellite.

shadow the shadings in an image caused by a light source. An element of image interpretation.

shape the form of objects. An element of image interpretation.

shapefile a series of files (with extensions such as .shp, .shx, and .dbf) that make up one vector data layer.

shortest path the route that corresponds to the lowest cumulative transit cost between stops in a network.

single frequency a GPS receiver that can pick up only the L1 frequency.

site and association the information referring the location of objects and their related attributes in an image. Used as elements of image interpretation.

Site Suitability the determination of the "useful" or "non-useful" locations based on a set of criteria.

size the physical dimensions (such as length and width) of objects. An element of image interpretation.

skewed when the unreferenced image is distorted or slanted during the transformation.

SKP file the file format and extension used by an object created in Google SketchUp.

SLC the Scan Line Corrector in the ETM+ sensor. Its failure in 2003 causes Landsat 7 ETM+ imagery to not contain all data from a scene.

slope a measurement of the rate of elevation change at a location, found by dividing the vertical height (the rise) by the horizontal length (the run).

slope aspect a determination of the direction that a slope is facing.

small-scale map a map with a lower value for its representative fraction. Such maps will usually show a large amount of geographic area.

SOS Science on a Sphere, a NOAA initiative used in projecting images of Earth onto a large sphere.

space segment one of the three segments of GPS, consisting of the satellites and the signals being broadcast from space.

spatial analysis examining the characteristics or features of spatial data, or how features spatially relate to each other.

spatial query selecting records or objects from one layer based upon their spatial relationships with other layers (rather than using attributes).

spatial reference the use of a real-world coordinate system for identifying locations.

spatial resolution the size of the area on the ground being represented by one pixel's worth of energy measurement.

SPCS the State Plane Coordinate System used for determining coordinates of locations within the United States.

SPCS zone one of the divisions of the United States set up by the SPCS.

spectral reflectance the percentage of the total incident energy that was reflected from that surface.

spectral resolution the bands and wavelengths measured by a sensor.

spectral signature a unique identifier for a particular item, generated by charting the percentage of reflected energy per wavelength against a value for that wavelength.

SPOT a satellite program operated by the French Space Agency and the Spot Image Corporation.

SQL the Structured Query Language—a formal setup for building queries.

SRTM the Shuttle Radar Topography Mission, flown in the year 2000, which mapped Earth's surface from orbit for the purpose of constructing digital elevation models of the planet.

Standard Deviation a data classification method that computes class break values by using the mean of the data values and the average distance a value is away from the mean.

standard false color composite an image arranged by placing the near-infrared band in the red color gun, the red band in the green color gun, and the green band in the blue color gun.

StateView the term used to describe each of the programs affiliated with AmericaView in each state.

stereoscopic 3D an effect that simulates the illusion of depth or immersion in an image by utilizing stereo imaging techniques.

stops destinations to visit on a network.

street centerline a file containing line segments representing roads.

Street View a component of Google Maps and Google Earth that allows the viewer to see 360-degree imagery around an area on a road.

Suitability Index a system whereby locations are "ranked" according to how well they fit a set of criteria.

Sun altitude the value between 0 and 90 used in constructing a hillshade to model the Sun's elevation above the terrain.

Sun azimuth the value between 0 and 360 used in constructing a hillshade to model the Sun's position in the sky to show the direction of the Sun's rays striking the surface.

Sun-synchronous orbit an orbital path set up so that the satellite crosses the same areas at the same local time.

swath width the width of the ground area the satellite is imaging.

symmetrical difference a type of GIS overlay that retains all features from both layers except for the features that they have in common.

temporal resolution a sensor's capability that determines how often it can view the same location on the ground.

Terra the flagship satellite of the EOS program.

TES the Tropospheric Emission Spectrometer instrument onboard Aura.

texture repeating tones in an image. An element of image interpretation.

textures graphics applied to 3D objects to create a more realistic appearance.

thematic map a map that displays a particular theme or feature.

three-dimensional (3D) model a model of the terrain that allows for multiple z-values to be assigned to each x/y coordinate location.

TIFF the Tagged Image File Format used for graphics or images.

TIGER/Line a file produced by the U.S. Census Bureau that contains (among other items) the line segments that correspond with roads across the United States.

time zones a method of measuring time around the world, found by dividing the world into subdivisions and relating the time in that division to the time in Greenwich, England.

TIN Triangulated Irregular Network. A terrain model that allows for non-equally spaced elevation points to be used in the creation of the surface.

TIR (thermal infrared) the portion of the electromagnetic spectrum with wavelengths between 3.0 and 14.0 micrometers.

TM the Thematic Mapper sensor onboard Landsat 4 and 5.

tone the grayscale levels (from black to white) or ranges of a color for objects present in an image. An element of image interpretation.

topographic map a map created by the USGS to show landscape and terrain, as well as location of features on the land.

topology how vector objects connect to each other (in terms of their adjacency, connectivity, and containment) independently of the objects' coordinates.

transit cost a value that represents how many units (of time or distance) are used in moving across a network edge.

translated when the unreferenced image is shifted during the transformation.

transmittance when light passes through a target.

trilateration finding a location in relation to three other points of reference.

true color composite an image arranged by placing the red band in the red color gun, the green band in the green color gun, and the blue band in the blue color gun.

true orthophoto an orthophoto where all objects look as if they're being seen from directly above.

two-and-a-half-dimensional (2.5D) model a model of the terrain that allows for a single z-value to be assigned to each x/y coordinate location.

type the lettering used on a map.

UAV unmanned aerial vehicle—a reconnaissance aircraft that is piloted from the ground via remote control.

Union the operation wherein the chosen features are all that meet the first criteria as well as all that meet the second criteria in the query.

union (overlay) a type of GIS overlay that combines all features from both layers.

Universal Transverse Mercator (UTM) the grid system of locating coordinates across the globe.

US Topo a digital topographic map series created by the USGS to allow multiple layers of data to be used on a map in GeoPDF file format.

user segment one of the three segments of GPS, consisting of the GPS receivers on the ground that pick up the signals from the satellites.

UTM zone one of the 60 divisions of the world set up by the UTM system, with each zone being 6 degrees of longitude wide.

UV (ultraviolet) the portion of the electromagnetic spectrum with wavelengths between 0.01 and 0.4 micrometers.

vector data model a conceptualization of representing spatial data with a series of vector objects (points, line, and polygons).

vector objects points, lines, and polygons that are used to model real-world phenomena using the vector data model.

vehicle navigation system a device used to plot the user's position on a map, using GPS technology to obtain the location.

vertical datum a baseline used as a starting point in measuring elevation values (either above or below this value).

vertical exaggeration a process whereby the z-values are artificially enhanced for terrain visualization purposes.

vertical photo an aerial photo in which the camera is looking down at a landscape.

VGI Volunteered Geographic Information, a term used to describe user-generated geospatial content and data.

viewshed a data layer that determines what an observer can see and cannot see from a particular location due to terrain.

virtual globe a software program that provides an interactive three-dimensional map of Earth.

Visible Earth a Website operated by NASA to distribute EOS images and animations of EOS satellites or datasets.

visible light the portion of the electromagnetic spectrum with wavelengths between 0.4 and 0.7 micrometers.

visual image interpretation the process of discerning information to identify objects in an aerial (or other remotely sensed) image.

WAAS the Wide Area Augmentation System—a satellite based augmentation system that covers the United States and other portions of North America.

wavelength the distance between the crests of two waves.

WGS84 the World Geodetic System of 1984 datum (used with the Global Positioning System).

wiki a database available so that anyone can edit it.

WorldView-1 a satellite launched in 2007 by the DigitalGlobe company, whose panchromatic sensor has 0.5-meter spatial resolution.

WorldView-2 a satellite launched in 2009 by the DigitalGlobe company, featuring an 8-band multispectral sensor with 1.84-meter resolution and a panchromatic sensor of 0.46-meter resolution.

Worldwide Reference System the global system of Paths and Rows that is used to identify what area on Earth's surface is present in which Landsat scene.

XOR the Boolean operator that corresponds with an Exclusive Or operation.

Y code an encrypted version of the P code.

z-value the elevation assigned to an x/y coordinate.

Index

f = figure; t = table